Titles in This Series

Volume

1 **Theory of singularities and its applications,**
V. I. Arnold, editor, ix + 333 pp., 1990

2 **Topics in representation theory,**
A. A. Kirillov, editor, xi + 247 pp., 1991

3 **Dynamical systems and statistical mechanics,**
Ya. G. Sinaĭ, editor, viii + 254 pp., 1991

4 **Algebraic K-theory,**
A. A. Suslin, editor, v + 170 pp., 1991

5 **Many-particle Hamiltonians: spectra and scattering,**
R. A. Minlos, editor, v + 194 pp., 1991

6 **Topological classification of integrable systems,**
A. T. Fomenko, editor, vi + 345 pp., 1991

7 **Estimates and asymptotics for discrete spectra of integral and differential equations,**
M. Sh. Birman, editor, x + 204 pp., 1991

Estimates and Asymptotics for Discrete Spectra of Integral and Differential Equations

Advances in
SOVIET
MATHEMATICS

Advances in
SOVIET
MATHEMATICS

Volume 7

Estimates and Asymptotics for Discrete Spectra of Integral and Differential Equations

M. Sh. Birman
Editor

American Mathematical Society
Providence, Rhode Island

Translation edited by A. B. SOSSINSKY

1991 *Mathematics Subject Classification.* Primary 34L20, 35J10, 35P20, 35S30, 45A05, 47A75, 47B99, 47G30; Secondary 35B40, 47A68.

Library of Congress Catalog Card Number: 91-640741

ISBN 0-8218-4106-8; ISSN 1051-8037

The paper used in this book is acid-free and falls within the guidelines
established to ensure permanence and durability. ♾

This publication was typeset using $\mathcal{AMS}$-TEX,
the American Mathematical Society's TEX macro system.

10 9 8 7 6 5 4 3 2 1 96 95 94 93 92 91

Contents

Editor's Preface

The papers of this collections contain adapted expositions of a series of talks on spectral theory given to the Leningrad Seminar on Mathematical Physics in the academic year of 1989–90.

A few words about the seminar itself. This is a city seminar (based on Leningrad University and LOMI, the Leningrad Branch of the Steklov Mathematical Institute of the Academy of Sciences of the USSR) on the theory of boundary value problems and related questions of analysis and mathematical physics. It was founded in 1947 by the late Professor V. I. Smirnov. Since his death, Professor O. A. Ladyzhenskaya has been directing the seminar. Many Leningrad mathematicians participate in its work, and often there are visitors. Topics of the talks are fairly diverse. Hence in selecting a collection for the Advances in Soviet Mathematics Series it is necessary to pick out expositions pertaining to some particular field. We have begun with spectral theory, though it cannot be considered as dominating in the variety of topics covered by the Seminar.

For the most part, the papers in this collection are devoted to investigations of the spectrum of the Schrödinger operator (or its generalizations) perturbed by some relatively compact operator. The discrete spectrum that emerges in the spectral gaps of the nonperturbed operator is studied. The corresponding estimates and asymptotic formulas for spectrum distribution functions are considered in the large-coupling-constant limit. The starting point here is the paper by Birman and Solomyak that opens the collection. It is devoted to the leading (but also simplest) case of a semi-infinite gap. The case of inner gaps is treated in the papers by Birman, Birman and Raĭkov, and Sobolev. In the paper by Birman, Karadzhov, and Solomyak, some related questions in the theory of functions are considered. A more detailed description of this group of papers can be found in the introduction to the Birman and Solomyak paper.

The large paper by Budylin and Buslaev is devoted to an integral equation with difference kernel on a finite interval. Asymptotic formulas for solutions with respect to the semiclassical parameter are investigated. The symbol is allowed to have singularities and zeros. A general geometric approach is developed (in the spirit of the Schwarz alternating method) that is applicable to similar problems.

The collection is concluded with Yafaev's paper. It is devoted to the classical problem of constructing scattering theory for the Schrödinger operator with potential decreasing faster than the Coulomb potential. This is ordinarily done on the basis of the theorem about decompositions into scattered plane waves. The author constructs a solution which is asymptotically the sum of an incoming and an outgoing spherical wave. A corresponding decomposition theorem is proved and a scattering theory is constructed. This is accompanied by the spectral analysis of the scattering matrix.

The collection was supposed to promptly give an account of current research in spectral theory by participants of the seminar. The technical stage of the work on it has taken more time than was expected. I hope nevertheless that, to a certain extent, the desired purpose has been achieved.

M. Sh. Birman

Translated by THE AUTHOR

ADVANCES IN SOVIET MATHEMATICS
Volume 7, 1991

Estimates for the Number of Negative Eigenvalues of the Schrödinger Operator and Its Generalizations

M. SH. BIRMAN AND M. Z. SOLOMYAK

ABSTRACT. Let $N(\alpha)$ be the number of negative eigenvalues of the operator

$$(-\Delta)^l + \varphi - \alpha V \qquad (*)$$

on $L_2(\mathbb{R}^d)$. Some growth estimates for $N(\alpha)$ (as $\alpha \to \infty$) are found, in terms of properties of φ and V. Principal attention is paid to the case when the growth is nonregular, i.e., $\alpha^{-d/2l} N(\alpha) \to \infty$. Various methods to derive nonregular estimates are developed. Several series of estimates are found in which both the rate of growth of $N(\alpha)$ and the corresponding conditions on φ and V vary, depending on relations between d and l. In some cases the estimates discussed are two-sided. We consider the inverse problem as well, i.e., the following one: to find the class of potentials guaranteeing the prescribed rate of growth of $N(\alpha)$. Examples of operators of the form $(*)$ are constructed with proper (power or power-logarithmic) asymptotic behavior of $N(\alpha)$. Some generalizations to the magnetic and the periodic Schrödinger operators are presented.

CONTENTS

1991 *Mathematics Subject Classification*. Primary 35J10, 35P20.

§0. Introduction

In this paper we discuss the discrete spectrum of the operator

$$A_l(\alpha V) = (-\Delta)^l - \alpha V, \qquad \alpha > 0, \tag{0.1}$$

with real potential V and "coupling constant" α, on the space $L_2(\mathbb{R}^d)$. Some other operators resembling (0.1) are also considered. Suppose that a selfadjoint operator A is bounded from below and its spectrum to the left of the point $\lambda \in \mathbb{R}$ is finite. Then by $N(\lambda; A)$ we denote the number of eigenvalues of A lying to the left of λ (and counted according to their multiplicities). We are concerned with characteristics of growth of $N(\lambda; A_l(\alpha V))$ as $\alpha \to \infty$, $\lambda \leqslant 0$ being fixed. Usually it is sufficient to consider nonnegative potentials V only, because by the variational principle we have

$$N(\lambda; A_l(\alpha V)) \leqslant N(\lambda; A_l(\alpha V_+)), \quad \text{where } 2V_\pm = |V| \pm V. \tag{0.2}$$

Estimates for the case $l = 1$ (i.e., for the Schrödinger operator) are of the greatest importance. The examination of the more general operator (0.1) (including the case when $l > 0$ is not an integer) allows one to better understand the problem and methods for solving it.

One associates with the operator (0.1) the so-called "semiclassical" exponent

$$\varkappa = \varkappa(d, l) = d/2l.$$

The order of growth of $N(\lambda; A_l(\alpha V))$ as $\alpha \to \infty$ cannot be smaller than $O(\alpha^\varkappa)$. Let us call *regular* the estimates guaranteeing precisely this order of growth. The best known regular estimate concerns the case $2l < d$ and reads as follows:

$$N(\lambda; A_l(\alpha V)) \leqslant C\alpha^\varkappa \int V_+^\varkappa \, dx, \qquad C = C(d, l), \quad 2l < d. \tag{0.3}$$

This estimate is sharp in the sense that for nonnegative potentials $V \in L_\varkappa(\mathbb{R}^d)$ the following asymptotic formula holds in the case $2l < d$:

$$N(\lambda; A_l(\alpha V)) \sim (2\pi)^{-d} v_d \alpha^\varkappa \int V^\varkappa \, dx, \qquad \alpha \to \infty, \quad \lambda \leqslant 0. \tag{0.4}$$

In (0.4) v_d denotes the volume of the unit ball in $\mathbb{R}^d$. For $2l \geqslant d$ (0.4) is also true, but under more restrictive assumptions about the potential.

In this paper we mainly deal with nonregular estimates. For example, in the case $2l < d$, we consider potentials with $V_+ \notin L_\varkappa$. We shall obtain a scale of nonregular estimates for $N(0; A_l(\alpha V))$ giving growth of order $O(\alpha^q)$ with $q > \varkappa$, as $\alpha \to \infty$. These estimates are sharp, each one for its particular class of potentials. A similar scale will be found for $N(\lambda; A_l(\alpha V))$ with $\lambda < 0$ as well. Among the things demonstrating sharpness, the explicit asymptotic formulas presented in §8 should be named. It turns out that for $2l > d$ and $V \geqslant 0$ our nonregular estimates are two-sided.

The problem of estimating the quantity $N(\lambda; A_l(\alpha V))$ provides a rather interesting task for an analyst. In this paper we propose two methods to obtain estimates: the interpolation method (§§4–6) and the "direct" one (§7). The first of them quickly leads to estimates in the framework of the power scale with respect to α. The direct method enables us to consider finer scales (e.g., power-logarithmic in α) and to attack the inverse problem: given a growth estimate for N, find the corresponding class of potentials. Under both approaches regular estimates are fundamental for getting nonregular ones.

A survey of earlier results on estimates of N can be found in [RS, BS6] (cf. also lecture [BS10]). Here we restrict ourselves to few remarks. The original estimates of Birman and Schwinger [B1, B2, Sch] for $l = 1$, $d = 3$ should be classified as nonregular. The regular estimate (0.3) has been obtained by Rozenblyum [R1, R2], who has refined the variational approach due to the authors of the present paper (cf. [BS2, BS3]). Other effective approaches to the proof of (0.3) have been given by Cwikel [C] and (for $l = 1$) by Lieb [L1]. The asymptotic formula (0.4) for $l = 1$, $d = 3$ appeared for the first time in [B2] (for problems in a bounded domain or, which is the same, for V with compact support). Estimate (0.3) together with the general statement about continuity of the asymptotic coefficient (cf. [BS3, BS5]) allows one to extend (0.4) to arbitrary $V \in L_\varkappa$, provided that this formula has been established for some set of potentials dense in $L_\varkappa$. We mention also [BB] in connection with regular estimates and the asymptotic formula (0.4) for $2l \geqslant d$. Finally, we want to quote the relatively recent papers [EK1–EK4] and [BSR], where a series of nonregular estimates has been obtained. These estimates give the order of $O(\alpha^q)$, $q > \varkappa$, for N but cannot be considered sharp in the power scale (see the commentary in remarks to Theorem 4.2). The authors are pleased to mention the fact that the publication of the articles [EK1–EK4] by Egorov and Kondrat′ev has significantly stimulated the work on this paper.

To present the main results of this article, we first need to discuss a good deal of background material of operator-theoretic and analytic nature. This will be done in §§1 and 2, 3. A more detailed discussion of the analytic background can be found in the authors' survey lectures [BS10]. A large part of the contents of the present paper corresponds to our notes [BS6–BS9]. They are short summaries of lectures given at several conferences in 1988–90. Some results are generalized or strengthened here. In [BS6–BS9] we mainly discussed the case $2l < d$; here, essentially, this restriction is dropped.

The main notions and notation will be introduced in §§1 and 2. By C, c we denote various constants. The constants depend throughout on the parameters d, l only, unless the opposite is explicitly stated. The notation for important constants is clear from the following example: $C_{4.4}$ is the constant from estimate (4.4). The symbol $\asymp$ stands for two-sided inequalities.

§1. Operators and forms

1. Let $\mathscr{H}$ be a separable Hilbert space. Let $(\cdot, \cdot)$ and $\|\cdot\|$ be the scalar product and the norm in it. By $\mathfrak{R} = \mathfrak{R}(\mathscr{H})$ we denote the algebra of all bounded linear operators on $\mathscr{H}$ and by $\mathfrak{S}_\infty = \mathfrak{S}_\infty(\mathscr{H})$ the ideal (in $\mathfrak{R}(\mathscr{H})$) of all compact operators.

Let A be a selfadjoint operator on $\mathscr{H}$ bounded from below, and let $a[u, v]$, $a[u] := a[u, u]$ be its sesquilinear and quadratic forms. For the latter we sometimes use the notation $a = QF(A)$. Next, let us denote by $d[a]$ the domain where a is well defined, and let m_a be the greatest lower bound of a. For $\gamma > -m_a$ we introduce the space $H_\gamma[a]$. This is $d[a]$ endowed with the metric generated by the form $a[u] + \gamma\|u\|^2$ (for different γ these metrics are equivalent). If $m_a > -1$ we write $H[a]$ instead of $H_1[a]$; in what follows $\|\cdot\|_a$ is the norm in $H[a]$. The form $QF[A]$ is always closed, i.e., the Hilbert space $H[a]$ is complete. For a semibounded selfadjoint operator A, let us introduce the spectrum distribution function $N(\lambda, A)$:

$$N(\lambda; A) := \dim E_A(-\infty, \lambda)\mathscr{H}, \qquad \lambda \in \mathbb{R},$$

where E_A is the spectral measure for A. The inequality $N(\lambda; A) < \infty$ means that to the left of λ the spectrum of A consists of a finite number of eigenvalues with finite multiplicities. If $N(\lambda; A) < \infty$ for every $\lambda < \lambda_0$ then the spectrum of A to the left of λ_0 is bounded from below and discrete.

The following statement ("Glazman lemma") is well known, cf. [BS5].

PROPOSITION 1.1. *Let A be a selfadjoint operator bounded from below, $a = QF(A)$. Then*

$$N(\lambda; A) = \sup\{\dim F : F \subset d[a],\ a[u] < \lambda\|u\|^2,\ 0 \neq u \in F\}. \tag{1.1}$$

For an arbitrary dense linear subspace $\mathscr{F} \subset d[a]$, it is sufficient to take the supremum in (1.1) over $F \subset \mathscr{F}$ only.

Now let $T = T^* \in \mathfrak{S}_\infty(\mathscr{H})$. Then one can introduce two distribution functions, separately for positive and negative eigenvalues:

$$n_-(\lambda; T) := N(-\lambda; T), \qquad n_+ := n_-(\lambda; -T), \qquad \lambda > 0. \tag{1.2}$$

The functions given by (1.2) make sense for arbitrary $T = T^* \in \mathfrak{R}(\mathscr{H})$ as well, but then it can happen that $n_+(\lambda; T) = \infty$ or $n_-(\lambda; T) = \infty$. Proposition 1.1 will take the following form.

PROPOSITION 1.2. *Let $T = T^* \in \mathfrak{R}(\mathscr{H})$. Then*

$$n_\pm(\lambda; T) = \sup\{\dim F : F \subset \mathscr{H},\ \pm(Tu, u) > \lambda\|u\|^2,\ 0 \neq u \in F\}. \tag{1.3}$$

Given a dense linear subspace $\mathscr{F}$ *in* $\mathscr{H}$, *the supremum in* (1.3) *can be taken over* $F \subset \mathscr{F}$ *only.*

Finally, let $\mathscr{H}_1$, $\mathscr{H}_2$ be two Hilbert spaces and $Q: \mathscr{H}_1 \to \mathscr{H}_2$ be a bounded linear operator. Its singular numbers (or s-numbers, cf. [GK]) are defined as the nonzero eigenvalues of the selfadjoint operator $(Q^*Q)^{1/2}$ (or, which is the same, those of $(QQ^*)^{1/2}$). For the s-numbers $\{s_n(Q)\}$ we introduce the distribution function

$$\nu(s; Q) = n_+(s^2; Q^*Q) = n_+(s^2; QQ^*), \qquad s > 0.$$

In particular, if $Q = Q^* \geqslant 0$, then $\nu(s; Q) = n_+(s; Q)$. It follows directly from (1.3) that

$$\begin{aligned} \nu(s; Q) &= \sup\{ \dim F : F \subset \mathscr{H}_1,\ \|Qu\| > s\|u\|,\ 0 \neq u \in F \} \\ &= \sup\{ \dim G : G \subset \mathscr{H}_2,\ \|Q^*v\| > s\|v\|,\ 0 \neq v \in G \}. \end{aligned} \tag{1.4}$$

It is sufficient here to take F (respectively, G) lying in some fixed dense linear subspace of $\mathscr{H}_1$ (of $\mathscr{H}_2$).

By using s-numbers, one can define symmetric ideals of compact operators (cf. [GK]). We need only the following ideals.

For $0 < p < \infty$, $\mathfrak{S}_p = \mathfrak{S}_p(\mathscr{H}_1, \mathscr{H}_2)$ is the class of all compact operators $Q: \mathscr{H}_1 \to \mathscr{H}_2$ for which the following functional is finite:

$$(\|Q\|_p)^p := \sum_n s_n^p(Q) = p \int_0^\infty s^{p-1} \nu(s; Q)\, ds. \tag{1.5}$$

The functional $\|\cdot\|_p$ is a norm for $p \geqslant 1$ and a quasinorm for $p < 1$.

Next, $\Sigma_p = \Sigma_p(\mathscr{H}_1, \mathscr{H}_2)$, $0 < p < \infty$, is the class of compact operators Q satisfying $s_n(Q) = O(n^{-1/p})$ or, which is the same, $\nu(s; Q) = O(s^{-p})$ as $s \to 0$. The class Σ_p is endowed with the quasinorm

$$|Q|_p = \left(\sup s^p \nu(s; Q)\right)^{1/p} = \sup_{n>0} n^{1/p} s_n(Q). \tag{1.6}$$

In contrast to $\mathfrak{S}_p$, the classes Σ_p are not separable. The subclasses

$$\Sigma_p^0 = \{ Q \in \Sigma_p : s_n(Q) = o(n^{-1/p}) \}$$

are separable. It is easy to see that

$$\mathfrak{S}_p \subset \Sigma_p^0, \qquad |Q|_p \leqslant \|Q\|_p, \qquad 0 < p < \infty. \tag{1.7}$$

2. Let a be a nonnegative quadratic form on $\mathscr{H}$ with $d[a]$ dense in $\mathscr{H}$. Suppose that a is *closed,* i.e., $d[a]$ is complete with respect to the metric form $a[u] + \gamma\|u\|^2$, $\gamma > 0$. Assume that a form b defined on $d[a]$ is bounded on the space $H[a]$:

$$|b[u]| \leqslant M(a[u] + \|u\|^2), \qquad u \in d[a]. \tag{1.8}$$

The form b generates a bounded linear operator (denoted $B_\gamma = B_\gamma(a, b)$) on $H_\gamma[a]$, $\forall\gamma > 0$. It is convenient to fix a dense linear subspace $\mathscr{F}$ in $H[a]$; when verifying (1.8), we can restrict ourselves to $u \in \mathscr{F}$. The equality $v = B_\gamma u$ means that

$$b[u, w] = a[v, w] + \gamma(v, w), \qquad \forall w \in \mathscr{F}.$$

The operator B is selfadjoint if and only if the form $b[u]$ is real-valued. Supposing that b is real-valued, let us consider the family of quadratic forms

$$a_\alpha[u] := a[u] - \alpha b[u], \qquad u \in d[a], \qquad \alpha > 0. \tag{1.9}$$

If for some $\alpha > 0$ the form given by (1.9) is bounded from below and closed on $d[a]$ (this holds automatically for $M\alpha < 1$), then it generates a semibounded selfadjoint operator A_α on $\mathscr{H}$. The negative spectrum of this operator will be the main object of our investigation. Let us formulate without proofs some elementary statements that we shall need (cf. [B2] and also [BS10]).

PROPOSITION 1.3. *Suppose that for some $\alpha_0 > 0$ the form a_{α_0} is bounded from below. Then a_α is semibounded and closed on $d[a]$ for every $\alpha < \alpha_0$.*

Let us now assume that b is represented as

$$b[u] = b^+[u] - b^-[u], \qquad u \in d[a], \tag{1.10}$$

where $b^\pm$ are nonnegative forms satisfying (1.8) with constants $M_\pm$ in place of M. Such representations always exist, and we suppose that one of them is fixed. We denote by $B^\pm$ the operators on $H[a]$ generated by the forms $b^\pm$ from (1.10).

PROPOSITION 1.4. *The form given by (1.9) is bounded from below and closed for every $\alpha > 0$ if (and in case $b = b^+$ also only if) the form b^+ has zero bound with respect to a:*

$$b^\pm[u] \leqslant \varepsilon a[u] + c(\varepsilon)\|u\|^2, \qquad u \in d[a], \quad \forall\varepsilon > 0. \tag{1.11}$$

It will be useful to supplement Proposition 1.4 by the following statement.

PROPOSITION 1.5. *If the operator B^+ is compact, then b^+ satisfies (1.11).*

3. Suppose now that $a[u] > 0$ for $u \neq 0$, and the form b is subject to the condition

$$|b[u]| \leqslant M_0 a[u], \qquad u \in d[a] \tag{1.12}$$

(this is stronger than (1.8)). To check this condition, we can restrict ourselves, as in the case of (1.8), to $u \in \mathscr{F}$. Let us consider the Hilbert space $\mathscr{H}[a]$—the completion of $d[a]$ (or, which is the same, of $\mathscr{F}$) in the metric generated by the form $a[u]$. If the form a is positive definite (and only in this case), we can interpret $\mathscr{H}[a]$ as $H_0[a]$. Conditions (1.8) and (1.12) are then equivalent. In general $\mathscr{H}[a]$ is not contained in $\mathscr{H}$.

Let $\hat{a}$ be the extension of a to $\mathscr{H}[a]$ by continuity. It follows from (1.12) that b also extends by continuity to $\mathscr{H}[a]$. The extended form $\hat{b}$ defines a bounded selfadjoint operator $\mathscr{B}$ on $\mathscr{H}[a]$. By Proposition 1.2, for this operator and every $\lambda > 0$ we have:

$$n_+(\lambda; \mathscr{B}) = \sup_{F \subset \mathscr{F}} \{ \dim F : b[u] > \lambda a[u],\ 0 \neq u \in F \}. \tag{1.13}$$

On the other hand, if for some $\alpha > 0$ the form a_α is semibounded and closed on $\mathscr{H}$, then, by (1.1)

$$N(0; A_\alpha) = \sup_{F \subset \mathscr{F}} \{ \dim F : a[u] < \alpha b[u],\ 0 \neq u \in F \}. \tag{1.14}$$

Comparing (1.13) and (1.14), we obtain the following result.

THEOREM 1.1. *Suppose that b satisfies (1.12) and the form defined in (1.9) is semibounded and closed on $d[a]$ for some $\alpha > 0$. Then*

$$N(0; A_\alpha) = n_+(\alpha^{-1}; \mathscr{B}). \tag{1.15}$$

Equation (1.15) allows one to deduce estimates for the number of eigenvalues of a semibounded selfadjoint operator A_α on $\mathscr{H}$ from estimates of the distribution function $n_+(\lambda; \mathscr{B})$ for some bounded operator $\mathscr{B}$ on another Hilbert space $\mathscr{H}[a]$. Technically this is quite convenient. The case when the form a_α is semibounded for all $\alpha > 0$ is of particular interest. Conditions sufficient for this are given by Propositions 1.4 and 1.5.

Replacing the form $a[u]$ by the form $a[u] + \gamma\|u\|^2$, $\gamma > 0$, in Theorem 1.1, we obtain the following statement.

THEOREM 1.2. *Suppose that b satisfies (1.8) and the form defined in (1.9) is semibounded and closed on $d[a]$ for some $\alpha > 0$. Then*

$$N(-\gamma; A_\alpha) = n_+(\alpha^{-1}; B_\gamma), \qquad \gamma > 0. \tag{1.16}$$

REMARK. Among other things, (1.15) and (1.16) imply that the function $N(-\gamma; A_\alpha)$, $\gamma > 0$, is nondecreasing in α.

§2. Function spaces

1. Let $(\mathscr{X}, \mu)$ be a measure space. Along with the function classes $L_p(\mathscr{X}, \mu)$, $0 < p \leqslant \infty$ (with the (quasi)norms $\|\cdot\|_p := \|\cdot\|_{L_p(\mathscr{X}, \mu)}$), we need "weak L_p-classes" $L_{p,w}(\mathscr{X}, \mu)$. Let us recall their definition. Let u be an arbitrary measurable and a.e. finite function on $\mathscr{X}$. We assign to it the distribution function

$$\mu_u(t) := \mu\{ x \in \mathscr{X} : |u(x)| > t \}, \qquad t > 0. \tag{2.1}$$

The class $L_{p,w} = L_{p,w}(\mathscr{X}, \mu)$, $0 < p < \infty$, consists of the functions u satisfying

$$|u|_p = |u|_{L_{p,w}(\mathscr{X}, \mu)} := \sup_{t>0} \left(t(\mu_u(t))^{1/p} \right). \tag{2.2}$$

The linear space $L_{p,w}$ is complete with respect to the quasinorm given by (2.2). If (and only if) $p > 1$, there exists an equivalent norm on $L_{p,w}$. The class $L_{p,w}$ is nonseparable. It has a distinguished separable subspace

$$L^0_{p,w}(\mathscr{X}, \mu) := \{ u \in L_{p,w}(\mathscr{X}, \mu) : \mu_u(t) = o(t^{-p}),\ t \to 0,\ t \to \infty \}.$$

Note that $L_p \subset L^0_{p,w}$, $|u|_p \leqslant \|u\|_p$.

As a particular case of the classes $L_{p,w}$, let us mention the "weak l_p-classes". Let K be a countable set and μ be the "counting measure" on K, i.e., $\mu(e) = \operatorname{card} e$, $\forall e \subset K$. Then $l_{p,w}(K) := L_{p,w}(K, \mu)$. The definition given by (2.2) takes the following form: a sequence $\tau = \{\tau_k\}_{k\in K}$ belongs to $l_{p,w}(K)$ if and only if

$$|\tau|_p = |\tau|_{l_{p,w}(K)} := \sup_{t>0} \left(t(\operatorname{card}\{ k \in K : |\tau_k| > t \})^{1/p}\right) < \infty.$$

If $K = \mathbb{N}$, the symbol K will be omitted in the notation.

Another special case of the spaces L_p, $L_{p,w}$ is represented by weighted classes on sets $X \subset \mathbb{R}^d$. Any measurable function $\psi \geqslant 0$ on X defines a measure $\mu^\psi(e) = \int_e \psi\, dx$. Let $L_p(X, \psi) := L_p(X, \mu^\psi)$, $L_{p,w}(X, \psi) := L_{p,w}(X, \mu^\psi)$. We omit the symbol X in the notation if $X = \mathbb{R}^d$; the same will be done with ψ if $\psi \equiv 1$. Clearly

$$|u|_{L_{p,w}(X,\psi)} = \sup_{t>0} \left\{ t \left(\int_{|u(x)|>t} \psi\, dx \right)^{1/p} \right\}.$$

Let us give some examples of functions in $L_{p,w}(\mathbb{R}^d)$.

EXAMPLE 2.1. Let $f_r(x) = |x|^{-r}$, $r > 0$. It follows directly from (2.2) that, for $p = d/r$, we have $f_r \in L_{p,w}(\mathbb{R}^d)$ and $|f|_r = v_d^{1/r}$.

EXAMPLE 2.2. Let $f(x) = \Phi(x/|x|)|x|^{-r}$, $r > 0$, where $\Phi \in L_p(\mathbb{S}^{d-1})$, $p = d/r$. Then $f \in L_{p,w}$ and $|f|_p = d^{-1/p}\|\Phi\|_p$.

EXAMPLE 2.3. The function $f(x) = \left(\sum_1^d |x_k|^{t_k}\right)^{-1}$, $t_k > 0$, belongs to $L_{p,w}$ for $p = \sum_1^d t_k^{-1}$.

We shall also need some function classes defined by using decompositions of $\mathbb{R}^d$ into cubes. Let $Q = [0, 1)^d$ and $Q_k = k + Q$, $k \in \mathbb{Z}^d$. To any function $u \in L_{r,\mathrm{loc}}(\mathbb{R}^d)$, we assign the function $\boldsymbol{\tau}(u, r) = \{\tau_k(u, r)\}$, $k \in \mathbb{Z}^d$, defined by $\tau_k(u, r) = \|u\|_{L_r(Q_k)}$ (here $0 < r < \infty$). The functions u satisfying $\boldsymbol{\tau}(u, r) \in l_q(\mathbb{Z}^d)$ constitute a linear space denoted by $l_q\{\mathbb{Z}^d; L_r(Q)\}$. The quasinorm in this space is given by $\|u\|_{l_q\{\mathbb{Z}^d; L_r(Q)\}} := \|\boldsymbol{\tau}(u, r)\|_{l_q(\mathbb{Z}^d)}$ (if $\min(r, q) \geqslant 1$, then this is a norm). One can define classes $l_{q,w}\{\mathbb{Z}^d; L_r(Q)\}$, etc., in a similar way.

Of course, it is possible to use other systems of sets instead of cubes. In §6, for example, some spaces connected with the decomposition of $\mathbb{R}^d$ into spherical shells will be introduced.

2. Below we systematically use several (known) versions of the Hardy inequality for functions on $\mathbb{R}^d$. Throughout $\int := \int_{\mathbb{R}^d}$. In what follows l (the index of smoothness) is positive but not necessarily an integer and $\varkappa := d/2l$. For $u \in C_0^\infty(\mathbb{R}^d)$ we set, by definition,

$$\int |\nabla_l u|^2 \, dx = \int |\xi|^{2l} |\hat{u}(\xi)|^2 \, d\xi, \tag{2.3}$$

where $\hat{u}$ is the Fourier transform of u.

PROPOSITION 2.1. *Let $2l < d$. Then for every function $\psi \in L_{\varkappa, w}(\mathbb{R}^d)$, $\psi \geqslant 0$ the following inequality holds*:

$$\int \psi |u|^2 \, dx \leqslant C |\psi|_\varkappa \int |\nabla_l u|^2 \, dx, \qquad u \in C_0^\infty(\mathbb{R}^d). \tag{2.4}$$

The classical Hardy inequality is a particular case of (2.4) (compare with Example 2.1):

$$\int |x|^{-2l} |u|^2 \, dx \leqslant c(d, l) \int |\nabla_l u|^2 \, dx, \qquad u \in C_0^\infty(\mathbb{R}^d), \qquad 2l < d. \tag{2.5}$$

If $2l \geqslant d$, then inequalities like (2.4) cannot hold for all $u \in C_0^\infty(\mathbb{R}^d)$, whatever be the function $\psi \not\equiv 0$. The situation is different for functions u vanishing in a neighborhood of a fixed point (e.g., the point $x = 0$). In what follows $\mathring{\mathbb{R}}^d := \mathbb{R}^d \setminus \{0\}$.

PROPOSITION 2.2. *Let $2l > d$, $l - d/2 \notin \mathbb{N}$. Then*

$$\int |x|^{-2l} |u|^2 \, dx \leqslant c(d, l) \int |\nabla_l u|^2 \, dx, \qquad u \in C_0^\infty(\mathring{\mathbb{R}}^d), \qquad 2l > d, \qquad l - d/2 \notin \mathbb{N}. \tag{2.6}$$

For $l - d/2 = 0, 1, \ldots$ certain variants of Hardy type inequalities are also valid. But they are not used in this article.

3. We shall also need another analytic object, namely, integral operators on $L_2(\mathbb{R}^d)$ of the form

$$(E_{ab} f)(x) = b(x) \int a(\xi) e^{i\xi x} f(\xi) \, d\xi. \tag{2.7}$$

Here a, b are measurable and a.e. finite functions on $\mathbb{R}^d$. Let us formulate two results. The first of them gives a boundedness condition for the operator defined by (2.7) and the second provides qualified estimates for its s-numbers.

PROPOSITION 2.3. *Let $a, b \in L_{p,w}$, $2 < p < \infty$. Then the operator defined by* (2.7) *is bounded and*

$$\|E_{a,b}\| \leqslant C |a|_p |b|_p, \qquad C = C(d, p). \tag{2.8}$$

Cf. [BKS] for the proof.

Estimate (2.8) can be considered as a very general form of the Hardy inequality. For example, choosing in it $a(\xi) = |\xi|^{-l}$, $b(x) = \psi^{1/2}(x)$ one can derive (2.5) with the constant $C_{2.5} = (2\pi)^{-d} v_d^{1/\varkappa} C_{2.8}^2(d, 2\varkappa)$.

PROPOSITION 2.4. *Let* $a \in L_{p,w}$, $b \in L_p$, $2 < p < \infty$. *Then* $E_{a,b} \in \Sigma_p$ *and*

$$|E_{a,b}|_p \leqslant C|a|_p \|b\|_p, \qquad C = C(d, p). \tag{2.9}$$

If $\mu_a(t) = o(t^{-p})$ as $t \to 0$, then $E_{a,b} \in \Sigma_p^0$.

Estimate (2.9) was formulated as a conjecture by Simon [S] and proved by Cwikel [C]. A somewhat different proof can be found in [BS4]. The above statement about the classes Σ_p^0 was observed in [BS8]. In [BS8, BS9, BKS] the results of Propositions 2.3, 2.4 have been generalized to wider classes of operators.

Proposition 2.4 can be supplemented by the following simple and well known result (cf., e.g., [S, BKS]).

PROPOSITION 2.5. *Let* $a, b \in L_p$, $2 \leqslant p < \infty$. *Then* $E_{a,b} \in \mathfrak{S}_p$ *and*

$$\|E_{a,b}\|_p \leqslant (2\pi)^{-(d/2)(1-2p^{-1})} \|a\|_p \|b\|_p.$$

4. In §§5 and 6 we shall need the Sobolev classes $H^l(X)$ on domains $X \subset \mathbb{R}^d$. Let us recall their description in terms of generalized derivatives (cf., e.g., [N, T]). Let the functional $I_l[u; X]$, $l > 0$, be defined as follows:

$$I_l[u; X] = \int_X \sum_{|\alpha|=l} \frac{l!}{\alpha!} |D^\alpha u|^2 \, dx, \qquad l \in \mathbb{N}; \tag{2.10}$$

$$I_l[u; X] = \int_X \int_X \frac{[l]!}{\alpha!} \frac{|D^\alpha u(x) - D^\alpha u(y)|^2}{|x-y|^{d+2(l-[l])}} \, dx dy, \quad l \notin \mathbb{N}. \tag{2.11}$$

Observe the following properties of the functionals given by (2.10), (2.11).

(1) *Superadditivity:* if domains X_ν constitute a decomposition of $\mathbb{R}^d$, then for $u \in C_0^\infty(\mathbb{R}^d)$ we have

$$\sum_\nu I_l[u; X_\nu] \leqslant I_l[u; \mathbb{R}^d]. \tag{2.12}$$

Here we get the equality when $l \in \mathbb{N}$.

(2) If $X = \mathbb{R}^d$, then there is an equivalent description in terms of the Fourier transform:

$$c \int |\xi|^{2l} |\hat{u}(\xi)|^2 \, d\xi \leqslant I_l[u; \mathbb{R}^d] \leqslant C \int |\xi|^{2l} |\hat{u}(\xi)|^2 \, d\xi. \tag{2.13}$$

In (2.13) $c = C = 1$ if $l \in \mathbb{N}$.

Now we recall the standard definition of the norm in $H^l(X)$:

$$(\|u\|_{H^l(X)})^2 = I_l[u; X] + \|u\|_{L_2(X)}^2. \tag{2.14}$$

For $X = \mathbb{R}^d$, $l \notin \mathbb{N}$, we shall also need the functional $I_{l,\varepsilon}[u]$ depending on the parameter $\varepsilon > 0$:

$$I_{l,\varepsilon}[u] = \int_{x,y\in\mathbb{R}^d : |x-y|<\varepsilon} \sum_{|\alpha|=[l]} \frac{|D^\alpha u(x) - D^\alpha u(y)|^2}{|x-y|^{d+2(l-[l])}} \, dxdy. \tag{2.15}$$

The following "equivalence inequality" holds (cf., e.g., [N, §2.5.1]):

$$\|u\|^2_{H^l(\mathbb{R}^d)} \leqslant C(\varepsilon)\left(I_{l,\varepsilon}[u] + \int |u|^2 \, dx\right). \tag{2.16}$$

§3. Formulation of the main problems

Our purpose is to investigate the function $N(\lambda; A_l(\alpha))$ for the operator

$$A_l(\alpha) = A_l(X, \varphi, \alpha V) = (-\Delta)^l + \varphi - \alpha V \tag{3.1}$$

in domains $X \subseteq \mathbb{R}^d$. We take $L_2(X)$ as the main Hilbert space; accordingly, $\|\cdot\| = \|\cdot\|_{L_2(X)}$. If $X \neq \mathbb{R}^d$, then Dirichlet conditions on the boundary ∂X are imposed. Introduction of the additional potential φ is useful to analyze the problem with $\varphi = 0$ as well.

Below we simplify the notation as follows: if $X = \mathbb{R}^d$ we omit X, if $\varphi \equiv 0$ we omit φ, and if $V \equiv 0$ we omit αV.

1. Suppose that a function $\varphi \in L_{1,\mathrm{loc}}(X)$, $\varphi \geqslant 0$, is fixed. For the initial quadratic form we take

$$a_l(X, \varphi)[u] := \int_X (|\nabla_l u|^2 + \varphi|u|^2) \, dx \tag{3.2}$$

with the domain

$$d[a_l(X, \varphi)] = \left\{ u \in \mathring{H}^l(X) : \int_X \varphi |u|^2 \, dx < \infty \right\}.$$

As usual, $\mathring{H}^l(X)$ is the completion of $C_0^\infty(X)$ with respect to the metric given by the integral in (3.2) with $\varphi \equiv 1$. The integral $\int_X |\nabla_l u|^2 \, dx$ for $u \in C_0^\infty(X)$ is defined by (2.3) (u is meant to be extended by 0 to $\mathbb{R}^d$). Supplying $d[a_l(X, \varphi)]$ with the metric form $a_l(X, \varphi)[u] + \gamma\|u\|^2$, we obtain a complete Hilbert space denoted by $\mathring{H}^l_\gamma(X, \varphi)$. Note that

$$H_\gamma[a_l(X, \varphi)] = \mathring{H}^l_\gamma(X, \varphi).$$

Thus the form defined by (3.2) is nonnegative and closed. Clearly $C_0^\infty(X)$ is dense in $H_\gamma[a_l(X, \varphi)]$. The selfadjoint operator on $L_2(X)$ generated by the form (3.2) is taken, *by definition,* as the realization of the operator $A_l(X, \varphi) = (-\Delta)^l + \varphi$.

Our next goal is to describe the corresponding class $\mathscr{H}[a_l(X, \varphi)]$. This description depends on the set X and the function φ. If $\varphi \not\equiv 0$, then

$\mathcal{H}[a_l(X, \varphi)]$ is embedded into the weighted space $L_2(X, \varphi)$ and, consequently, is *a fortiori* a space of (equivalence classes of) functions on X. In particular, if $\varphi \geqslant c > 0$, then $\mathcal{H}[a_l(X, \varphi)]$ coincides with $H_0[a_l(X, \varphi)]$. If $\varphi \equiv 0$, then $\mathcal{H}[a_l(X)]$ is not necessarily a space of functions: this depends on the fact of whether or not Hardy type inequalities hold for $C_0^\infty(X)$. For $\varphi \equiv 0$, we restrict ourselves to the case when $\mathcal{H}[a_l(X)]$ can be described as a function space. Note that it then turns out that for an arbitrary $\varphi \not\equiv 0$

$$\mathcal{H}[a_l(X, \varphi)] = \mathcal{H}[a_l(X)] \cap L_2(X, \varphi).$$

Let us single out two principal cases when $\mathcal{H}[a_l(X)]$ is a function space.

PROPOSITION 3.1. *Let $2l < d$. Then the class $\mathcal{H}[a_l]$ coincides (algebraically and topologically) with the homogeneous Sobolev space*

$$\mathcal{H}^l(\mathbb{R}^d) := \left\{ u \in H^l_{\mathrm{loc}}(\mathbb{R}^d) : \int (|\nabla_l u|^2 + |x|^{-2l}|u|^2)\, dx < \infty \right\}, \quad 2l < d. \tag{3.3}$$

PROPOSITION 3.2. *Let $2l > d$, $l - d/2 \notin \mathbb{N}$. Then the class $\mathcal{H}[a_l(\dot{\mathbb{R}}^d)]$ coincides (algebraically and topologically) with the space*

$$\begin{aligned}\mathcal{H}^l(\dot{\mathbb{R}}^d) := \Big\{ u \in H^l_{\mathrm{loc}}(\dot{\mathbb{R}}^d) : \int (|\nabla_l u|^2 + |x|^{-2l}|u|^2)\, dx < \infty, \\ (D^\alpha u)(0) = 0,\ |\alpha| < l - d/2 \Big\}, \\ 2l > d, \quad l - d/2 \notin \mathbb{N}.\end{aligned} \tag{3.4}$$

Both statements follow directly from (2.5) and (2.6).

Note that for $2l \geqslant d$ the class $\mathcal{H}[a_l]$ cannot be realized as a space of functions on $\mathbb{R}^d$.

2. For a function $V \in L_{1,\mathrm{loc}}(X)$ set

$$b[u] = b_V[u] = \int_X V|u|^2\, dx, \qquad u \in L_2(X; |V|). \tag{3.5}$$

Conditions (1.8), (1.11) for the form defined by (3.5) turn into

$$\left|\int_X V|u|^2\, dx\right| \leqslant M \int_X (|\nabla_l u|^2 + (\varphi + 1)|u|^2)\, dx; \tag{3.6}$$

$$\left|\int_X V|u|^2\, dx\right| \leqslant M_0 \int_X (|\nabla_l u|^2 + \varphi|u|^2)\, dx. \tag{3.7}$$

When checking these estimates, it is sufficient to take $u \in C_0^\infty(X)$.

If (3.6) is valid, then the form b_V generates a bounded operator on $H_\gamma[a_l(\varphi, V)]$; let us denote it by $B_{l,\gamma}(X, \varphi, V)$. If (3.7) holds, then b_V generates an operator $\mathcal{B}_l(X, \varphi, V)$ on the space $\mathcal{H}[a_l(X, \varphi)]$ as well. Clearly

$$B_{l,\gamma}(X, \varphi, V) = \mathcal{B}_l(X, \varphi + \gamma, V).$$

We omit φ in the notation if $\varphi \equiv 0$.

For the operators $\mathscr{B}_l$, $B_{l,\gamma}$ to be well defined, V must not necessarily be real-valued. These operators are selfadjoint if and only if $V = \overline{V}$; for $V \geqslant 0$ they are positive. To decompose the form defined by (3.5) as in (1.10), we always take

$$b^{\pm}[u] = \int V_{\pm}|u|^2\,dx, \qquad 2V_{\pm} = |V| \pm V.$$

3. Suppose that $V = \overline{V}$ and (3.7) holds. On $d[a_l(X, \varphi)]$ consider the quadratic form

$$a_l(X, \varphi, \alpha V)[u] = \int_X (|\nabla_l u|^2 + \varphi|u|^2 - \alpha V|u|^2)\,dx. \tag{3.8}$$

If for some $\alpha > 0$ this form is semibounded and closed, then it defines a selfadjoint operator $A_l(X, \varphi, \alpha V)$ on $L_2(X)$. It is natural to consider this operator as a realization of the differential operator (3.1). Equation (1.15) turns into

$$N(0; A_l(X, \varphi, \alpha V)) = n_+(\alpha^{-1}; \mathscr{B}_l(X, \varphi, V)). \tag{3.9}$$

If the operator $B_{l,1}$ is compact (*a fortiori,* if $\mathscr{B}_l(X, \varphi, V)$ is compact) then the form defined by (3.8) is semibounded and closed for every $\alpha > 0$. Hence (3.9) is also true for all $\alpha > 0$. Note that for $V \geqslant 0$ compactness of $\mathscr{B}_l(X, \varphi, V)$ is equivalent to compactness of the embedding of the space $\mathscr{H}[a_l(X, \varphi)]$ into the weighted space $L_2(X, V)$.

If instead of (3.7) the weaker condition (3.6) is fulfilled and, moreover, the form given by (3.8) is semibounded and closed on $d[a_l(X, \varphi)]$, then instead of (3.9) we have the following equality, implied by (1.16):

$$N(-\gamma; A_l(X, \varphi, \alpha V)) = n_+(\alpha^{-1}; B_{l,\gamma}(X, \varphi, \alpha V)), \qquad \gamma > 0. \tag{3.10}$$

If the operator $B_{l,\gamma}(\cdot)$ is compact (this property does not depend on $\gamma > 0$) then (3.10) holds for all $\alpha, \gamma > 0$.

4. In view of equations (3.9), (3.10), the question about the estimates for $N(-\gamma; A_l(X, \varphi, \alpha V))$, $\gamma \geqslant 0$, can be reduced to the problem of estimating the eigenvalues of the bounded operators $B_{l,\gamma}(\cdot)$, $\mathscr{B}_l(\cdot)$. If $\varphi \equiv 0$, $V \geqslant 0$, $X = \mathbb{R}^d$, $\gamma \geqslant 0$ (for $2l < d$) or $\gamma > 0$ (for $2l \geqslant d$), then a further reduction is possible, to estimates of singular numbers of the following integral operator on $L_2(\mathbb{R}^d)$:

$$S_{l,\gamma}(V)\colon f(\xi) \longmapsto (2\pi)^{-d/2} V^{1/2}(x) \int (|\xi|^{2l} + \gamma)^{-1/2} f(\xi) e^{i\xi x}\,d\xi. \tag{3.11}$$

PROPOSITION 3.3. *Suppose that for some V, γ the operator* (3.11) *is bounded on $L_2(\mathbb{R}^d)$. Then the corresponding operator $B_{l,\gamma}(V)$ is bounded on $H[a_l]$*

(if $\gamma > 0$), the operator $\mathscr{B}_l(V)$ is bounded on $\mathscr{H}[a_l]$ (if $\gamma = 0$), and the following properties hold:

$$N(-\gamma; A_l(\alpha V)) = n_+(\alpha^{-1}; B_{l,\gamma}(V)) = \nu(\alpha^{-1/2}; S_{l,\gamma}(V)), \qquad \gamma > 0; \tag{3.12}$$

$$N(0; A_l(\alpha V)) = n_+(\alpha^{-1}; \mathscr{B}_l(V)) = \nu(\alpha^{-1/2}; S_{l,0}(V)). \tag{3.13}$$

PROOF. Both statements (for $\gamma > 0$ and for $\gamma = 0$) are proved in the same way; we consider the case $\gamma = 0$ only. Let Φ, Φ^{-1} be the direct and inverse Fourier transformations for $\mathbb{R}^d$. The operator

$$U_l\colon f \longmapsto \Phi^{-1}_{\xi\mapsto x}(|\xi|^{-l} f(\xi))$$

is a unitary mapping of $L_2(\mathbb{R}^d)$ onto $\mathscr{H}^l(\mathbb{R}^d)$, $2l < d$. For $V \geqslant 0$, we have $b_V[u] = \|V^{1/2}u\|^2$. Thus the quadratic form of the operator $U_l^{-1}\mathscr{B}_l(V)U_l$ (acting on $L_2(\mathbb{R}^d)$) is $\|V^{1/2}U_l f\|^2$. Clearly $S_{l,0}(V)\colon f \mapsto V^{1/2}U_l f$. Thus $\mathscr{B}_l(V)$ is bounded if and only if $S_{l,0}(V)$ is. The equality

$$(\mathscr{B}_l(V)u, u)_{\mathscr{H}[a_l]} = \|S_{l,0}(V)f\|^2, \qquad f(\xi) = |\xi|^l \hat{u}(\xi),$$

shows that the eigenvalues of $\mathscr{B}_l(V)$ are equal to the squares of the s-numbers of $S_{l,0}(V)$. This implies (3.13). □

§4. The case $2l < d$: fundamental estimates and interpolation

We shall present two ways of obtaining estimates for $N(-\gamma; A_l(\alpha V))$. Both stem from the same basic results. One of them is a Hardy type inequality and the other gives a sharp condition guaranteeing regular order of growth of $N(\cdot)$ as $\alpha \to \infty$. Both ways lead, for $2l < d$, to estimates involving some function as a parameter. In all other respects they are quite different. In this section we use the relationships with the operators given by (2.7) and then interpolate. Another way (cf. §7, Subsection 3) is based on an artificial incorporation of a function-parameter into the operator, followed by an application of fundamental estimates.

1. An analytic description of the classes $H[a_l]$, $\mathscr{H}[a_l]$ was given in §3. Namely, $H[a_l] = H^l(\mathbb{R}^d)$ (the Sobolev space), $\mathscr{H}[a_l] = \mathscr{H}^l(\mathbb{R}^d)$ (the homogeneous Sobolev space, see (3.3)). It is convenient to describe in axiomatic form the potentials V for which (3.6) or (3.7) hold with $\varphi \equiv 0$.

Let us introduce the classes of "Hardy weights" $\mathscr{W}(d, l)$, $W(d, l)$. Suppose that $\psi \in L_{1,\mathrm{loc}}(\mathbb{R}^d)$, $\psi \geqslant 0$. Then $\psi \in \mathscr{W}(d, l)$ if

$$\int \psi|u|^2\,dx \leqslant M_0(\psi)\int |\nabla_l u|^2\,dx, \qquad u \in C_0^\infty(\mathbb{R}^d), \tag{4.1}$$

and $\psi \in W(d, l)$ if

$$\int \psi |u|^2 dx \leqslant M(\psi) \int (|\nabla_l u|^2 + |u|^2) dx, \qquad u \in C_0^\infty(\mathbb{R}^d). \tag{4.2}$$

It is assumed in the sequel that $M_0(\psi)$, $M(\psi)$ are the best possible constants in (4.1), (4.2).

The next statement (formulated without proof) gives sufficient conditions for ψ to belong to $\mathscr{W}(d, l)$ or $W(d, l)$ in the case $2l < d$.

PROPOSITION 4.1. *Let $2l < d$ and $\psi \in L_{1,\mathrm{loc}}(\mathbb{R}^d)$, $\psi \geqslant 0$. Then*

(1) *$\psi \in L_{\varkappa, w}$ implies $\psi \in \mathscr{W}(d, l)$ and $M_0(\psi) \leqslant c_{2.4} |\psi|_\varkappa$;*

(2) *$\psi \in l_\infty(\mathbb{Z}^d; L_{\varkappa, w}(Q))$ implies $\psi \in W(d, l)$ and*

$$M(\psi) \leqslant c \sup_{k \in \mathbb{Z}^d} |\psi|_{L_{\varkappa, w}}(Q_k). \tag{4.3}$$

A complete description of the classes $\mathscr{W}(d, l)$, $W(d, l)$ in terms of capacity is known, see [MS]. Proposition 4.1 can be considered as a consequence of this description. See [BKS] for an elementary proof based on the "exact" form of Young's theorem on convolutions. Clearly statement (1) coincides with Proposition 2.1. If $2l \geqslant d$, then the class $\mathscr{W}(d, l)$ is trivial: it contains only the function $\psi \equiv 0$.

2. The fundamental spectrum estimate for $A_l(\alpha V)$ reads as follows.

PROPOSITION 4.2. *Let $2l < d$, $V = \overline{V} = L_{1,\mathrm{loc}}(\mathbb{R}^d)$, and $V_+ \in L_\varkappa$. Then*

$$\begin{gathered} N(-\gamma; A_l(\alpha V)) \leqslant C\alpha^\varkappa \int V_+^\varkappa dx, \\ \gamma \geqslant 0, \quad C = C(d, l) = (2\pi)^{-d\varkappa} v_d (C_{2.9}(d, 2\varkappa))^{2\varkappa}. \end{gathered} \tag{4.4}$$

PROOF. By (0.2) it is sufficient to consider the case $V \geqslant 0$. Then we can apply Proposition 3.3. The operator $S_{l,\gamma}(V)$ defined in (3.11) is of the form (2.7), and (see Example 2.1)

$$a(\xi) = (|\xi|^{2l} + \gamma)^{-1/2} \in L_{2\varkappa, w}, \qquad |a|_{2\varkappa} = v_d^{1/d}, \quad \forall \gamma \geqslant 0.$$

If $V \in L_\varkappa$, then the hypotheses of Proposition 2.4 are fulfilled, and so

$$|S_{l,\gamma}(V)|_{2\varkappa} \leqslant C_{2.9} (2\pi)^{-d/2} v_d^{l/d} \|V\|_\varkappa^{1/2}. \tag{4.5}$$

In other words (see (1.6)),

$$\nu(s; S_{l,\gamma}(V)) \leqslant C_{2.9}^{2\varkappa} (2\pi)^{-d\varkappa} v_d \|V\|_\varkappa^\varkappa s^{-2\varkappa}.$$

By (3.12) and (3.13) this is equivalent to (4.4). □

Estimate (4.4) was obtained (by another method) by Rozenblyum [R1] for $l \in \mathbb{N}$ and extended to $l \in \mathbb{R}_+$ by the authors [BS3]. The proof based on Cwikel's result (Proposition 2.4) now seems to be the simplest. It can be generalized to a certain class of nonelliptic operators, see (4.21) below.

For $l = 1$ yet another proof of (4.4) was suggested by Lieb [L1, L2]. It is based on functional integration and gives the best known value of $C_{4.4}$, at least for $d = 3$.

3. Considering inequalities (4.1), (4.2), and estimate (4.4) as starting points for interpolation, one can obtain various intermediate results. This leads to estimates of the quantities $\nu(s;\cdot)$ for the operators $\mathscr{B}_l(V)$ and $B_{l,\gamma}(V)$. To pass to $N(-\gamma; A_l(\alpha V))$, $\gamma \geqslant 0$, one can then simply use (3.12) and (3.13). In general, constants in the estimates for $\nu(s; B_{l,\gamma}(V))$ grow as $\gamma \to 0$. In all other respects these estimates are equivalent for all $\gamma > 0$, and we state them, as a rule, for $\gamma = 1$.

We begin with the case $\gamma = 0$.

THEOREM 4.1. *Let $2l < d$ and $\psi \in \mathscr{W}(d, l)$, $M_0(\psi) \leqslant 1$. Then the following estimates hold for $\varkappa < q < \infty$ (the constants in them do not depend on ψ):*

$$(1) \quad \|\mathscr{B}_l(V)\|_q \leqslant C\|V/\psi\|_{L_q(\psi^\varkappa)}, \qquad C = C(d, l, q), \tag{4.6}$$

$$(2) \quad |\mathscr{B}_l(V)|_q \leqslant C|V/\psi|_{L_{q,w}(\psi^\varkappa)}, \qquad C = C(d, l, q). \tag{4.7}$$

Moreover, if $V/\psi \in L^0_{q,w}(\psi^\varkappa)$, then $\mathscr{B}_l(V) \in \Sigma^0_q$.

PROOF. Consider the mapping

$$\Pi = \Pi_l(\psi)\colon V/\psi \longmapsto \mathscr{B}_l(V). \tag{4.8}$$

It follows from the definition of $\mathscr{W}(d, l)$ and from the evident inequality $\|\mathscr{B}_l(V)\| \leqslant \|\mathscr{B}_l(|V|)\|$ that

$$\Pi\colon L_\infty(\mathbb{R}^d) = L_\infty(\mathbb{R}^d; \psi^\varkappa) \longrightarrow \mathfrak{R}(\mathscr{H}^l(\mathbb{R}^d)), \qquad \|\Pi\| \leqslant 1. \tag{4.9}$$

Further, (4.5) with $\gamma = 0$ means, by (3.13), that for $V \geqslant 0$

$$|\mathscr{B}_l(V)|_\varkappa \leqslant C_{4.4}^{1/\varkappa}\|V\|_\varkappa = C_{4.4}^{1/\varkappa}\|V/\psi\|_{L_\varkappa(\psi^\varkappa)}. \tag{4.10}$$

This is true for complex V as well, because $|\mathscr{B}_l(V)|_\varkappa \leqslant |\mathscr{B}_l(|V|)|_\varkappa$. Estimate (4.10) means that

$$\Pi : L_\varkappa(\mathbb{R}^d, \psi^\varkappa) \longrightarrow \Sigma_\varkappa(\mathscr{H}^l(\mathbb{R}^d)), \qquad \|\Pi\| \leqslant C_{4.4}^{1/\varkappa}. \tag{4.11}$$

Now an arbitrary interpolation functor can be applied to (4.9) and (4.11) (cf. [BL, T]). In particular, using the Lions-Peetre functors $K_{\theta,q}$ and $K_{\theta,\infty}$ with $\theta = \varkappa/q$ one can get (4.6), (4.7). The statement about Σ^0_q follows from (1.7) and the fact that $L_q(\psi^\varkappa)$ is dense in $L^0_{q,w}(\psi^\varkappa)$. □

REMARK. (4.7) can be rewritten in the form

$$\sup_{s>0}(s^q \nu(s; \mathscr{B}_l(V))) \leqslant C_{4.7}^q \sup_{t>0}\left(t^q \int_{V>t\psi} \psi^\varkappa\, dx\right).$$

This can be supplemented by the inequality

$$\limsup_{s\to 0}(s^q\nu(s;\mathscr{B}_l(V)))\leqslant C_{4.7}^q \limsup_{t+t^{-1}\to\infty}\left(t^q\int_{V>t\psi}\psi^\varkappa\,dx\right). \tag{4.12}$$

Let us prove (4.12). Theorem 4.1 states that for $q>\varkappa$ the mapping $\Pi_\psi : L_{q,w}(\psi^\varkappa)\longrightarrow\Sigma_q$ is continuous and $\Pi_\psi : L^0_{q,w}(\psi^\varkappa)\longrightarrow\Sigma^0_q$. Consequently, the following quotient mapping is well defined:

$$\check{\Pi}_\psi : L_{q,w}(\psi^\varkappa)/L^0_{q,w}(\psi^\varkappa)\longrightarrow\Sigma_q/\Sigma^0_q.$$

For the corresponding quasinorms, we have $\|\check{\Pi}_\psi\|\leqslant\|\Pi_\psi\|$. It is well known (cf. [BS5, §11.6]) that the quasinorm in Σ_q/Σ_q^0 of the quotient class generated by an operator $T\in\Sigma_q$ is equal to $\limsup_{s\to 0}(s\nu^{1/q}(s;T))$. Similarly, the quasinorm in $L_{q,w}/L^0_{q,w}$ of the quotient class generated by a function $u\in L_{q,w}(\mathscr{X},\mu)$ equals

$$\max\left(\limsup_{t\to\infty}(t\mu_u^{1/q}(t));\limsup_{t\to 0}(t\mu_n^{1/q}(t))\right)=\limsup_{t+t^{-1}\to\infty}(t\mu_u^{1/q}(t)).$$

In view of these remarks, (4.7) implies (4.12).

It is clear that the use of other interpolation functors would widen the conclusion of Theorem 4.1. For example, an application of the functor $K_{\varkappa/q,r}$, $0<r<\infty$, gives estimates in terms of Lorentz spaces.

Under real interpolation, the constants in estimates are hard to trace, and for this reason we have not expressed explicitly the constants in (4.6), (4.7) in terms of the basic constant $C_{4.4}$. Another, more direct proof of these estimates to be given in §7, Subsection 3, will enable us to obtain such expressions.

Theorem 4.1 implies the main results about estimates of $N(0;A_l(\alpha V))$ for $2l<d$.

THEOREM 4.2. *Let $2l<d$ and $\psi\in\mathscr{W}(d,l)$, $M_0(\psi)\leqslant 1$. Suppose that $V=\overline{V}\in L_{1,\mathrm{loc}}(\mathbb{R}^d)$. Then for $\varkappa<q<\infty$*

(1) *$V_+/\psi\in L_q(\psi^\varkappa)$ implies*

$$\int\alpha^{-q-1}N(0;A_l(\alpha V))\,d\alpha\leqslant q^{-1}C_{4.6}^q\int V_+^q\psi^{\varkappa-q}\,dx; \tag{4.13}$$

(2) *$V_+/\psi\in L_{q,w}(\psi^\varkappa)$ implies*

$$\sup_{\alpha>0}(\alpha^{-q}N(0;A_l(\alpha V)))\leqslant C_{4.7}^q\sup\left(t^q\int_{V(x)>t\psi(x)}\psi^\varkappa\,dx\right); \tag{4.14}$$

$$\begin{aligned}&\limsup_{\alpha\to\infty}(\alpha^{-q}N(0;A_l(\alpha V)))\\&\leqslant C_{4.7}^q\limsup_{t+t^{-1}\to\infty}\left(t^q\int_{V(x)>t\psi(x)}\psi^\varkappa\,dx\right).\end{aligned} \tag{4.15}$$

PROOF. Inequality (0.2) allows one to consider the case $V \geqslant 0$ only. In view of the first equality in (3.13), all the assertions of Theorem 4.2 are equivalent, for $V \geqslant 0$, to the corresponding assertions of Theorem 4.1 and the Remark to it. When deriving (4.13), one uses (1.5). □

REMARKS. 1) Applying complex interpolation (cf. [T]) to (4.8), (4.9), we obtain the following estimate:

$$N(0;\, A_l(\alpha V)) \leqslant C_{4.4}\alpha^q \int V_+^q \psi^{\varkappa-q}\, dx, \qquad q > \varkappa. \tag{4.16}$$

This follows also from (1.7), (4.13) with a worse constant. Estimate (4.16) is not sharp in the sense that if the integral on the right is finite, then in fact $N(0;\, A_l(\alpha V)) = o(\alpha^q)$.

2) By Proposition 2.1, under the hypotheses of Theorems 4.1, 4.2, every nonnegative function of class $L_{\varkappa,w}$ can be taken for ψ after appropriate normalization. In particular, taking $\psi(x) = (C_{2.4}v_d^{1/\varkappa})^{-1}|x|^{-2l}$ in (4.16), we obtain

$$\begin{gathered} N(0;\, A_l(\alpha V)) \leqslant C\alpha^q \int V_+^q |x|^{2ql-d}\, dx, \\ C = C_{4.4}(C_{2.4}v_d^{1/\varkappa})^{q-\varkappa}, \quad q > \varkappa. \end{gathered} \tag{4.17}$$

For $l \in \mathbb{N}$, this was obtained in [EK2, EK3], and also in [BSR], for $l = 1$. In [BSR] a numerical value for $C_{4.17}$ was given, while the authors of [EK2, EK3] had no similar objectives.

3) Specifying $\psi(x) = (C_{2.4}v_d^{1/\varkappa})^{-1}|x|^{-2l}$ in (4.14) leads to

$$N(0;\, A_l(\alpha V)) \leqslant C\alpha^q \sup_{t>0}\left(t^q \int_{|x|^{2l}V(x)>t} |x|^{-d}\, dx\right), \qquad q > \varkappa. \tag{4.18}$$

This is a direct strengthening of (4.17) (if the constant is disregarded).

4) Estimates (4.17), (4.18) are but particular cases of estimates (4.13), (4.14) that contain the function ψ as a parameter. If V behaves "anisotropically" near ∞, the weights from Examples 2.2, 2.3 may turn out to be more convenient.

5) Estimate (4.14) (unlike (4.16)) gives a sharp order. We present an example in support of this statement.

EXAMPLE 4.1. Let $\psi \in L_{\varkappa,w}$, $\psi > 0$. Then it is easily seen that

$$|\ln \psi|^{-\delta} \in L_{q,w}(\psi^\varkappa), \qquad q\delta = 1.$$

For the potential $V = \psi|\ln\psi|^{-\delta}$, $\delta\varkappa < 1$, (4.14) implies $N(0;\, A_l(\alpha V)) = O(\alpha^q)$.

Let, in particular, $\psi(x) = |x|^{-2l}$ and $V(x) = |x|^{-2l}|\ln|x||^{-\delta}$. Then

$$\int_{V/\psi>t} \psi^\varkappa\, dx = 2dv_d t^{-q},$$

so that the above indication of the class $L_{q,w}(\psi^\varkappa)$ is sharp. On the other hand, for the potential in question it is not difficult to obtain (provided

$\delta\varkappa < 1$) an estimate of $N(0; A_l(\alpha V))$ from below having the same order $O(\alpha^q)$. See also §8, Subsection 3, for an explicit asymptotic formula for $N(0; A_1(\alpha V))$, as $\alpha \to \infty$.

4. We pass now to estimates for $N(-1; A_l(\alpha V))$. The same scheme as in Subsection 3 will be applied. The proofs can easily be reconstructed and we omit them. Let us start at once with an analog of Theorem 4.2.

THEOREM 4.3. *Let $2l < d$ and $\psi \in W(d, l)$, $M(\psi) \leqslant 1$. Suppose that $V = \overline{V} \in L_{1,\mathrm{loc}}(\mathbb{R}^d)$. Then for $\varkappa < q < \infty$ the statements of Theorem* 4.2 *hold with $N(0; A_l(\alpha V))$ replaced by $N(-1; A_l(\alpha V))$.*

Even the simplest choice $\psi \equiv 1$ leads to important results. The change of variable $x \mapsto \gamma^{-1/2l}x$ allows one to include the parameter γ into the estimates here.

COROLLARY 4.1. *Let $2l < d$ and $V = \overline{V} \in L_{1,\mathrm{loc}}(\mathbb{R}^d)$. Then for $\varkappa < q < \infty$ the following estimates hold*:

(1) $V_+ \in L_q$ *implies*

$$\alpha^{-q-1}N(-\gamma; A_l(\alpha V))d\alpha \leqslant q^{-1}C_{4.6}^q\gamma^{\varkappa-q}\int V_+^q\,dx; \tag{4.19}$$

(2) $V_+ \in L_{q,w}$ *implies*

$$\begin{aligned}&\sup_{\alpha>0}(\alpha^{-q}N(-\gamma; A_l(\alpha V)))\\ &\qquad\leqslant C_{4.7}^q\gamma^{\varkappa-q}\sup_{t>0}(t^q\,\mathrm{meas}\{x : V(x) > t\});\end{aligned} \tag{4.20}$$

$$\begin{aligned}&\limsup_{\alpha\to\infty}(\alpha^{-q}N(-\gamma; A_l(\alpha V)))\\ &\qquad\leqslant C_{4.7}^q\gamma^{\varkappa-q}\limsup_{t+t^{-1}\to\infty}(t^q\,\mathrm{meas}\{x : V(x) > t\}).\end{aligned} \tag{4.21}$$

In §8 we shall give explicit asymptotic formulas for $N(-\gamma; A_l(\alpha V))$ as $\alpha \to \infty$ for a certain class of potentials V (cf. Corollary 8.1). This will show that (4.20) is sharp.

5. Let us compare the results of Theorems 4.2, 4.3 with the regular estimate (4.4). The latter is uniform in $\gamma \geqslant 0$. The coefficient in the asymptotic formula (0.4) (confirming sharpness of (4.4)) does not contain γ. This means that, as α becomes large, the main part of the eigenvalues of $A_l(\alpha V)$ tends to be disposed to the left of any fixed point $-\gamma < 0$.

The situation in the nonregular case ($V_+ \notin L_\varkappa$) is different. This is transparent from estimates (4.19), (4.20) that are not uniform in γ. The order of growth of $N(-\gamma; A_l(\alpha V))$ as $\alpha \to \infty$ is in general smaller for $\gamma > 0$ than for $\gamma = 0$. It may happen that $N(-\gamma; A_l(\alpha V)) < \infty$ for $\gamma > 0$ and

$N(0;\, A_l(\alpha V)) = \infty$. This is the case, say, for the potentials $V(x) = |x|^{-r}$, $0 < r < 2l$. A more interesting example is provided by the potential

$$V(x) = \min\left(1\,,\, |x|^{-2l}(1 + |\ln|x||)^{-\delta}\right), \qquad \delta\varkappa < 1. \tag{4.22}$$

It follows from results of §7 (see Table (7.11), line 6) that here

$$N(-\gamma;\, A_l(\alpha V)) = O(\alpha^{\varkappa}(\ln\alpha)^{1-\varkappa\delta}) \quad \text{for } \gamma > 0\,.$$

On the other hand, the estimate $N(0;\, A_l(\alpha V)) = O(\alpha^{1/\delta})$ (implied by (4.14)) cannot be improved for the potential given by (4.22).

It is clear from formulas (4.15) and (4.21) that the asymptotic behavior of $N(-\gamma;\, A_l(\alpha V))$ as $\alpha \to \infty$ depends in fact on the quasinorm of the equivalence class generated by the potential V_+ in the quotient space $L_{q,w}(\psi^{\varkappa})/L^0_{q,w}(\psi^{\varkappa})$. In other words, it depends only on the behavior of V_+ near its principal singularities (on the behavior of V_+ near ∞, if $V_+ \in L_\infty$).

6. The results presented above can be generalized to a certain class of nonelliptic operators. Let $P(\xi)$ be a positive function on $\mathbb{R}^d$ with $1/P \in L_{\varkappa,w}$ for some $\varkappa > 1$. This is true, for example, for $P(\xi) = \sum_1^d |\xi_k|^{t_k}$ with $\varkappa = \sum_1^d t_k^{-1} > 1$ (cf. Example 2.3). Consider the quadratic form

$$a_P[u] = \int P(\xi)|\hat{u}(\xi)|^2\, d\xi$$

on its natural domain of definition $\{\, u \in L_2 : a_P[u] < \infty \,\}$.

Proposition 3.3 implies the following Hardy type inequality:

$$\int \psi|u|^2\, dx \leqslant C|\psi|_{\varkappa} a_P[u], \qquad u \in C_0^\infty(\mathbb{R}^d), \quad \psi > 0, \quad \psi \in L_{\varkappa,w};$$

$$C = (2\pi)^{-d} C_{2.8}^2(d\,,\, 2\varkappa)|1/P|_{\varkappa}. \tag{4.23}$$

In particular, we can take $\psi = 1/P$ in (4.23). It follows from (4.23) that $\mathscr{H}[a_P]$ can be realized as a function space embedded in $L_2(\mathbb{R}^d\,,\, \psi)$.

If $V \in L_{\varkappa,w}$, then, again by (4.23), the quadratic form $\int V|u|^2\, dx$ is bounded on $\mathscr{H}[a_P]$ and hence it generates a bounded operator $\mathscr{B}_P(V)$ on $\mathscr{H}[a_P]$. Now the form

$$a_P(\alpha V)[u] = a_P[u] - \alpha\int V|u|^2\, dx\,, \qquad u \in d[a_P],$$

is semibounded and closed on $L_2(\mathbb{R}^d)$, at least for sufficiently small $\alpha > 0$. The selfadjoint operator $A_P(\alpha V)$ generated by this form can be considered as a natural realization of $P(\mathscr{D}) - \alpha V$. In view of (1.15),

$$N(0;\, A_P(\alpha V)) = n_+(\alpha^{-1};\, \mathscr{B}_P(V)).$$

For $V \geqslant 0$ consider the integral operator

$$S_P(V)\colon f(\xi) \longmapsto (2\pi)^{-d/2} V^{1/2}(x)\int (P(\xi))^{-1/2} f(\xi) e^{i\xi x}\, d\xi.$$

In view of a result similar to Proposition 3.3,

$$n_+(s;\mathscr{B}_P(V)) = \nu(s^{1/2};S_P(V)), \qquad s>0.$$

Proposition 2.4 implies the following "regular" estimate:

$$N(0;A_P(\alpha V)) \leqslant C\alpha^{\varkappa}\int V_+^{\varkappa}\,dx. \tag{4.24}$$

The subsequent interpolation between (4.23) and (4.24) is carried out in the same way as in the proof of Theorem 4.1. The result that follows generalizes Theorem 4.2 (however, we restrict ourselves to $\psi\in L_{\varkappa,w}$ when presenting its statement).

THEOREM 4.4. *Let $P(\xi)$, $\xi\in\mathbb{R}^d$, be a positive function with $1/P\in L_{\varkappa,w}$, $\varkappa>1$. Let $\psi\in L_{\varkappa,w}$, $\psi>0$, $|\psi|_{\varkappa}\leqslant C_{4.23}^{-1}$. Let $V=\overline{V}\in L_{1,\mathrm{loc}}(\mathbb{R}^d)$. Then for $\varkappa<q<\infty$ both statements of Theorem 4.2 hold with $A_P(\alpha V)$ substituted for $A_l(\alpha V)$, the constants in the estimates depending on d, q, and P.*

Theorem 4.3 can be generalized in a similar way. Here, however, there are no effective conditions on ψ that could play the role of (4.3).

§5. The case $2l>d$. Spectrum estimates to the left of $\lambda=-1$

Unlike the case $2l<d$, for $2l\geqslant d$ there exist no estimates for $N(0;A_l(\alpha V))$ that are homogeneous in α (as (4.4)). Indeed, a homogeneous estimate would imply that $N(0;\cdot)<1$ for small α and thus $N(0;\cdot)=0$. But it is known that for nonnegative V with $V\not\equiv 0$ and $2l\geqslant d$ we always have $N(0;A_l(\alpha V))>0$, $\forall\alpha>0$. Using physical terminology, one can say that if $2l\geqslant d$, then the operator $(-\Delta)^l$ has resonance at the point $\lambda=0$.

For this reason the estimates of $N(-\gamma;A_l(\alpha V))$ for $\gamma>0$ and for $\gamma=0$ are of different character. In this section we consider the first case, taking $\gamma=1$ to be definite. The estimates we shall obtain do not contain any function-parameter. In fact this leads to no losses, for $2l>d$ at least: it turns out that for $V\geqslant 0$ the estimates in Theorem 5.1 are two-sided.

1. For $V\geqslant 0$ the next statement gives a boundedness criterion for the operator $B_{l,1}(V)$ in the case $2l>d$.

PROPOSITION 5.1. *Let $2l>d$ and $V\in L_{1,\mathrm{loc}}(\mathbb{R}^d)$, $V\geqslant 0$. The inequality*

$$\int V|u|^2\,dx\leqslant M(V)\int(|\nabla_l u|^2+|u|^2)\,dx, \qquad u\in H^l(\mathbb{R}^d), \tag{5.1}$$

holds if and only if $V\in l_\infty(\mathbb{Z}^d;L_1(Q))$. Supposing $M(V)$ to be the best possible constant in (5.1), we have

$$c\|V\|_{l_\infty(\mathbb{Z}^d;L_1(Q))}\leqslant M(V)\leqslant C\|V\|_{l_\infty(\mathbb{Z}^d;L_1(Q))}. \tag{5.2}$$

PROOF. Let $I_l[u; X]$ be the functional given by (2.10) (if $l \in \mathbb{N}$) or by (2.11) (if $l \notin \mathbb{N}$). For $2l > d$, we have

$$\max_{Q_k} |u(x)|^2 \leqslant C \left(I_l[u; Q_k] + \int_{Q_k} |u|^2 \, dx \right), \tag{5.3}$$

the constant being independent of $k \in \mathbb{Z}^d$. Summing over k and taking into account (2.12), (2.13), we obtain (5.1) with

$$M(V) = C_{2.13} C_{5.3} \sup_k \left(\int_{Q_k} V \, dx \right).$$

Conversely, fix a function $u_0 \in C_0^\infty(\mathbb{R}^d)$ such that $u_0 = 1$ on Q_0 and set $u_k(x) = u_0(x-k)$, $k \in \mathbb{Z}^d$. Denote $\sigma := \int (|\nabla_l u_0|^2 + |u_0|^2) \, dx$. Applying (5.1) to u_k, we find

$$\int_{Q_k} V \, dx \leqslant M(V) \int (|\nabla_l u_k|^2 + |u_k|^2) \, dx = \sigma M(V). \quad \square$$

The fundamental estimate for $N(-1; A_l(\alpha V))$ is essentially taken from [BB]. Here it is presented in a refined form. Below $\tau_k(V) := \tau_k(V, 1) = \int_{Q_k} |V| \, dx$.

PROPOSITION 5.2. *Let* $2l > d$ *and* $V \in L_{1,\mathrm{loc}}(\mathbb{R}^d)$. *Then*

$$\nu(s; B_{l,1}(V)) \leqslant C s^{-\varkappa} \sum_{k \in \mathbb{Z}^d \,:\, \tau_k(V) > cs} \tau_k^\varkappa(V). \tag{5.4}$$

PROOF. Let $B(V; X)$ be the operator on $H^l(X)$ generated by the form $\int_X V|u|^2 \, dx$ (the metric on $H^l(X)$ being given by (2.14)). In view of the variational principle and the superadditivity property (2.12), we get

$$\nu(c_1 s; B_{l,1}(V)) \leqslant \nu(s; B(V; \mathbb{R}^d)) \leqslant \sum_{k \in \mathbb{Z}^d} (s; B(V; Q_k)),$$

$$c_1 = \max(1, C_{2.13}).$$

The estimate $\nu(s; B(V; Q_k)) \leqslant c_2 s^{-\varkappa} (\int_{Q_k} |V| \, dx)^\varkappa$ for each particular cube has been obtained (in terms of approximation) in [BS1, Theorem 3.3], see also [BS3, Theorem 2.3]. Let us also take into account (in view of the embedding theorem) that $\|B(V; Q_k)\| \leqslant c_3 \int_{Q_k} |V| \, dx$, and $\nu(s; B) = 0$ for $s \geqslant \|B\|$. Collecting all these estimates, we get (5.4) with $C_{5.4} = c_1^\varkappa c_2$, $c_{5.4} = (c_1 c_3)^{-1}$. $\square$

Note that $c_1 = 1$ if $l \in \mathbb{N}$.

COROLLARY 5.1. *Let* $2l > d$ *and* $V \in l_\varkappa(\mathbb{Z}^d; L_1(Q))$. *Then* $B_{l,1}(V) \in \Sigma_\varkappa$ *and*

$$|B_{l,1}(V)|_\varkappa \leqslant C \|V\|_{l_\varkappa(\mathbb{Z}^d; L_1(Q))}, \qquad C = C_{5.3}^{1/\varkappa}. \tag{5.5}$$

This estimate is a coarsening of (5.4). For $l \in \mathbb{N}$ it was proved in [BB] (as well as the estimate (5.17) below in the case $2l = d$). (5.5) is sufficient for most applications. In §7, Subsection 5, we shall demonstrate some additional possibilities given by the sharper estimate (5.4).

2. Let us interpolate between the results of Proposition 5.1 and Corollary 5.1. In fact we interpolate the mapping

$$\Pi\colon V \longmapsto B_{l,1}(V). \tag{5.6}$$

There will be some technical differences comparative to §4. First , we have to interpolate some spaces of vector-valued sequences. Besides, since $\varkappa < 1$, we shall meet some nonnormed spaces. But general interpolation theory still contains the required results, cf. [BL, T].

THEOREM 5.1. *Let* $2l > d$, $V = \overline{V} \in L_{1,\mathrm{loc}}(\mathbb{R}^d)$. *Then for* $\varkappa < q < \infty$

(1) $V_+ \in l_q(\mathbb{Z}^d; L_1(Q))$ *implies*

$$\int_0^\infty \alpha^{-q-1} N(-1; A_l(\alpha V))\, d\alpha \leqslant C \sum_{k \in \mathbb{Z}^d} \left(\int_{Q_k} V_+\, dx \right)^q, \qquad C = C(d, l, q); \tag{5.7}$$

(2) $V_+ \in l_{q,w}(\mathbb{Z}^d; L_1(Q))$ *implies*

$$\sup_{\alpha>0} (\alpha^{-q} N(-1; A_l(\alpha V))) \leqslant C \sup_{t>0} \left(t^q \operatorname{card} \left\{ k \in \mathbb{Z}^d : \int_{Q_k} V_+\, dx > t \right\} \right), \qquad C = C(d, l, q), \tag{5.8}$$

$$\limsup_{\alpha \to \infty} (\alpha^{-q} A_l(\alpha V)) \leqslant C_{5.8} \limsup_{t \to 0} \left(t^q \operatorname{card} \left\{ k \in \mathbb{Z}^d : \int_{Q_k} V_+\, dx > t \right\} \right); \tag{5.9}$$

(3) *if* $V \geqslant 0$, *then the inequalities opposite to* (5.7)–(5.9) *are valid.*

PROOF. By Proposition 5.6, for the mapping defined in (5.6), we have

$$\Pi\colon l_\infty(\mathbb{Z}^d; L_1(Q)) \longrightarrow \mathfrak{R}(H^l(\mathbb{R}^d)), \qquad \|\Pi\| \leqslant C_{5.2}.$$

Estimate (5.5) means that

$$\Pi\colon l_\varkappa(\mathbb{Z}^d; L_1(Q)) \longrightarrow \Sigma_\varkappa(H^l(\mathbb{R}^d)), \qquad \|\Pi\| \leqslant C_{5.5}.$$

Applying the interpolation functors $K_{\theta,q}$, $K_{\theta,\infty}$, $\theta = \varkappa/q$, $\varkappa < q < \infty$, we obtain

$$\Pi\colon l_q(\mathbb{Z}^d; L_1(Q)) \longrightarrow \mathfrak{S}_q,$$
$$\Pi\colon l_{q,w}(\mathbb{Z}^d; L_1(Q)) \longrightarrow \Sigma_q.$$

These relations mean that

$$\|B_{l,1}(V)\|_q \leqslant c\|V\|_{l_q(\mathbb{Z}^d; L_1(Q))},$$
$$|B_{l,1}(V)|_q \leqslant c|V|_{l_{q,w}(\mathbb{Z}^d; L_1(Q))}.$$

This is equivalent to (5.7), (5.8) for $V \geqslant 0$ in view of (1.5) and (3.12). To pass to arbitrary V, one should use (0.2). As to (5.9), this can be proved on the basis of the fact that $\Pi : l^0_{q,w}(\mathbb{Z}^d; L_1(Q)) \longrightarrow \Sigma^0_q$, by an argument similar to that in the proof of Theorem 4.2.

To verify (3), we use the sequence $\{u_k\}$ constructed in the proof of Proposition 5.1. Assume that, in addition, $u_0(x) = 0$ for $\operatorname{dist}(x, Q_0) > 1/3$. Denote by $\tilde{H}$ the closure in $H^l(\mathbb{R}^d)$ of the linear hull of the functions $\{u_k\}$, $k \in (2\mathbb{Z})^d$. Let $V \in l_\infty(\mathbb{Z}^d, L_1(Q))$, $V \geqslant 0$, and let $B(V)$ be the operator on $\tilde{H}$ generated by the form $\int V|u|^2\,dx$. In view of the variational principle, $\lambda_j(\tilde{B}(V)) \leqslant \lambda_j(B_{l,1}(V))$, $\forall j$. In the formula $u = \sum c_k u_k$ (k runs over $(2\mathbb{Z})^d$), the summands on the right are disjointly supported, whence

$$\int V|u|^2\,dx \geqslant \sum_k |c_k|^2 \int_{Q_k} V\,dx.$$

We consider first the simpler case $l \in \mathbb{N}$. Then

$$\int (|\nabla_l u|^2 + |u|^2)\,dx = \sigma \sum_k |c_k|^2. \tag{5.10}$$

Hence $\lambda_j(\tilde{B}(V)) \geqslant \sigma^{-1} w_j$, where $\{w_j\}$, $j \in \mathbb{N}$, is the nonincreasing rearrangement of the sequence $\{\int_{Q_k} V\,dx\}$, $k \in (2\mathbb{Z})^d$. Thus

$$\sum_{k\in(2\mathbb{Z})^d} \left(\int_{Q_k} V\,dx\right) \leqslant \sigma^q \|\tilde{B}(V)\|_q^q \leqslant \sigma^q \|B_{l,1}(V)\|_q^q$$

and similarly for the quasinorms in $l_{q,w}$ and Σ_q. Doing the same for the "shifted" lattices and summing the results, for $B = B_{l,1}(V)$ we obtain

$$\|V\|_{l_q(\mathbb{Z}^d; L_1(Q))} \leqslant 2^{d/q}\|B\|_q,$$
$$|V|_{l_{q,w}(\mathbb{Z}^d; L_1(Q))} \leqslant 2^{d/q}|B|_q.$$

This proves the inequalities opposite to (5.7), (5.8) provided $l \in \mathbb{N}$. The reverse of (5.9) can be proved by the same method.

If $l \notin \mathbb{N}$, then (5.5) does not hold. But for $\varepsilon \leqslant 1/3$, for the functional defined in (2.15), we have

$$I_{l,\varepsilon}[u] + \|u\|^2 = \sigma(\varepsilon) \sum_k |c_k|^2, \qquad \sigma(\varepsilon) = I_{l,\varepsilon}[u_0] + \|u_0\|^2.$$

In view of (2.16) and (2.13), the functional $I_{l,\varepsilon}$ defines an equivalent norm on $H^l(\mathbb{R}^d)$. The rest does not differ from the case $l \in \mathbb{N}$. □

REMARKS. 1) Using the substitution $x \mapsto \gamma^{-1/2l} x$, we can derive estimates for $N(-\gamma; A_l(\alpha V))$ from (5.7)–(5.9) for all $\gamma > 0$. We shall not write them out explicitly; we only note that they involve (quasi)norms of the potential V connected with partitions of $\mathbb{R}^d$ into cubes with edge length $\gamma^{-1/2l}$.

2) Taking into account (1.5), (1.7), from (5.7) with $q = 1$, we get:

$$\begin{aligned} \alpha^{-1} N(-1; A_l(\alpha V)) &\leqslant \int_0^\infty \alpha^{-2} N(-1; A_l(\alpha V))\, dx \\ &\leqslant C_{5.7}(d, l, 1) \int V_+\, dx. \end{aligned} \tag{5.11}$$

This estimate is more transparent than the general inequalities (5.7), (5.8) because it involves no partition of $\mathbb{R}^d$ into cubes. There is also a direct way to prove (5.11) that, besides, gives the sharp constant. Indeed, let $V \geqslant 0$ and let $S_{l,\gamma}(V)$ be the operator defined by (3.11). By using (3.12) and (1.5), we get

$$\begin{gathered} \int \alpha^2 N(-1; A_l(\alpha V))\, dx = \|B_{l,1}(V)\|_1 = \|S_{l,1}(V)\|_2^2 = C \int V\, dx, \\ C = (2\pi)^{-d} \int (|\xi|^{2l} + 1)^{-1}\, d\xi. \end{gathered} \tag{5.12}$$

It is clear that $C_{5.7}$ can be replaced by $C_{5.12}$ in (5.11).

3) Interpolating between (5.11) and the evident inequality $\|B_{l,1}(V)\| \leqslant \|V\|_\infty$ leads to the following useful and simple estimates:

$$\int_0^\infty \alpha^{-q-1} N(-1; A_l(\alpha V))\, dx \leqslant c \int V_+^q\, dx, \qquad q \geqslant 1; \tag{5.13}$$

$$N(-1; A_l(\alpha V)) \leqslant c\alpha^q |V_+|_q^q, \qquad q > 1. \tag{5.14}$$

Of course, they can be derived directly from (5.7), (5.8). For example, (5.13) follows from (5.7) by the Hölder inequality.

4) Suppose that a potential $V > 0$ satisfies the following condition with constant $K > 0$:

$$V(x) \leqslant K V(y) \quad \text{if } |x - y| \leqslant 1. \tag{5.15}$$

Then, for every q, we have two-sided estimates

$$\begin{gathered} \|V\|_{l_q(\mathbb{Z}^d; L_1(Q))} \asymp \|V\|_q, \\ |V|_{l_{q,w}(\mathbb{Z}^d; L_1(Q))} \asymp |V|_q. \end{gathered}$$

Consequently, for the potentials satisfying (5.15), one can rewrite (5.7), (5.8) in the equivalent forms (5.13), (5.14) for all $q > \varkappa$. But the constants in (5.13) for $q < 1$ and in (5.14) for $q \leqslant 1$ depend on K.

Estimates (5.13) and (5.14) seemingly coincide with (4.19), (4.20), but have to do with another range of d, l.

3. Now we consider the border case $2l = d$. We have not succeeded in obtaining sharp results in this situation. So the statements presented below are rough.

PROPOSITION 5.3. *Let* $2l = d$, $V \geqslant 0$, *and* $V \in l_\infty(\mathbb{Z}^d; L_r(Q))$ *for some* $r > 1$. *Then*

$$\int V|u|^2\,dx \leqslant C\|V\|_{l_\infty(\mathbb{Z}^d;L_r(Q))}\int (|\nabla_l u|^2 + |u|^2)\,dx, \qquad u \in H^l(\mathbb{R}^d), \quad C = C(d, r). \tag{5.16}$$

PROOF. Let $r' = (1-r^{-1})^{-1}$. By the Hölder inequality and the embedding $H^l(Q_k) \to L_{2r'}(Q_k)$ (holding for $2l = d$), we have

$$\begin{aligned}\int_{Q_k} V|u|^2\,dx &\leqslant \left(\int_{Q_k} V^r\,dx\right)^{1/r}\left(\int_{Q_k}|u|^{2r'}\,dx\right)^{1/r'}\\ &\leqslant c\left(\int_{Q_k} V^r\,dx\right)^{1/r}\left(I_l[u;Q_k] + \int_{Q_k}|u|^2\,dx\right).\end{aligned}$$

The constant c does not depend on $k \in \mathbb{Z}^d$. Summing over k and taking into account (2.12), (2.13), we obtain (5.16). □

PROPOSITION 5.4. *Let* $2l = d$ *and* $V \in l_1(\mathbb{Z}^d; L_r(Q))$ *for some* $r > 1$. *Then* $B_{l,1}(V) \in \Sigma_1$ *and*

$$\nu(s; B_{l,1}(V)) \leqslant Cs^{-1}\|V\|_{l_1(\mathbb{Z}^d;L_r(Q))}, \qquad C = C(d, r). \tag{5.17}$$

THEOREM 5.2. *Let* $2l = d$, $V = \overline{V} \in L_{1,\mathrm{loc}}(\mathbb{R}^d)$, *and let us fix* $r > 1$. *Then for* $1 < q < \infty$:

(1) $V_+ \in l_q(\mathbb{Z}^d; L_r(Q))$ *implies*

$$\int_0^\infty \alpha^{-q-1}N(-1; A_l(\alpha V))\,dx \leqslant c(d, r, q)\sum_{k\in\mathbb{Z}^d}\left(\int_{Q_k} V_+^r\,dx\right)^{q/r}. \tag{5.18}$$

(2) $V_+ \in l_{q,w}(\mathbb{Z}^d; L_r(Q))$ *implies*

$$\begin{aligned}&\sup_{\alpha>0}(\alpha^{-q}N(-1; A_l(\alpha V)))\\ &\qquad\leqslant c(d, r, q)\sup_{t>0}\left(t^q\,\mathrm{card}\left\{k \in \mathbb{Z}^d : \int_{Q_k} V_+^r\,dx > t^r\right\}\right).\end{aligned} \tag{5.19}$$

This estimate still holds if one replaces "sup" *on the left by* $\limsup_{\alpha\to\infty}$ *and* "sup" *on the right by* $\limsup_{t\to 0}$.

We omit the proofs of Proposition 5.4 and Theorem 5.2: they are quite similar to the proofs of the corresponding assertions for $2l > d$.

Taking $r = q$ in (5.18), we see that (5.13) remains true for $2l = d$ and $q > 1$. From this one can derive (5.14) by using interpolation. After the substitution $x \mapsto \gamma^{-1/2l}x$ these estimates will look like (4.9), (4.20). Thus it turns out that, whatever be the relations between d and l, we have similar type estimates for $N(-\gamma; A_l(\alpha V))$, $\gamma > 0$, though they have been proved in different ways. We are going to present a simple and unified proof of these estimates which does not depend on the general Theorems 4.3, 5.1, 5.2.

PROPOSITION 5.5. *Let $l > 0$ and $V = \overline{V} \in L_{1,\mathrm{loc}}(\mathbb{R}^d)$. Then for $\gamma > 0$ the following estimates hold.*

(1) *If $V_+ \in L_q$, then for $\varkappa < q < \infty$ (in case $2l \leqslant d$) and for $1 \leqslant q < \infty$ (in case $2l > d$)*

$$\int_0^\infty \alpha^{-q-1} N(-\gamma; A_l(\alpha V))\, d\alpha \leqslant c\gamma^{\varkappa - q} \int V_+^q\, dx, \qquad c = c(d, l, q). \tag{5.20}$$

(2) *If $V_+ \in L_{q,w}$, then for $\max(1, \varkappa) < q < \infty$*

$$\sup_{\alpha>0}\left(\alpha^{-q} N(-\gamma; A_l(\alpha V))\right) \leqslant c\gamma^{\varkappa-q} \sup_{t>0}\left(t^q \operatorname{meas}\{x : V(x) > t\}\right), \qquad c = c(d, l, q). \tag{5.21}$$

The last estimate remains true with "sup" *replaced by* "lim sup" (*on the left as $\alpha \to \infty$, on the right as $t + t^{-1} \to \infty$*).

PROOF. As usual, it is sufficient to consider the case $V \geqslant 0$. For such a V we apply (3.12). The function $(|\xi|^{2l} + \gamma)^{-1/2}$ is in L_p for $p > d/l$ and, moreover,

$$\int (|\xi|^{2l} + \gamma)^{-p/2}\, d\xi = c\gamma^{-p/2+\varkappa}, \qquad c = c(d, l, p).$$

Thus for $p > 2$ Proposition 2.4 can be applied to the operator given by (3.11) (we take $a = V^{1/2}$, $b = (|\xi|^{2l} + \gamma)^{-1/2}$ in this Proposition). This leads to (5.21) with $q = p/2$. One can pass to the estimate for lim sup along the lines of the Remark to Theorem 4.1. In the same way, for $p \geqslant 2$ Proposition 2.5 is applicable to the operator given by (3.11). This leads to (5.20). □

§6. The case $2l > d$. Spectrum estimates to the left of $\lambda = 0$

In this section we consider the operator $(-\Delta)^l + |x|^{-2l} - \alpha V$ for the domain $\mathring{\mathbb{R}}^d$ and the operator $(-\Delta)^l - \alpha V$ for $\mathbb{R}^d$. When studying the former, we use the same methods as in §5, with the only difference that the partition into cubes will be replaced by a partition into spherical shells. The investigation of the latter will be reduced to that of the former by means of the Hardy inequality (2.6). This leads to the additional restriction $l - d/2 \notin \mathbb{N}$.

1. In what follows Ω_n, $n \in \mathbb{Z}$, denotes the spherical shell $2^{n-1} < |x| < 2^n$. Let us assign to a function $V \in L_{1,\mathrm{loc}}(\mathring{\mathbb{R}}^d)$ the sequence $\boldsymbol{\eta}(V, l) = \{\eta_n(V, l)\}$, where

$$\eta_n(V, l) = \int_{\Omega_n} |x|^{2l-d} |V(x)|\, dx, \qquad n \in \mathbb{Z}. \tag{6.1}$$

Now define the function spaces

$$G(\mathring{\mathbb{R}}^d, l) = \{ V \in L_{1,\mathrm{loc}}(\mathring{\mathbb{R}}^d) : \boldsymbol{\eta}(V, l) \in l_p(\mathbb{Z}) \}, \qquad 0 < p \leqslant \infty, \tag{6.2}$$

$$G_{p,w}(\mathring{\mathbb{R}}^d, l) = \{ V \in L_{1,\mathrm{loc}}(\mathring{\mathbb{R}}^d) : \boldsymbol{\eta}(V, l) \in l_{p,w}(\mathbb{Z}) \}, \quad 0 < p < \infty \tag{6.3}$$

with the natural (quasi)norms $\|V\|_{G_p(\mathring{\mathbb{R}}^d, l)} = \|\boldsymbol{\eta}(V, l)\|_p$, $|V|_{G_{p,w}(\mathring{\mathbb{R}}^d, l)} = |\boldsymbol{\eta}(V, l)|_p$. We also introduce in a natural way the class $G^0_{p,w}(\mathring{\mathbb{R}}^d, l)$.

The class $\mathscr{H}[a_l(\mathring{\mathbb{R}}^d, |x|^{-2l})]$ clearly coincides with the space $\mathscr{H}^l(\mathring{\mathbb{R}}^d)$ described in (3.4), the restriction $l - d/2 \notin \mathbb{N}$ being in fact irrelevant here. The next statement is similar to Proposition 5.1.

PROPOSITION 6.1. *Let $2l > d$ and $V \in L_{1,\mathrm{loc}}(\mathring{\mathbb{R}}^d)$, $V \geqslant 0$. The inequality*

$$\int V|u|^2\, dx \leqslant \mathscr{M}(V) \int (|\nabla_l u|^2 + |x|^{-2l}|u|^2)\, dx, \qquad u \in \mathscr{H}(\mathring{\mathbb{R}}^d), \tag{6.4}$$

is valid if and only if $\boldsymbol{\eta}(V, l) \in l_\infty(\mathbb{Z})$. The best possible constant in (6.4) satisfies

$$c\|\boldsymbol{\eta}(V, l)\|_\infty \leqslant \mathscr{M}(V) \leqslant C\|\boldsymbol{\eta}(V, l)\|_\infty. \tag{6.5}$$

PROOF. For $u \in H^l(\Omega_0)$ and $2l > d$ we have, in view of the embedding $H^l \to C(\overline{\Omega}_0)$,

$$\sup_{\Omega_0} |u(x)|^2 \leqslant c \left(I_l[u, \Omega_0] + \int_{\Omega_0} |x|^{-2l}|u|^2\, dx \right). \tag{6.6}$$

Substituting $2^{-n}x$ for x, we see that for $u \in H^l(\Omega_n)$

$$\sup_{\Omega_n} |u(x)|^2 \leqslant 2^{n(2l-d)} C_{6.6} \left(I_l[u, \Omega_n] + \int_{\Omega_n} |x|^{-2l}|u|^2\, dx \right).$$

Consequently,

$$\begin{aligned}\int_{\Omega_n} V|u|^2\, dx &\leqslant \eta_n(V, l) \sup_{\Omega_n} (|x|^{d-2l}|u(x)|^2) \\ &\leqslant 2^{2l-d} C_{6.6} \eta_n(V, l) \left(I_l[u, \Omega_n] + \int_{\Omega_n} |x|^{-2l}|u|^2\, dx \right).\end{aligned}$$

Summing over $n \in \mathbb{Z}$ and taking into account (2.12), (2.13), we obtain (6.4) with the constant $\mathscr{M}(V) = 2^{2l-d} C_{6.6} \max(1, C_{2.13}) \|\boldsymbol{\eta}(V, l)\|_\infty$.

Conversely, fix a function $u_0 \in C_0(\mathring{\mathbb{R}}^d)$, $u_0|_{\Omega_0} \equiv 1$, and set $u_n(x) = u_0(2^{-n}x)$, $n \in \mathbb{Z}$. Denote $\int (|\nabla_l u_0|^2 + |x|^{-2l}|u_0|^2)\, dx = \tau$, then

$$\int (|\nabla_l u_n|^2 + |x|^{-2l}|u_n|^2)\, dx = 2^{n(d-2l)}\tau.$$

If (6.4) is true, then

$$\eta_n(V, l) \leqslant 2^{n(2l-d)} \int V|u_n|^2\,dx$$
$$\leqslant 2^{n(2l-d)} \mathcal{M}(V) \int (|\nabla_l u_n|^2 + |x|^{-2l}|u_n|^2)\,dx = \tau\mathcal{M}(V).$$

This implies the right-hand side inequality in (6.5) with $C_{6.5} = \tau^{-1}$. □

PROPOSITION 6.2. *Let $2l > d$ and $V \in L_{1,\mathrm{loc}}(\mathring{\mathbb{R}}^d)$. Then*

$$\nu(s; \mathcal{B}_l(\mathring{\mathbb{R}}^d, |x|^{-2l}, V)) \leqslant Cs^{-\varkappa} \sum_{n\in\mathbb{Z}\,:\,\eta_n(V,l)>cs} \eta_n^{\varkappa}(V, l). \qquad (6.7)$$

PROOF. We write $\mathcal{B}$ instead of $\mathcal{B}_l(\mathring{\mathbb{R}}^d, |x|^{-2l}, V)$ for brevity. Let us endow $H^l(\Omega_n)$ with the metric $I_l[u, \Omega_n] + \int_{\Omega_n} |x|^{-2l}|u|^2\,dx$, and let $\mathcal{B}_n$ be the operator on $H^l(\Omega_n)$ generated by the form $\int_{\Omega_n} V|u|^2\,dx$. By the variational principle and (2.12),

$$\nu(c_1 s; \mathcal{B}) \leqslant \sum_{n\in\mathbb{Z}} \nu(s; \mathcal{B}_n), \qquad c_1 = \max(1, C_{2.13}). \qquad (6.8)$$

It follows from [BS1] (compare with the proof of Proposition 5.2) that

$$\nu(s; \mathcal{B}_0) \leqslant c_2 s^{-\varkappa} \eta_0^{\varkappa}(V, l).$$

Since $\|\mathcal{B}_0\| \leqslant c_3\eta_0(V, l)$, we have $\nu(s; \mathcal{B}_0) = 0$ for $s \geqslant c_3\eta_0(V, l)$. The substitution $x \mapsto 2^{-n}x$ sends the shell Ω_n onto Ω_0 and gives

$$\begin{aligned} &\nu(s; \mathcal{B}_n) \leqslant c_2 s^{-\varkappa} \eta_n^{\varkappa}(V, l), \\ &\nu(s; \mathcal{B}_n) = 0 \quad \text{for } s \geqslant c_3\eta_n(V, l). \end{aligned} \qquad (6.9)$$

Comparing (6.8) and (6.9), we obtain (6.7) with $C_{6.7} = c_1^{\varkappa} c_2$ and $c_{6.7} = (c_1 c_3)^{-1}$. □

As in Proposition 5.2, $c_1 = 1$ for $l \in \mathbb{N}$.

COROLLARY 6.1. *Let $2l > d$ and $V \in G_{\varkappa}(\mathring{\mathbb{R}}^d, l)$. Then*

$$\mathcal{B} := \mathcal{B}_l(\mathring{\mathbb{R}}^d, |x|^{-2l}, V) \in \Sigma_{\varkappa}$$

and

$$|\mathcal{B}|_{\varkappa}^{\varkappa} = \sup_{s>0}(s^{\varkappa}\nu(s; \mathcal{B})) \leqslant C_{6.7} \sum_{n\in\mathbb{Z}} \eta_n^{\varkappa}(V, l). \qquad (6.10)$$

2. THEOREM 6.1. *Let $2l > d$, $V = \overline{V} \in L_{1,\mathrm{loc}}(\mathring{\mathbb{R}}^d)$. Then for the operator $A(\alpha) = A_l(\mathring{\mathbb{R}}^d, |x|^{-2l}, \alpha V)$ and $\varkappa < q < \infty$ the following statements hold.*

(1) *If $V_+ \in G_q(\mathring{\mathbb{R}}^d, l)$, then*

$$\int_0^{\infty} \alpha^{-q-1} N(0; A(\alpha))\,d\alpha \leqslant C \sum_{n\in\mathbb{Z}} \eta_n^q(V_+, l), \quad C = C(d, l, q). \qquad (6.11)$$

(2) *If* $V_+ \in G_{q,w}(\mathring{\mathbb{R}}^d, l)$, *then*

$$\sup_{\alpha>0}(\alpha^{-q} N(0; A(\alpha))) \leqslant C \sup_{t>0}(t^q \operatorname{card}\{n \in \mathbb{Z} : \eta_n(V_+, l) > t\}), \quad C = C(d, l, q); \tag{6.12}$$

$$\limsup_{\alpha\to\infty}(\alpha^{-q} N(0; A(\alpha))) \leqslant C_{6.12} \limsup_{t\to 0}(t^q \operatorname{card}\{n \in \mathbb{Z} : \eta_n(V_+, l) > t\}). \tag{6.13}$$

(3) *If* $V \geqslant 0$ *and* $l \in \mathbb{N}$, *then inequalities* (6.11)–(6.13) *can be reversed.*

PROOF. To prove (1) and (2) we interpolate the mapping

$$\Pi: |x|^{2l-d} V \longmapsto \mathscr{B} := \mathscr{B}_l(\mathring{\mathbb{R}}^d, |x|^{-2l}, V).$$

We identify the function $|x|^{2l-d} V$ with the set of its restrictions to the shells Ω_n. Proposition 6.1 implies

$$\Pi: G_\infty(\mathring{\mathbb{R}}^d, l) \longrightarrow \mathfrak{R}(\mathscr{H}^l(\mathring{\mathbb{R}}^d)), \qquad \|\Pi\| \leqslant C_{6.5}.$$

Estimate (6.10) means that

$$\Pi: G_\varkappa(\mathring{\mathbb{R}}^d, l) \longrightarrow \Sigma_\varkappa(\mathscr{H}^l(\mathring{\mathbb{R}}^d)), \qquad \|\Pi\| \leqslant C_{6.7}.$$

Further steps do not differ from the proof of statements (1), (2) of Theorem 5.1.

When proving statement (3), we shall also follow the pattern of Theorem 5.1. Let u_n, $n \in \mathbb{Z}$, be the sequence of functions constructed in the proof of Proposition 6.1. Assume in addition that $u_0(x) = 0$ for $|x| \notin (2^{-4/3}, 2^{1/3})$. Assume $\tilde{\mathscr{H}}$ be the closure (in $\mathscr{H}^l(\mathbb{R}^d)$) of the linear hull of the functions $\{u_{2n}\}$, $n \in \mathbb{Z}$. Let $V \geqslant 0$, $V \in G_\infty(\mathring{\mathbb{R}}^d, l)$, and let $\tilde{\mathscr{B}}$ be the operator on $\tilde{\mathscr{H}}$ generated by the form $\int V|u|^2\,dx$. Then $\lambda_j(\tilde{\mathscr{B}}) \leqslant \lambda_j(\mathscr{B})$, $\forall j$. The summands in $\sum_n c_n u_{2n} = u$ have disjoint supports, whence

$$\int V|u|^2\,dx \geqslant \sum_n |c_n|^2 \int_{\Omega_{2n}} V\,dx \geqslant \sum_n 2^{-2n(2l-d)} |c_n|^2 \eta_{2n}(V, l),$$

$$\int (|\nabla_l u|^2 + |x|^{-2l}|u|^2)\,dx = \tau \sum_n 2^{-2n(2l-d)} |c_n|^2.$$

It follows that $\lambda_j(\tilde{\mathscr{B}}) \geqslant \tau^{-1} w_j$, where $\{w_j\}$, $j \in \mathbb{N}$, is the nonincreasing rearrangement of the sequence $\{\eta_{2n}(V, l)\}$. Thus

$$\|\{\eta_{2n}(V, l)\}\|_q \leqslant \tau \|\tilde{\mathscr{B}}\|_q \leqslant \tau \|\mathscr{B}\|_q,$$

and similarly for the quasinorms in $l_{q,w}$ and Σ_q. The same inequality holds for the sequence $\{\eta_{2n+1}(V, l)\}$. Consequently

$$\|V\|_{G_q(\mathring{\mathbb{R}}^d, l)} \leqslant 2^{1/q} \tau \|\mathscr{B}\|_q,$$

$$|V|_{G_{q,w}(\mathring{\mathbb{R}}^d, l)} \leqslant 2^{1/q} \tau |\mathscr{B}|_q.$$

This proves the inequalities opposite to (6.11), (6.12). The reverse of (6.13) can be proved by the same method. □

3. The problem of estimating the quantity $N(0; A_l(\alpha V))$, $2l > d$, for $l - d/2 \notin \mathbb{N}$ can easily be reduced to the problem examined in Subsections 1, 2. First consider the operator $A_l(\mathring{\mathbb{R}}^d, \alpha V)$. The class $\mathscr{H}[a_l(\mathring{\mathbb{R}}^d)]$ was described in Proposition 3.2. It coincides with the class $\mathscr{H}^l(\mathring{\mathbb{R}}^d)$ and the corresponding metric forms are equivalent, i.e.,

$$\int (|\nabla_l u|^2 + |x|^{-2l}|u|^2)\,dx \leqslant C \int |\nabla_l u|^2\,dx, \qquad u \in C_0^\infty(\mathring{\mathbb{R}}^d). \tag{6.14}$$

Hence for every potential V

$$\nu(s; \mathscr{B}_l(\mathring{\mathbb{R}}^d, V)) \leqslant (C_{6.14}^{-1}s; \mathscr{B}_l(\mathring{\mathbb{R}}^d, |x|^{-2l}, V)).$$

It follows by (3.9) that

$$N(0; A_l(\mathring{\mathbb{R}}^d, \alpha V)) \leqslant N(0; A_l(\mathring{\mathbb{R}}^d; |x|^{-2l}, C_{6.14}\alpha V)). \tag{6.15}$$

Thus the results of Propositions 6.1, 6.2 and Theorem 6.1 can be transferred automatically to the operator $A_l(\mathring{\mathbb{R}}^d, \alpha V)$ with evident changes of constants in the estimates.

Now the quadratic form $a_l(\mathring{\mathbb{R}}^d)$ differs from the form $a_l = a_l(\mathbb{R}^d)$ only by the fact that the functions $u \in \mathscr{H}[a_l(\mathring{\mathbb{R}}^d)]$ satisfy, at $x = 0$, the additional conditions indicated in (3.4). Below, $m = m(d, l)$ is the standard notation for the number of these conditions. In view of the variational principle,

$$\begin{aligned} N(\lambda; A_l(\mathring{\mathbb{R}}^d, \alpha V)) &\leqslant N(\lambda; A_l(\alpha V)) \\ &\leqslant m + N(\lambda; A_l(\mathring{\mathbb{R}}^d, \alpha V)), \quad \forall \lambda \in \mathbb{R}. \end{aligned} \tag{6.16}$$

This inequality enables one to transfer the preceding estimates to the operator $A_l(\alpha V)$. It is also worth noting that now $V \in L_{1,\mathrm{loc}}(\mathbb{R}^d)$ (and not $V \in L_{1,\mathrm{loc}}(\mathring{\mathbb{R}}^d)$ as in Subsections 1, 2). In view of the inequalities

$$\int_{\Omega_n} |x|^{2l-d}|V(x)|\,dx \leqslant \int_{|x|\leqslant 1} |V|\,dx, \qquad n \leqslant 0,$$

$$\sum_{n\leqslant 0} \left(\int_{\Omega_n} |x|^{2l-d}|V(x)|\,dx \right)^{\varkappa} \leqslant (1 - 2^{-\varkappa(2l-d)})^{-1} \int_{|x|<1} |V|\,dx,$$

estimates (6.5), (6.7) along with their consequences (6.10)–(6.13) can be simplified by excluding the integrals over the shells Ω_n with $n \leqslant 0$. To make the formulation that will follow more transparent, we define a sequence $\hat{\boldsymbol{\eta}}(V, l) = \{\hat{\eta}_n(V, l)\}$, $n \in \mathbb{Z}_+$ (compare with (6.1)):

$$\begin{aligned} \hat{\eta}_0(V, l) &= \int_{|x|<1} |V|\,dx; \\ \hat{\eta}_n(V, l) &= \eta_n(V, l) = \int_{\Omega_n} |x|^{2l-d}|V(x)|\,dx, \qquad n \in \mathbb{N}. \end{aligned} \tag{6.17}$$

Along with the classes defined in (6.2), (6.3), we introduce the following function spaces (with the natural (quasi)norms)

$$\hat{G}_p(\mathbb{R}^d, l) = \{ V \in L_{1,\mathrm{loc}}(\mathbb{R}^d) : \hat{\boldsymbol{\eta}}(V, l) \in l_p(\mathbb{Z}_+) \}, \qquad 0 < p < \infty;$$
$$\hat{G}_{p,w}(\mathbb{R}^d, l) = \{ V \in L_{1,\mathrm{loc}}(\mathbb{R}^d) : \hat{\boldsymbol{\eta}}(V, l) \in l_{p,w}(\mathbb{Z}_+) \}, \qquad 0 < p < \infty.$$

The theorem we are going to formulate collects the results corresponding to Proposition 6.2, Corollary 6.1, and statements (2), (3) of Theorem 6.1. The integral estimate similar to (6.11) is omitted.

THEOREM 6.2. *Let* $2l > d$, $l - d/2 \notin \mathbb{N}$, $V = \overline{V} \in L_{1,\mathrm{loc}}(\mathbb{R}^d)$. *Then*
(1) *We have*

$$N(0; A_l(\alpha V)) \leqslant m + C\alpha^{\varkappa} \sum_{n \geqslant 0 \,:\, \hat{\eta}_n(V_+, l) > cs} \hat{\eta}_n^{\varkappa}(V_+, l) \tag{6.18}$$

and consequently,

$$N(0; A_l(\alpha V)) \leqslant m + C_{6.18}\alpha^{\varkappa} \sum_{n \geqslant 0} \hat{\eta}_n^{\varkappa}(V_+, l). \tag{6.19}$$

(2) *If* $V_+ \in \hat{G}_{q,w}(\mathbb{R}^d, l)$ *with* $\varkappa < q < \infty$, *then*

$$\begin{gathered} N(0; A_l(\alpha V)) \leqslant m + C\alpha^q \sup_{t>0}\big(t^q \operatorname{card}\{ n \geqslant 0 : \hat{\eta}_n(V_+, l) > t \}\big), \\ C = C(d, q, l); \end{gathered} \tag{6.20}$$

$$\begin{aligned} &\limsup_{\alpha \to \infty}\big(\alpha^{-q} N(0; A_l(\alpha V))\big) \\ &\qquad \leqslant C_{6.20} \limsup_{t \to 0}\big(t^q \operatorname{card}\{ n \geqslant 0 : \hat{\eta}_n(V_+, l) > t \}\big). \end{aligned} \tag{6.21}$$

(3) *If* $V \geqslant 0$, *then for* $\varkappa < q < \infty$ *we have the two-sided estimate*

$$\sup_{\alpha > 0}\big(\alpha^{-q} N(0; A_l(\alpha V)) - m\big)_+ \asymp |V_+|_{\hat{G}_{q,w}(\mathbb{R}^d, l)}$$

and also the opposite of (6.21).

PROOF. In view of the preceding remarks, statements (1) and (2) follow from the corresponding statements about the operator $A_l(\mathbb{R}^d, |x|^{-2l}, \alpha V)$.

To prove statement (3), let us use the same functions u_n that were used in the proofs of Proposition 6.1 and Theorem 6.1, but for $n > 0$ only. We supplement the sequence $\{u_n\}$, $n \in \mathbb{N}$, by a function $\tilde{u}_0 \in C_0^\infty(\mathbb{R}^d)$ satisfying $\tilde{u}_0(x) = 1$ for $|x| < 1$, $\tilde{u}_0(x) = 0$ for $|x| > 4/3$. With the help of the sequence constructed, we can prove statement (3) for $l \in \mathbb{N}$ in the same way as in Theorem 6.1.

If $l \notin \mathbb{N}$, then the functional given by (2.15) with $\varepsilon \leqslant 1/3$ should be used. The reader can recover the details on the basis of the analogy with Theorem 5.1. □

Under the hypotheses of Theorem 6.1, a similar trick does not work because the distances between the supports of the functions u_n tend to 0 as $n \to -\infty$.

REMARKS. 1) Sometimes it is more convenient to use the initial sequence (6.1) instead of (6.17) in the above estimates. We point out, for example, an estimate implied by (6.11), (6.16) (in view of (1.7)):

$$N(0; A_l(\alpha V)) \leqslant m + qC_{6.11}\alpha^q \sum_{n\in\mathbb{Z}} \eta_n^q(V_+, l), \qquad \varkappa < q < \infty. \tag{6.22}$$

By the Hölder inequality, it follows from (6.22) that for $q \geqslant 1$

$$\begin{gathered} N(0; A_l(\alpha V)) \leqslant m + C\alpha^q \int |x|^{2lq-d} V_+^q \, dx, \\ 1 \leqslant q < \infty, \quad C = qC_{6.11}(v_d d \ln 2)^{q-1}. \end{gathered} \tag{6.23}$$

Here the right-hand side contains no integrals over spherical shells. Estimate (6.23) for $l \in \mathbb{N}$ was obtained in [EK2, EK3] by a direct method. It is much rougher than the sharp estimate (6.18) and, moreover, is no longer true for $\varkappa < q < 1$.

2) For $d = 1$, $l \in \mathbb{N}$, the results of this section clearly embrace also the operator $(-1)^l u^{(2l)} - \alpha V u$ on the half-axis.

Concluding the section, we consider an example.

EXAMPLE 6.1. For the potential given by (4.19), we have $\hat{\eta}_0 = v_d$, $\hat{\eta}_n \asymp n^{-\delta}$, $n \in \mathbb{N}$. Consequently, if $2l > d$, $l - d/2 \notin \mathbb{N}$, then $N(0; A_l(\alpha V)) = O(\alpha^q)$, $q = \delta^{-1}$, for this potential (the order is sharp; compare with Example 4.1).

§7. A noninterpolation approach

The interpolation approach is convenient in many respects, but the domain of its applications is restricted. For example, it is evidently inappropriate to estimate the growth of $N(-\gamma; A_l(\alpha V))$ in terms of membership of the operator $B_{l,\gamma}(V)$ ($\mathscr{B}_l(V)$ if $\gamma = 0$) in some operator ideals noninterpolatory with respect to the pair $(\mathfrak{R}, \Sigma_\varkappa)$.

Here we consider another approach. It is based on a direct application of some fundamental estimates. Estimates containing some "purely integral" norm of the potential, such as (4.4), (5.13), (6.23), are the most convenient for this purpose. Inequalities like (5.5), (5.10) also can serve as fundamental estimates, though results derived from them look somewhat heavy. We also illustrate applications of the initial estimates (5.4), (6.7).

The approach discussed leads, among other things, to new proofs of the estimates from Theorems 4.2, 4.3, 5.1, and 6.1. We do not consider the case $2l = d$ here, its fundamental estimate (5.17) being *a fortiori* rough.

1. First suppose that $2l < d$ and let $\varphi \in L_{1,\mathrm{loc}}(\mathbb{R}^d)$, $\varphi > 0$. Let $V = \overline{V} \in L_{1,\mathrm{loc}}(\mathbb{R}^d)$ and assume that $(\alpha V - \varphi)_+ \in L_\varkappa$ for some $\alpha > 0$. In the case $X = \mathbb{R}^d$ the operator defined in (3.1) can be represented in the form

$$A_l(\varphi, \alpha V) = A_l(\alpha V - \varphi). \tag{7.1}$$

To the latter (4.4) can be applied. This gives

$$N(0; A_l(\varphi, \alpha V)) = N(0; A_l(\alpha V - \varphi)) \leqslant C_{4.4} \int (\alpha V - \varphi)_+^{\varkappa} \, dx. \tag{7.2}$$

There are many consequences of (7.2). Even the first one (which is simply a coarsening of the estimate) is interesting enough.

THEOREM 7.1. *Let* $2l < d$, φ, $V \in L_{1,\mathrm{loc}}(\mathbb{R}^d)$, $V = \overline{V}$, *and* $\varphi > 0$. *Then*

$$N(0; A_l(\varphi, \alpha V)) \leqslant C_{4.4} \alpha^{\varkappa} \int_{\alpha V > \varphi} V^{\varkappa} \, dx, \qquad \alpha > 0. \tag{7.3}$$

We single out the case $\varphi \equiv \gamma > 0$ that corresponds to the problem of estimating the spectrum of $A_l(\alpha V)$ to the left of the point $(-\gamma)$.

COROLLARY 7.1. *Let* $2l < d$, $V = \overline{V} \in L_{1,\mathrm{loc}}(\mathbb{R}^d)$. *Then for* $\gamma > 0$

$$N(-\gamma; A_l(\alpha V)) \leqslant C_{4.4} \alpha^{\varkappa} \int_{V > \alpha^{-1}\gamma} V^{\varkappa} \, dx, \qquad 2l < d. \tag{7.4}$$

EXAMPLE 7.1. Let $V(x) = 0$ for $|x| < 1$ and $V(x) = \Phi(\theta)|x|^{-2l}$, $\theta = x/|x| \in \mathbb{S}^{d-1}$ for $|x| \geqslant 1$. Computing the integral in (7.4), we find

$$N(-\gamma; A_l(\alpha V)) \leqslant (2l)^{-1} C_{4.4} \alpha^{\varkappa} \int_{\mathbb{S}^{d-1}} \Phi_+^{\varkappa} (\ln(\alpha \gamma^{-1} \Phi_+))_+ \, d\theta.$$

A slightly weaker estimate:

$$N(-\gamma; A_l(\alpha V)) \leqslant c\alpha^{\varkappa} \left(\ln(\alpha\gamma^{-1}) \int_{\mathbb{S}^{d-1}} \Phi_+^{\varkappa} \, d\theta + \int_{\mathbb{S}^{d-1}} \Phi_+^{\varkappa} \ln(1 + \Phi_+) \, d\theta \right), \qquad \alpha > \gamma.$$

To justify the above calculations, it is sufficient to suppose that $|\Phi| + \Phi_+^{\varkappa} \ln(1 + \Phi_+) \in L_1(\mathbb{S}^{d-1})$.

EXAMPLE 7.2. Let $V(x) = 0$ for $|x| < e$, $V(x) = (\ln|x|)^{-1}$ for $|x| \geqslant e$. Then by (7.4)

$$N(-\gamma; A_l(\alpha V)) \leqslant C_{4.4} \alpha^{\varkappa} v_d d \int_{\alpha^{-1}\gamma}^{1} t^{\varkappa - 2} e^{d/t} \, dt, \qquad \alpha > \gamma.$$

Here $N(\cdot)$ grows exponentially.

EXAMPLE 7.3. Consider the following operator on $L_2(\mathbb{R}^d)$:

$$A_{l,\alpha} = (-\Delta)^l + \Phi(\theta)|x|^{\beta} - \alpha \Psi(\theta)|x|^{\delta}, \qquad -2l < \delta < \beta < \infty. \tag{7.5}$$

Here $\varphi(x) = \Phi(\theta)|x|^{\beta}$, $V(x) = \Psi(\theta)|x|^{\delta}$, $\theta = x/|x| \in \mathbb{S}^{d-1}$. If $2l < d$ and $\Phi < 0$, $\Psi = \overline{\Psi}$, then (7.3) can be applied. Computing the integral in (7.3), we find

$$N(0;\, A_{l,\delta}) \leqslant c\alpha^q \int_{\mathbb{S}^{d-1}} \Psi_+^q \Phi^{-p}\, d\theta\,, \tag{7.6}$$

$$p = \varkappa \frac{2l+\delta}{\beta-\delta}\,, \qquad q = \varkappa \frac{2l+\beta}{\beta-\delta}. \tag{7.7}$$

In §8, Subsection 1, there is an asymptotic formula for $N(0;\, A_{l,\alpha})$ under some additional assumptions about Φ and Ψ. It shows that not only the order with respect to α in (7.6) is sharp, but so is the coefficient expressed by the integral.

2. Let us introduce the distribution function of V/φ with respect to the weight $\varphi^\varkappa$:

$$\mu_{V,\varphi}(t) := \int_{V>t\varphi} \varphi^\varkappa\, dx\,, \qquad t>0.$$

Estimate (7.2) can be rewritten as follows:

$$\begin{aligned} N(0;\, A_l(\varphi\,,\alpha V)) &\leqslant -C_{1.1}\alpha^\varkappa \int_{\alpha^{-1}}^\infty (t-\alpha^{-1})^\varkappa\, d\mu_{V,\varphi}(t) \\ &= C_{4.4}\varkappa\alpha^\varkappa \int_{\alpha^{-1}}^\infty (t-\alpha^{-1})^{\varkappa-1} \mu_{V,\varphi}(t)\, dt. \end{aligned} \tag{7.8}$$

The first integral in (7.8) is finite if and only if $\int^\infty t^\varkappa\, d\mu_{V,\varphi}(t) < \infty$ or, which is the same, if and only if the integral in (7.3) is finite. The same conditions justify integration by parts in (7.8).

To systematize consequences of estimate (7.8), let us introduce two integral transformations. The transformation $\Gamma_\varkappa$, $\varkappa > 1$, is defined on the set of nonincreasing functions $\mu(t)$, $t>0$, with $-\int^\infty t^\varkappa\, d\mu(t) < \infty$, as follows:

$$\begin{aligned} \Gamma_\varkappa\colon \mu \longmapsto g\,, \qquad g(\lambda) &= -\lambda^{-\varkappa} \int_\lambda^\infty (t-\lambda)^\varkappa\, d\mu(t) \\ &= \varkappa\lambda^{-\varkappa} \int_\lambda^\infty (t-\lambda)^{\varkappa-1} \mu(t)\, dt. \end{aligned} \tag{7.9}$$

The function g is nonincreasing along with μ.

Next we define the dual transformation $\Gamma_\varkappa^+$ on the set of nondecreasing functions $\rho(\lambda)$, $\lambda \geqslant 0$, satisfying $\rho(0) = \rho(+0) = 0$, $\int_0 \lambda^{-\varkappa}\, d\rho(\lambda) < \infty$:

$$\Gamma_\varkappa^+\colon \rho \longmapsto F\,, \qquad F(t) = \int_0^t \lambda^{-\varkappa}(t-\lambda)^\varkappa\, d\rho(\lambda). \tag{7.10}$$

F is also nondecreasing, $F(0) = F(+0) = 0$. Note that $\Gamma_\varkappa^+ = R\Gamma_\varkappa R$, where R denotes the transformation $f(s) \mapsto f(s^{-1})$, $s \in \mathbb{R}_+$.

The table below contains "typical" functions μ, ρ and their transforms $g = \Gamma_\varkappa \mu$, $F = \Gamma_\varkappa^+ \rho$. Except for the first two lines, the expressions for μ, ρ

TABLE 1

$\mu(t) = t^{-q}$,	$q > \varkappa$	$g(\lambda) = \varkappa B(\varkappa, q-\varkappa)\lambda^{-q}$
$\rho(\lambda) = \lambda^q$,	$q > \varkappa$	$F(t) = \varkappa B(\varkappa, q-\varkappa)t^q$
$\mu(t) \sim t^{-\varkappa}$		$g(\lambda) \sim c\mu(\lambda)\ln(1/\lambda)$
$\mu(t) \sim t^{-q}(\ln 1/t)^r$,	$q > \varkappa$, $r \in \mathbb{R}$	$g(\lambda) \sim c\mu(\lambda)$
$\mu(t) \sim t^{-\varkappa}(\ln 1/t)^r$,	$r > -1$	$g(\lambda) \sim c\mu(\lambda)\ln 1/\lambda$
$\mu(t) \sim t^{-\varkappa}(\ln 1/t)^{-1}$		$g(\lambda) \sim c\lambda^{-\varkappa}\ln(\ln 1/\lambda)$
$\rho(\lambda) \sim \lambda^q(\ln 1/\lambda)^r$,	$\varkappa$, $r \in \mathbb{R}$	$F(t) \sim c\rho(t)$
$\rho(\lambda) \sim \lambda^{\varkappa}(\ln 1/\lambda)^r$,	$r > -1$	$F(t) \sim c\rho(t)\ln 1/t$

(7.11)

are asymptotic as $t, \lambda \to 0$. Similarly, for g, F the asymptotics near 0 is given (without explicit indication of the constant c).

In the sequel a reference of the form (7.11_k) means that we refer to the kth line in Table 1. In (7.11_1), (7.11_2) B is the Euler beta-function.

Now let us turn back to (7.8) and rewrite it as follows:

$$N(0; A_l(\varphi, \alpha V)) \leqslant C_{4.4} g(\alpha^{-1}), \qquad g = \Gamma_\varkappa \mu_{V,\varphi}. \tag{7.12}$$

This directly implies the next statement.

THEOREM 7.2. *Let $2l < d$ and $\varphi \in L_{1,\mathrm{loc}}(\mathbb{R}^d)$, $\varphi > 0$. Suppose that a nonincreasing function $G(t)$, $t > 0$, is fixed, with $\int^\infty t^{\varkappa-1}G(t)\,dt < \infty$. Then for the potentials $V = \overline{V} \in L_{1,\mathrm{loc}}(\mathbb{R}^d)$ satisfying*

$$m(V) := \sup_{t>0}(\mu_{\varphi,V}(t)/G(t)) < \infty$$

the following estimate holds:

$$N(0; A_l(\varphi, \alpha V)) \leqslant C_{4.4} m(V) H(\alpha^{-1}), \qquad H = \Gamma_\varkappa G.$$

The next result gives an integral estimate for $N(0; A_l(\varphi, \alpha V))$.

THEOREM 7.3. *Let $2l < d$ and $\varphi \in L_{1,\mathrm{loc}}(\mathbb{R}^d)$, $\varphi > 0$. Let $\rho(\lambda)$, $\lambda > 0$, be a nondecreasing function satisfying $\rho(0) = \rho(+0) = 0$ and $\int_0 \lambda^{-\varkappa}\,d\rho(\lambda) < \infty$. Then for $V = \overline{V} \in L_{1,\mathrm{loc}}(\mathbb{R}^d)$ we have*

$$\int_0^\infty N(0; A_l(\varphi, \lambda^{-1}V))\,d\rho(\lambda) \leqslant C_{4.4}\int F(V_+/\varphi)\varphi^\varkappa\,dx, \qquad F = \Gamma_\varkappa^+ \rho. \tag{7.13}$$

PROOF. By (7.8)

$$N(\lambda^{-1}) := N(0; A_l(\varphi, \lambda^{-1}V)) \leqslant -C_{4.4}\lambda^{-\varkappa}\int_\lambda^\infty (t-\lambda)^\varkappa\,d\mu_{V,\varphi}(t).$$

Consequently

$$\begin{aligned}\int_0^\infty N(\lambda^{-1})\,d\rho(\lambda) &\leqslant -C_{4.4}\int_0^\infty \lambda^{-\varkappa}\int_\lambda^\infty (t-\lambda)^\varkappa\,d\mu_{V,\varphi}(t)d\rho(\lambda)\\ &= -C_{4.4}\int_0^\infty F(t)\,d\mu_{V,\varphi}(t)\\ &= C_{4.4}\int F(V_+/\varphi)\varphi^\varkappa\,dx. \quad\square\end{aligned}$$

It should be noted that for $V \geqslant 0$ the integral on the left-hand side of (7.13) equals

$$\operatorname{Tr}\rho(\mathscr{B}_l(\varphi, V)) = \sum_n \rho(\lambda_n(\mathscr{B}_l(\varphi, V))).$$

Note also that the monotonicity of $N(\cdot)$ (see the Remark to Theorem 1.2) implies

$$\rho(\alpha^{-1})N(0;A_l(\varphi, \alpha V)) \leqslant C_{4.4}\int F(V/\varphi)\varphi^\varkappa\,dx$$

and

$$\rho(\alpha^{-1})N(0;A_l(\varphi, \alpha V)) \longrightarrow 0, \qquad \alpha\to\infty.$$

The transformations defined by (7.9), (7.10) are invertible (as mappings onto their images). To find the inverses, one can differentiate and, for $\varkappa \notin \mathbb{N}$, solve the Abel integral equation. The inversion formulas are quite evident and we do not write them out. Their existence allows one to solve, on the basis of Theorems 7.2, 7.3, the "inverse problem": for a fixed φ, to determine the class of potentials V ensuring a given rate of growth (in α) of $N(0;A_l(\varphi, \alpha V))$.

When applying Theorems 7.2, 7.3, it is useful to keep in mind Table 1. For example, it follows from (7.11_1), (7.11_2) that for $\varkappa < q < \infty$

$$N(0;A_l(\varphi, \alpha V)) \leqslant C_{4.4}\varkappa B(\varkappa, q-\varkappa)\alpha^q \sup_{t>0}(t^q\mu_{\varphi,V}(t)); \tag{7.14}$$

$$\begin{aligned}\int_0^\infty \alpha^{-q-1}&N(0;A_l(\varphi, \alpha V))\,d\alpha\\ &\leqslant C_{4.4}B(\varkappa+1, q-\varkappa)\int V_+^q\varphi^{\varkappa-q}\,dx.\end{aligned} \tag{7.15}$$

Let, e.g., $\varphi(x)\equiv\gamma>0$. Then (7.14), (7.15) imply

$$N(-\gamma;A_l(\alpha V)) \leqslant C_{4.4}\varkappa B(\varkappa, q-\varkappa)\gamma^{\varkappa-q}\alpha^q\sup_{t>0}(t^q \operatorname{meas}\{x: V(x)>t\}),$$

$$\int \alpha^{-q-1}N(-\gamma;A_l(\alpha V))\,d\alpha \leqslant C_{4.4}B(\varkappa+1, q-\varkappa)\gamma^{\varkappa-q}\int V_+^q\,dx.$$

Thus we have recovered (4.18), (4.17). But now we have simple expressions for the constants in terms of the basic constant $C_{4.4}$.

3. Estimates for $N(0; A_l(\alpha V))$ can be reduced to estimates from Subsections 1, 2 in the following way. Fix a function $\psi \in \mathscr{W}(d, l)$. According to (4.1), the following operator inequality holds:

$$(-\Delta)^l + \psi - (1+M_0)\alpha V \leqslant (1+M_0)((-\Delta)^l - \alpha\Delta), \qquad M_0 = M_0(\psi).$$

By the variational principle, it implies the following statement.

PROPOSITION 7.1. *Let* $2l < d$, $V = \overline{V} \in L_{1,\mathrm{loc}}(\mathbb{R}^d)$. *Then for every function* $\psi \in \mathscr{W}(d, l)$ *we have*

$$N(0; A_l(\alpha V)) \leqslant N(0; A_l(\psi, (1+M_0)\alpha V)), \qquad M_0 = M_0(\psi), \tag{7.16}$$

where $M_0(\psi)$ *is the best possible constant in* (4.1).

It is convenient to apply Proposition 7.1 in combination with Theorems 7.1–7.3. For example, Theorem 7.1 implies the following statement.

THEOREM 7.4. *Let* $2l < d$, $V = \overline{V} \in L_{1,\mathrm{loc}}(\mathbb{R}^d)$. *Then*

$$N(0; A_l(\alpha V)) \leqslant C_{4.4}\alpha^{\varkappa} \inf_{\psi \in \mathscr{W}(d,l)} (1+M_0)^{\varkappa} \int_{(1+M_0)\alpha V > \psi} V^{\varkappa}\, dx, \tag{7.17}$$
$$M_0 = M_0(\psi).$$

EXAMPLE 7.4 (compare with Example 4.1). Let $\psi \in L_{\varkappa, w}$, $\psi > 0$, and $C_{2.5}|\psi|_{\varkappa} \leqslant 1$ (so that $M_0(\psi) \leqslant 1$). Let $V = \psi(1 + |\ln \psi|)^{-1/\varkappa}$. According to (7.17),

$$N(0; A_l(\alpha V)) \leqslant c\alpha^{\varkappa} \int_{2\alpha(1+|\ln\psi|)^{-1/\varkappa} > 1} \psi^{\varkappa}(1+|\ln\psi|)^{-1}\, dx$$
$$= -c\alpha^{\varkappa} \int_{1+|\ln t| < (2\alpha)^{-\varkappa}} t^{\varkappa}(1+|\ln t|)^{-1}\, d\mu_{\psi}(t)$$

(the function $\mu_{(\cdot)}(t)$ is defined in (2.1)). Since $\mu_{\psi}(t) \leqslant (|\psi|_{\varkappa} t^{-1})^{\varkappa}$, it is easy to estimate the integral (after integration by parts), and we obtain $N(0; A_l(\alpha V)) \leqslant c\alpha^{\varkappa}\ln(2\alpha)$, $\alpha > 1/2$. This order of growth is sharp, e.g., for the potential $V = |x|^{-2l}(1 + |\ln|x||)^{-1/\varkappa}$.

With the help of Theorems 7.2 and 7.3, it is possible to obtain a new proof of the estimates in Theorem 4.2. Let $\psi \in \mathscr{W}(d, l)$, $M_0(\psi) \leqslant 1$. Set $\psi_h = h\psi$ ($h > 0$ will be specified later). By (7.16), (7.14)

$$N(0; A_l(\alpha V)) \leqslant C\alpha^q(1+hM_0)^q \sup_{t>0}\left(t^q h^{\varkappa} \int_{V > th\psi} \psi^{\varkappa}\, dx\right)$$
$$= C\alpha^q(1+hM_0)^q h^{\varkappa - q} \sup_{t>0}\left(t^q \int_{V > t\psi} \psi^{\varkappa}\, dx\right),$$
$$C = C_{4.4}\varkappa B(\varkappa, q-\varkappa).$$

Minimizing over h, we arrive at (4.14) with the constant

$$C = C_{4.4}\varkappa B(\varkappa, q-\varkappa) q^q \varkappa^{-q}(q-\varkappa)^{-(q-\varkappa)}. \tag{7.18}$$

One can derive (4.13) from (7.16) and (7.15) with the constant $q^{-1}C_{7.18}$ in a similar way.

4. The approach discussed can be applied to the case $2l > d$ as well. For example, (7.1) and (5.5) imply

$$N(-1; A_l(\varphi, \alpha V)) \leqslant C_{5.3} \sum_{k\in\mathbb{Z}^d} \left(\int_{Q_k} (\alpha V - \varphi)_+ \, dx \right)^{\varkappa}; \qquad (7.19)$$

if $l - d/2 \notin \mathbb{N}$, then (7.1) and (6.19) imply

$$N(0; A_l(\varphi, \alpha V)) \leqslant m + C_{6.18} \sum_{n\geqslant 0} \hat{\eta}_n^{\varkappa}((\alpha V - \varphi)_+, l). \qquad (7.20)$$

Here the $\hat{\eta}_n$ are defined in (6.17). It is assumed in (7.19), (7.20) that φ, $V \in L_{1,\mathrm{loc}}(\mathbb{R}^d)$. If only the weaker condition φ, $V \in L_{1,\mathrm{loc}}(\mathring{\mathbb{R}}^d)$ is satisfied, then in view of (7.1), (6.10), and (6.15) we have

$$N(0; A_l(\mathring{\mathbb{R}}^d, \varphi, \alpha V)) \leqslant C \sum_{n\in\mathbb{Z}} \eta_n^{\varkappa}((\alpha V - \varphi)_+, l) \qquad (7.21)$$

(the η_n are defined in (6.1)).

EXAMPLE 7.5. For $2l > d$, $l - d/2 \notin \mathbb{N}$, consider the following operator on $L_2(\mathbb{R}^d)$ (compare with (7.5)):

$$A_{l,\alpha} = (-\Delta)^l + |x|^\beta - \alpha|x|^\delta, \qquad -2l < \delta < \beta. \qquad (7.22)$$

Here $(\alpha V - \varphi)_+ = 0$ for $x > |\alpha|^{1/(\beta-\delta)}$. Therefore summation in (7.21) extends only to those $n \in \mathbb{Z}$ for which $2^{n-1} < \alpha^{1/(\beta-\delta)}$. For these n replace φ by 0 in (7.21). Then it is easily seen that $N(0; A_{l,\alpha}) = O(\alpha^q)$, where q is defined in (7.7). If $\delta > -d$, then N for the operator on $L_2(\mathbb{R}^d)$ given by (7.22) is of the same order of growth. Compare with Example 7.3 (we had $2l < d$ there) and with §8, Subsection 1.

EXAMPLE 7.6. Consider the following operator on $L_2(\mathbb{R}^d)$ for $2l > d$, $l - d/2 \notin \mathbb{N}$:

$$\tilde{A}_{l,\alpha} = (-\Delta)^l + |x|^\beta - \alpha|x|^\beta(1 + \ln^2|x|)^{-\rho}, \qquad \beta > -2l, \quad \rho > 0.$$

Calculations in the spirit of Example 7.5 lead to the estimate

$$N(0; \tilde{A}_{l,\alpha}) \leqslant c\alpha^{\varkappa} \sum_{n<\alpha^{1/2\rho}+1} |n|^{-\varkappa\rho} 2^{\varkappa n(\beta+2l)}.$$

That is, we have exponential growth in α.

The next result can be proved in the same way as Theorem 7.1. Instead of (4.4), we use (5.11) and (6.23) for $q = 1$.

THEOREM 7.5. *Let $2l > d$, φ, $V \in L_{1,\mathrm{loc}}(\mathbb{R}^d)$, $V = \overline{V}$, $\varphi > 0$. Then for $\alpha > 0$*

$$N(-1; A_l(\varphi, \alpha V)) \leqslant C_{5.12} \int (\alpha V - \varphi)_+ \, dx \leqslant C_{5.12}\alpha \int_{\alpha V > \varphi} V \, dx \qquad (7.23)$$

and, if $l - d/2 \notin \mathbb{N}$,

$$N(0;\, A_l(\varphi\,,\,\alpha V)) \leqslant m + C\int |x|^{2l-d}(\alpha V - \varphi)_+\,dx$$
$$\leqslant m + C\alpha \int_{x\,:\,\alpha V>\varphi} |x|^{2l-d}V\,dx\,, \quad C = C_{6.11}|_{q=1}. \quad (7.24)$$

Estimates (7.23), (7.24) are simpler than (7.19), (7.20), because they involve no partitions of $\mathbb{R}^d$ into cubes or spherical shells. On the other hand, they cannot give for $N(-\gamma;\, A_l(\varphi\,,\,\alpha V))$ an order of growth smaller than $O(\alpha)$. This is a consequence of the fact that here we proceed from the "intermediate" estimates (5.11), (6.23). This approach may cause other losses as well. For example, (7.24) applied to the operator defined by (7.22) gives an exaggerated order $N(0;\, A_{l\,,\,\alpha}) = O(\alpha^{1/\varkappa})$.

5. Concluding the section, let us illustrate applications of (5.4), (6.18). Recall that these estimates concern the case $2l > d$.

Let us introduce the distribution functions for some sequences connected with the potential $V \in L_{1\,,\,\mathrm{loc}}(\mathbb{R}^d)$ (more precisely, with its positive part V_+):

$$\sigma_V(s) = \operatorname{card}\left\{ k \in \mathbb{Z}^d : \int_{Q_k} V_+\,dx > s \right\}, \quad (7.25)$$

$$\hat{\sigma}_{V\,,\,l}(s) = \operatorname{card}\{\, n \geqslant 0 : \hat{\eta}_n(V_+\,,\,l) > s \,\}. \quad (7.26)$$

The sequence $\{\hat{\eta}_n(V\,,\,l)\}$ was defined in (6.17). With the help of these functions, one can rewrite (5.4), (6.18) in a form more convenient for applications. Taking into account (3.12), (3.13), and (0.2), we shall do this directly for the operator $A_l(\alpha V)$ rather than for the corresponding operators $B_{l\,,\,1}$ and $\mathscr{B}_l$.

PROPOSITION 7.2. *Let* $2l > d$ *and* $V = \overline{V} \in L_{1\,,\,\mathrm{loc}}(\mathbb{R}^d)$. *Then for* $\alpha > 0$

$$N(-1;\, A_l(\alpha V)) \leqslant C\left(\sigma_V(c/\alpha) + \int_c^\infty t^{\varkappa-1}\sigma_V(t/\alpha)\,dt\right). \quad (7.27)$$

If $l - d/2 \notin \mathbb{N}$, *then*

$$N(0;\, A_l(\alpha V)) \leqslant m + C\left(\hat{\sigma}_{V\,,\,l}(c/\alpha) + \int_c^\infty t^{\varkappa-1}\hat{\sigma}_{V\,,\,l}(t/\alpha)\,dt\right). \quad (7.28)$$

PROOF. Estimate (5.4) for $V \geqslant 0$ can be rewritten as

$$n_+(s;\, B_l(\alpha V)) \leqslant -C_{5.4}s^{-\varkappa}\int_{t>C_{5.4}s} t^{\varkappa}\,d\sigma_V(t)$$
$$= C_{5.4}c_{5.4}^{\varkappa}\sigma_V(c_{5.4}s) + C_{5.4}\varkappa s^{-\varkappa}\int_{t>c_{5.4}s} t^{\varkappa-1}\sigma_V(t)\,dt.$$

Changing the variable $t \mapsto st$ and taking into account (3.12), we obtain (7.27) with $C = C_{5.4}\max(\varkappa\,,\,c_{5.4}^{\varkappa})$, $c = c_{5.4}$. To drop the assumption $V \geqslant 0$ one can use (0.2). Similarly, (7.28) can be derived from (6.18). □

It is very easy to derive (5.8), (6.20) from (7.27), (7.28). Indeed, let

$$\boldsymbol{\tau}(V_+) = \left\{\int_{Q_k} V_+\,dx\right\} \in l_{q,w}(\mathbb{Z}^d), \qquad q > \varkappa.$$

Then $\sigma_V(s) \leqslant s^{-q}\|\boldsymbol{\tau}(V_+)\|_q^q$ and, in view of (7.27),

$$N(-1\,;A_l(\alpha V)) \leqslant C(c^{-q} + (q-\varkappa)^{-1}c^{\varkappa-q})\|\boldsymbol{\tau}(V_+)\|_q^q\alpha^q.$$

This coincides with (5.8). The same argument shows that (7.27) implies (6.20).

Not only power-type estimates for $N(\cdot)$ can be derived from (7.27), (7.28). Let us turn again to the potentials from Examples 7.1, 7.2, but this time for $2l > d$.

EXAMPLE 7.7. Let

$$V(x) = 0 \qquad \text{for } |x| < 1,$$
$$V(x) = |x|^{-2l} \qquad \text{for } |x| \geqslant 1.$$

Then

$$\int_{Q_k} V\,dx \leqslant 1, \qquad \int_{Q_k} V\,dx \asymp (1+|k|)^{-2l}.$$

Therefore

$$\sigma_V(s) = 0 \qquad \text{for } s \geqslant 1,$$
$$\sigma_V(s) \asymp s^{-\varkappa} \qquad \text{for } s < 1.$$

According to (7.27)

$$N(-1\,;A_l(\alpha V)) \leqslant \text{const}\big(\alpha^{\varkappa} + \alpha^{\varkappa}\ln(\alpha/c)\big).$$

EXAMPLE 7.8. For V from Example 7.2, we have

$$\int_{Q_k} V\,dx \leqslant 1 \quad \text{and} \quad \int_{Q_k} V\,dx \asymp (\ln|k|)^{-1}, \qquad k \geqslant 3;$$

therefore $\sigma_V(s) = 0$ for $s \geqslant 1$, and $\sigma_V(s) \leqslant e^{\rho/s}$, where ρ is some constant. Thus

$$N(-1\,;A_l(\alpha V)) \leqslant C\left(e^{\alpha\rho/c} + \int_c^{\alpha} t^{\varkappa-1}e^{\alpha\rho/t}\,dt\right).$$

Both examples are difficult to study on the basis of estimates (5.5), (6.19) and their consequences.

The next example is slightly more general.

EXAMPLE 7.9. Let $V \in L_{1,\mathrm{loc}}(\mathbb{R}^d)$, $2l > d$, $l - d/2 \notin \mathbb{N}$, and

$$V(x) = |x|^{-2l}w(|x|)\Phi(x/|x|), \qquad |x| \geqslant 1, \tag{7.29}$$

where $\Phi \in L_1(\mathbb{S}^{d-1})$, $w \geqslant 0$. For the potential (7.29), we have

$$\hat{\eta}_n(V\,,l) = \|\Phi_+\|_1\sigma_n(w), \qquad \sigma_n(w) = \int_{2^{n-1}}^{2^n} w(r)r^{-1}\,dr, \quad n \in \mathbb{N}.$$

By (7.28) the growth of $N(0; A_l(\alpha V))$ is determined by the behavior of the sequence $\{\sigma_n(w)\}$. In particular, if this sequence tends to 0, then $N(0; A_l(\alpha V))$ is finite for all $\alpha > 0$. If $w(r)$ is monotone, then the sequence in question can be replaced by $\{w(2^n)\}$.

§8. Examples of explicit asymptotic formulas

The main purpose that we pursue in the examples below is to show that it is possible to give explicit asymptotic formulas for $N(\cdot)$ in nonregular cases. As a rule, our calculations are based on estimates proved above, the formulas obtained demonstrating their sharpness. In this section we suppose that $l \in \mathbb{N}$.

1. First of all we study the operator given by (7.5). The numbers p and q in the statement below have been defined in (7.7).

THEOREM 8.1. (1) *Let $2l < d$ and suppose that in* (7.5) $\Phi > 0$, $\Phi + \Phi^{-1} \in L_\infty(\mathbb{S}^{d-1})$, $\Psi = \overline{\Psi} \in L_q(\mathbb{S}^{d-1})$. *Then for the operator defined by* (7.5) *the following asymptotic formula is true*:

$$N(0; A_{l,\alpha}) \sim C\alpha^q \int_{\mathbb{S}^{d-1}} \Psi_+^q \Phi^{-p}\, d\theta, \qquad \alpha \to \infty,$$
$$C = C(d, l, \beta, \delta) = v_d (2\pi)^{-d} \int_0^1 (t^\delta - t^\beta)^\varkappa t^{d-1}\, dt, \quad \varkappa = d/2l. \tag{8.1}$$

(2) *If $2l \geqslant d$, $\delta > -d$ then statement* (1) *holds for $\Psi \in L_\infty(\mathbb{S}^{d-1})$.*

PROOF. (1) According to (1.1), $N(0; A_{l,\alpha})$ is equal to the maximal dimension of subspaces in $C_0^\infty(\mathbb{R}^d)$ on which the following inequality holds:

$$\int |\nabla_l u|^2\, dx - \int (\alpha\Psi(\theta)|x|^\delta - \Phi(\theta)|x|^\beta)|u|^2\, dx < 0, \qquad u \not\equiv 0.$$

The substitution $x \mapsto \alpha^{1/(\beta-\delta)} x$ transforms this inequality into

$$\lambda \int |\nabla_l u|^2\, dx - \int (\Psi(\theta)|x|^\delta - \Phi(\theta)|x|^\beta)|u|^2\, dx < 0,$$
$$\lambda = \lambda(\alpha) := \alpha^{-(2l+\beta)/(\beta-\delta)} = \alpha^{-q/\varkappa}. \tag{8.2}$$

By (1.3), the maximal dimension of subspaces in $C_0^\infty(\mathbb{R}^d)$ on which (8.2) is satisfied is equal to $n_+(\lambda; \mathscr{B}_l(W))$, where

$$W(x) = \Psi(\theta)|x|^\delta - \Phi(\theta)|x|^\beta. \tag{8.3}$$

Comparing these facts, we obtain

$$N(0; A_{l,\alpha}) = n_+(\lambda; \mathscr{B}_l(W)), \qquad \lambda = \lambda(\alpha) = \alpha^{-q/\varkappa}. \tag{8.4}$$

Now we use the following result in [BS3, Theorem 4.14 and Remark 4.16]. Let $V = \overline{V} \in L_{1,\mathrm{loc}}(\mathbb{R}^d)$, $V_+ \in L_\varkappa(\mathbb{R}^d)$, and let an open set $\Omega \subset \mathbb{R}^d$ exist with $V \geqslant 0$ on Ω and $V \leqslant 0$ off Ω. Then

$$n_+(\lambda; \mathscr{B}_l(V)) \sim v_d(2\pi)^{-d}\lambda^{-\varkappa}\int V_+^{\varkappa}\,dx, \qquad \lambda \to +0. \tag{8.5}$$

For the potential given by (8.3), we have $W \in L_{1,\mathrm{loc}}$, $W_+ \in L_\varkappa$. The supplementary condition on the existence of Ω is automatically true provided $g := \Psi/\Phi \in C(\mathbb{S}^{d-1})$. In this case, applying (8.5) to the operator $\mathscr{B}_l(W)$ and taking into account (8.4), we obtain (8.1) after elementary calculations.

To drop the assumption $g \in C(\mathbb{S}^{d-1})$ let us use a special case of (3.9):

$$\begin{gathered} N(0; A_{l,\alpha}) = n_+(\alpha^{-1}; \tilde{\mathscr{B}}_l(\Phi, \Psi)), \\ \tilde{\mathscr{B}}_l(\Phi, \Psi) := \mathscr{B}_l(\Phi(\theta)|x|^\beta, \Psi(\theta)|x|^\delta). \end{gathered} \tag{8.6}$$

We recall that $\tilde{\mathscr{B}}_l$ is an operator on $\mathscr{H}[a_l(\varphi)]$, $\varphi(x) = \Phi(\theta)|x|^\beta$. It follows from (8.6), (8.1) that for $g \in C(\mathbb{S}^{d-1})$, we have

$$n_+(\lambda; \tilde{\mathscr{B}}_l(\Phi, \Psi)) \sim C_{8.1}\lambda^{-q}\int_{\mathbb{S}^{d-1}} \Psi_+^q\Phi^{-p}\,d\theta. \tag{8.7}$$

Besides, applying (7.6), (8.6) to $\tilde{\mathscr{B}}_l(\Phi, \pm\Psi)$ we arrive at

$$\nu(s; \tilde{\mathscr{B}}_l(\Phi, \Psi)) \leqslant c\alpha^q\int_{\mathbb{S}^{d-1}} |\Psi|^q\Phi^{-p}\,d\theta, \qquad \Psi \in L_q(\mathbb{S}^{d-1}). \tag{8.8}$$

Fix $\varepsilon > 0$ and find $g_\varepsilon = \overline{g}_\varepsilon \in C(\mathbb{S}^{d-1})$ with $\|g_\varepsilon - g\|_{L_q} < \varepsilon$. Let $\Phi_\varepsilon = g_\varepsilon\Phi$. For the operator $\tilde{\mathscr{B}}_l(\Phi, \Psi_\varepsilon)$ the asymptotic formula (8.7) holds and, by (8.8),

$$|\tilde{\mathscr{B}}_l(\Phi, \Psi_\varepsilon) - \tilde{\mathscr{B}}_l(\Phi, \Psi)|_q \leqslant c\varepsilon.$$

By the lemma on continuity of asymptotic coefficients (Lemma 1.18 in [BS3]), (8.7) (and thus (8.1)) extends to arbitrary $\Psi = \overline{\Psi} \in L_q(\mathbb{S}^{d-1})$.

(2) As in the proof of statement (1), the study of $N(0; A_{l,\alpha})$ reduces to an examination of (8.2). But for $2l \geqslant d$ more detailed considerations are needed, for no direct reference to [BS3] is possible. Indeed, Theorem 3.6 in [BS3] (that allows $2l \geqslant d$) concerns operators in bounded domains only. Besides, for $2l \geqslant d$ the operator $\mathscr{B}_l(W)$ is not well defined.

Under the hypotheses of statement (2), there exists an $R > 0$ such that the function given by (8.3) is negative for $2|x| > R$. Let $X = X(R)$ be the ball $|x| < R$. Consider the operator $\mathscr{B}_l(X, W)$ generated by the form $\int_X W|u|^2\,dx$ on the space $\mathring{H}^l(X)$ (with the metric $\int_X |\nabla_l u|^2\,dx$). In view of the variational principle, (8.2) implies

$$N(0; A_{l,\alpha}) \geqslant n_+(\lambda; \mathscr{B}_l(\lambda, W)), \qquad \lambda = \lambda(\alpha) > 0. \tag{8.9}$$

To obtain the upper estimate, fix τ, $0 < \tau < \inf\Phi$. For $u \in C_0^\infty(\mathbb{R}^d)$, consider the inequality

$$\begin{gathered} \lambda\int(|\nabla_l u|^2 + |x|^\beta|u|^2)\,dx - \int W_\tau|u|^2\,dx < 0, \\ W_\tau(x) := \Psi(\theta)|x|^\delta - (\Phi(\theta) - \tau)|x|^\beta. \end{gathered} \tag{8.10}$$

For $\lambda \in (0, \tau)$ the set $\{u \in C_0^\infty : (8.10) \text{ holds}\}$ is wider than the similar set for (8.2). If τ is sufficiently small, then automatically $W_\tau(x) < 0$ outside the ball X.

Let $\hat{\mathscr{B}}_l = \hat{\mathscr{B}}_l(X, W_\tau)$ be the operator generated by the form $\int_X W_\tau|u|^2\,dx$ on the space $H^l(X)$ endowed with the metric $\int_X(|\nabla_l u|^2 + |x|^\beta|u|^2)\,dx$. By (1.3), the maximal dimension of subspaces in $H^l(X)$ on which (8.10) holds equals $n_+(\lambda; \hat{\mathscr{B}}_l)$. In view of the variational principle, (8.10) implies the upper estimate

$$N(0; A_{l,\alpha}) \leqslant n_+(\lambda; \hat{\mathscr{B}}_l(X, W_\tau)), \qquad 0 < \lambda = \lambda(\alpha) < \tau. \tag{8.11}$$

Next, let $\check{\mathscr{B}}_l = \check{\mathscr{B}}_l(X, W_\tau)$ be the operator generated by the same form $\int_X W_\tau|u|^2\,dx$ on the space $H^l(X)$ endowed this time with the standard metric $\int_X(|\nabla_l u|^2 + |u|^2)\,dx$. Deriving the asymptotic formula for $n_+(\lambda)$, we can replace $\hat{\mathscr{B}}_l$ by $\check{\mathscr{B}}_l$. Indeed, the "distortion of the metric" given by

$$\int (1 - |x|^\beta)|u|^2\,dx$$

is a compact form on $H^l(X)$ for $\beta > -d$. Therefore (cf. [BS3, Lemma 1.16]) the asymptotic coefficient remains unchanged. The spectral asymptotics for $\mathscr{B}_l(X, W)$, $\check{\mathscr{B}}_l(X, W)$ is known: by Theorem 4.6 of [BS3] it can be calculated according to formula (8.5). Thus

$$\begin{aligned} \lim_{\lambda\to+0} \lambda^\varkappa n_+(\lambda; \mathscr{B}_l(X, W)) &= C_{8.1}\int_{\mathbb{S}^{d-1}} \Psi_+^q \Phi^{-p}\,d\theta, \\ \lim_{\lambda\to+0} \lambda^\varkappa n_+(\lambda; \hat{\mathscr{B}}_l(X, W_\tau)) &= \lim_{\lambda\to+0} \lambda^\varkappa n_+(\lambda, \check{\mathscr{B}}_l) \\ &= C_{8.1}\int_{\mathbb{S}^{d-1}} \Psi_+^q(\Phi-\tau)^{-p}\,d\theta. \end{aligned}$$

Together with (8.9) and (8.11), this implies

$$\begin{aligned} C_{8.1}\int \Psi_+^q\Phi^{-p}\,d\theta &\leqslant \liminf_{\alpha\to\infty}(\alpha^{-q}N(0; A_{l,\alpha})) \\ &\leqslant \limsup_{\alpha\to\infty}(\alpha^{-q}N(0; A_{l,\alpha})) \\ &\leqslant C_{8.1}\int \Psi_+^q(\Phi-\tau)^{-p}\,d\theta. \end{aligned}$$

Letting $\tau \to 0$, we arrive at (8.1). □

REMARKS. 1) Statement (2) of Theorem 8.1 holds for $\delta \in (-2l, -d]$ as well, but this time the operator given by (7.5) must be considered on $L_2(\dot{\mathbb{R}}^d)$. The scheme of the proof remains the same, but one needs to know the generalization of Theorem 4.6 of [BS3] to potentials $V \in L_{1,\mathrm{loc}}(X \setminus \{0\})$ possessing a nonsummable singularity of order $O(|x|^\delta)$ near $x = 0$. For odd d, such a generalization can be obtained with the help of the Hardy inequality (2.6). For even d, the required estimate also can be obtained by using an appropriate version of the Hardy inequality. We shall not enter into details here.

2) Under the hypotheses of statement (1) of Theorem 8.1, the asymptotic formula (8.1) extends to operators $A_l(\varphi, \alpha V)$, where the potentials φ, $V \in L_{1,\mathrm{loc}}(\mathbb{R}^d)$ satisfy

$$|\varphi(x) - \Phi(\theta)|x|^\beta| \leqslant \mu_1(|x|), \qquad \mu_1(r) = o(r^\beta), \quad r \to \infty; \tag{8.12}$$

$$|V(x) - \Psi(\theta)|x|^\delta| \leqslant \mu_2(|x|)\Psi_0(\theta), \quad \mu_2(r) = o(r^\delta), \quad r \to \infty; \tag{8.13}$$

here $\Psi_0 \in L_q(\mathbb{S}^{d-1})$. The first condition corresponds to a compact distortion of the metric in $\mathscr{H}[a_l(\varphi)]$, the second one corresponds to the well-known Weyl lemma (cf. Lemmas 1.16, 1.17 in [BS3]).

The same is true under the hypotheses of statement (2), provided that (8.12) and (8.13) hold with $\Psi_0 \equiv 1$. In the situation of Remark 1, it should be assumed that similar estimates hold also as $r \to 0$.

3) If in Theorem 8.1 $\Phi \equiv \gamma = \mathrm{const}$, $\beta = 0$, then we obtain an asymptotic formula for $N(-\gamma; A_l(\alpha V))$. For the sake of convenience we interchange the roles of δ and $-\delta$.

Proposition 8.1. *Let $l \in \mathbb{N}$ and $V_\Psi(x) = \Psi(x/|x|)|x|^{-\delta}$, $0 < \delta < \min(d, 2l)$, where Ψ satisfies the hypotheses of Theorem* 8.1. *Then for $\gamma > 0$*

$$N(-\gamma; A_l(\alpha V)) \sim C\alpha^q \gamma^{\varkappa - q} \int_{\mathbb{S}^{d-1}} \Psi_+^q \, d\theta, \qquad \alpha \to \infty, \quad q = d/\delta, \quad C = C_{8.1}(d, l, 0, -\delta). \tag{8.14}$$

The result extends to arbitrary potentials V with $N(-\gamma; A_l(\alpha|V - V_\Psi|)) = o(\alpha^q)$. Concrete conditions on the perturbation $V - V_\Psi$ can be derived from Remark 2 *and Theorems* 4.3, 5.1, *and* 5.2.

Taking into account Remark 1 *above, it is possible to extend all this to $\delta \in [d, 2l)$.*

4) For $d = 1$ the results of Remark 3 also hold for operators on the half-axis $\mathbb{R}_+$ under arbitrary selfadjoint boundary conditions at $x = 0$. In (8.14) the term containing the integral should be replaced by $\Psi_+^q(1)$.

5) For $l = 1$, $\beta = 2$, the statement of Theorem 8.1 can be applied to the harmonic oscillator perturbed by a homogeneous potential of order $\delta \in (-2, 2)$. If $d = 1$, $\delta \leqslant -1$, then one should also keep in mind Remark 1.

2. Here we consider the potentials of Example 7.1. The asymptotic formula (8.15) (for $l = 1$) was announced in [BS8]. However it was assumed there that $\Phi \in L_\varkappa(\mathbb{S}^{d-1})$, which is less than satisfactory.

Theorem 8.2. *Let $l \in \mathbb{N}$, $2l < d$ and $V(x) = 0$ for $|x| < 1$, $V(x) = \overline{V}(x) = \Phi(x/|x|)|x|^{-2l}$ for $|x| \geqslant 1$. Let*

$$\int_{\mathbb{S}^{d-1}} |\Phi|^\varkappa \ln(1 + |\Phi|) \, d\theta < \infty.$$

Then we have the following asymptotic formula:

$$N(-1;\, A_l(\alpha V)) \sim \frac{1}{2l} v_d (2\pi)^{-d} \alpha^{\varkappa} \ln\alpha \int_{\mathbb{S}^{d-1}} \Phi_+^{\varkappa}\, d\theta, \qquad \alpha\to\infty. \tag{8.15}$$

PROOF. It is more convenient to consider the equivalent problem about the asymptotics as $\lambda\to 0$ of the function $n_+(\lambda;\, B)$, $B=B_l(V)$, $\lambda=\alpha^{-1}$. We shall need the following remark. Let

$$X_k=\{x\in\mathbb{R}^d : e^{k-1}<|x|<e^k\}, \qquad k\in\mathbb{N},$$

and let $B(k)$ denote the operator on the space $H^l(X_k)$ (with the metric $\int_X(|\nabla_l u|^2+|x|^{-2l}+|u|^2)\,dx$) generated by the quadratic form

$$\int_{X_k}\Phi|x|^{-2l}|u|^2\,dx. \tag{8.16}$$

The substitution $x=e^{k-1}y$ shows that all operators $B(k)$ are unitarily equivalent to the "standard" operator $B(1)$. A similar fact is valid for the operators $B_0(k)$ on the space $\mathring{H}^l(X_k)$ (with metric $\int_{X_k}|\nabla_l u|^2\,dx$) that are generated by the form (8.16).

Suppose first that $\Phi\in L_\infty(\mathbb{S}^{d-1})$. For $\lambda<\|\Phi\|_\infty$ set*

$$k(\lambda)=\mathrm{Entier}[(1/2l)\ln(\|\Phi\|_\infty\lambda^{-1})]+1.$$

Let us estimate $n_+(\lambda;\, B)$ from above. Let $\hat{B}(k)$ be the operator on the space $H^l(X_k)$ with the standard metric $\int_{X_k}(|\nabla_l u|^2+|u|^2)\,dx$ generated by the form (8.16). Then

$$n_+(\lambda;\, B)\leqslant\sum_{k\in\mathbb{N}} n(\lambda;\, \hat{B}(k)),$$

the summands with numbers $k>k(\lambda)$ being in fact equal to 0. Therefore

$$n_+(\lambda;\, B)\leqslant\sum_{k\leqslant k(\lambda)} n(\lambda;\, \hat{B}(k))\leqslant\sum_{k\leqslant k(\lambda)} n(\lambda;\, B(k))=k(\lambda)n(\lambda;\, B(1)).$$

Now we find, by using the known asymptotic formula for $n_+(\lambda;\, B(1))$,

$$\begin{aligned}\limsup_{\lambda\to+0}(\lambda^{\varkappa}(\ln\lambda^{-1}))^{-1}n_+(\lambda;\, B)&\leqslant\frac{1}{2l}v_d(2\pi)^{-d}\int_{X_1}|x|^{-d}\Phi_+^{\varkappa}\,dx\\ &=\frac{1}{2l}v_d(2\pi)^{-d}\int_{\mathbb{S}^{d-1}}\Phi_+^{\varkappa}\,d\theta.\end{aligned} \tag{8.17}$$

Let us pass to the lower estimate. The quantity $n_+(\lambda;\, B)$ is equal to the maximal dimension of subspaces in $C_0^\infty(\mathbb{R}^d)$ on which the following inequality holds:

$$\int|\nabla_l u|^2\,dx\leqslant\lambda^{-1}\int_{|x|>1}\Phi|x|^{-2l}|u|^2\,dx-\int|u|^2\,dx.$$

* *Editor's note.* The French "Entier" here means "integral part of".

Let $\mathscr{X}(\lambda) = \{ x \in \mathbb{R}^d : |x| < e^{k(\lambda)} \}$. Consider three operators $B^{(1)}$, $B^{(2)}$, and $B^{(3)} = \lambda(B^{(1)} + B^{(2)})$ on the space $\mathring{H}^l(\mathscr{X}(\lambda))$ endowed with the metric $\int_{\mathscr{X}(\lambda)} |\nabla_l u|^2\,dx$: let $B^{(1)}$ be generated by the form

$$\lambda^{-1} \int_{1<|x|<e^{k(\lambda)}} \Phi |x|^{-2l} |u|^2\,dx - \int_{\mathscr{X}(\lambda)} |u|^2\,dx$$

and $B^{(2)}$ generated by the form $\int_{\mathscr{X}(\lambda)} |u|^2\,dx$. According to the Weyl inequality (eq. (1.31) in [BS3]) for $\mu > 0$, $0 < \vartheta < 1$ we have:

$$\begin{aligned} n_+(\mu; B^{(3)}) &= n_+(\mu\lambda^{-1}; B^{(1)} + B^{(2)}) \\ &\leqslant n_+(\mu\lambda^{-1}(1-\vartheta); B^{(1)}) + n_+(\mu\lambda^{-1}\vartheta; B^{(2)}). \end{aligned} \tag{8.18}$$

In view of the variational principle, $n_+(\lambda; B) \geqslant n_+(1; B^{(1)})$ and

$$n_+(\mu; B^{(3)}) \geqslant \sum_{k \leqslant k(\lambda)} n_+(\mu; B_0(k)) = k(\lambda) n_+(\mu; B_0(1)).$$

Therefore, setting $\mu = \lambda(1-\vartheta)^{-1}$ in (8.18), we find

$$\begin{aligned} n_+(\lambda, B) &\geqslant n_+(\lambda(1-\vartheta)^{-1}; B^{(3)}) - n_+(\vartheta(1-\vartheta)^{-1}; B^{(2)}) \\ &\geqslant k(\lambda) n_+(\lambda(1-\vartheta)^{-1}; B_0(1)) - n_+(\vartheta(1-\vartheta)^{-1}; B^{(2)}). \end{aligned} \tag{8.19}$$

By (4.4),

$$\begin{aligned} n_+(\vartheta(1-\vartheta)^{-1}; B^{(2)}) &\leqslant C_{4.4} (\vartheta(1-\vartheta)^{-1})^{\varkappa} v_d e^{k(\lambda)d} \\ &\leqslant c(\vartheta^{-1} - 1)^{\varkappa} (\|\Phi\|_\infty \lambda^{-1})^{\varkappa}. \end{aligned}$$

Now it follows from (8.19) and Weyl's asymptotic formula for $n(\lambda; B_0(1))$ that

$$\liminf_{\lambda \to +0} (\lambda^{\varkappa} (\ln \lambda^{-1})^{-1} n_+(\lambda; B)) \geqslant \frac{1}{2l} v_d (2\pi)^{-d} (1-\vartheta)^{\varkappa} \int_{\mathbb{S}^{d-1}} \Phi_+^{\varkappa}\,d\theta.$$

Letting $\vartheta \to 0$ and taking into account (8.17), we obtain

$$n_+(\lambda; B) \sim \frac{1}{2l} v_d (2\pi)^{-d} \lambda^{-\varkappa} \ln(\lambda^{-1}) \int_{\mathbb{S}^{d-1}} \Phi_+^{\varkappa}\,d\theta,$$

which is equivalent to (8.15).

To get rid of the restriction $\Phi \in L_\infty$, we use the estimate of Example 7.1. Applying it to the potentials $\pm\Phi(\theta)|x|^{-2l}$, we find that for $\lambda < e$

$$\nu(\lambda; B) \leqslant c\lambda^{\varkappa} \left(\ln \frac{1}{\lambda} \int |\Phi|^{\varkappa}\,d\theta + \int |\Phi|^{\varkappa} \ln(1 + |\Phi|)\,d\theta \right).$$

It remains to refer to Lemma 2.2 of [VS] on the continuity of the asymptotic coefficient in "power-logarithmic" asymptotics. □

3. Here we compute the asymptotics of $N(0; A_l(\alpha V))$ for the spherically symmetric potential of Example 4.1. For $l = 1$ the condition $2l < d$ leads to the restriction $d \geqslant 3$. Let $Y = \{x \in \mathbb{R}^d : |x| > 1\}$. On $L_2(Y)$ consider the operators

$$A^{(\varepsilon)}(\alpha) = -\Delta - \alpha|x|^{-2}(\ln|x|)^{-1/q}, \tag{8.20}$$

ε standing for $\mathscr{D}$ or $\mathscr{N}$ in accordance with the boundary condition on ∂Y that we impose (Dirichlet or Neumann).

PROPOSITION 8.2. *If $d < 2q < \infty$, then for the operators defined by* (8.20) *we have the following asymptotic formula*:

$$N(0; A^{(\varepsilon)}(\alpha)) \sim \alpha^q \frac{1}{2\sqrt{\pi}} \frac{\Gamma(q-\frac{1}{2})}{\Gamma(q)} \sum_{j=0}^{\infty} k_j \left(\Lambda_j + \left(\frac{d-2}{d}\right)^2\right)^{\frac{1}{2}-q}, \qquad \alpha \to \infty; \tag{8.21}$$

here Λ_j are the eigenvalues of the Laplace operator on the sphere $\mathbb{S}^{d-1}$ and k_j are their multiplicities.

We recall (cf. [SW]) that

$$\Lambda_j = j(j+d-2), \qquad k_j = \binom{d+j-2}{d-2} + \binom{d+j-3}{d-2}.$$

The condition $2q > d$ guarantees that the series in (8.21) converges.

With the help of a standard variational argument, one obtains the following statement from (8.21).

THEOREM 8.3. *Let $2q > d \geqslant 3$, and $V = \overline{V} \in L_{d/2,\mathrm{loc}}(\mathbb{R}^d)$, $V(x) \sim |x|^{-2}(\ln|x|)^{-1/q}$ as $|x| \to \infty$. Then the asymptotic formula* (8.21) *holds for $A_l(\alpha V)$.*

The proof of Proposition 8.2 is based on an elementary separation of variables. $N(0; A^{(\varepsilon)}(\alpha))$ is equal to the maximal dimension of subspaces F on which

$$\int_Y |\nabla u|^2\, dx < \alpha \int_Y |x|^{-2}(\ln|x|)^{-\delta}|u|^2\, dx, \qquad q\delta = 1. \tag{8.22}$$

It is sufficient to take $F \subset C_0^\infty(X)$ (in the case $\varepsilon = \mathscr{D}$) or $F \in \{u \in C^\infty(\overline{Y}) : \operatorname{supp} u \text{ is bounded}\}$ (in the case $\varepsilon = \mathscr{N}$). Passing to spherical coordinates and substituting $r = e^t$, $u = v\exp(-t(d-2)/2)$, we turn (8.22) into

$$\begin{gathered}\int_0^\infty \int_{\mathbb{S}^{d-1}} \left(|v_t|^2 + |\nabla_\theta v|^2 + \left(\frac{d-2}{2}\right)^2 |v|^2\right) dt\, d\theta \\ < \alpha \int_0^\infty \int_{\mathbb{S}^{d-1}} t^{-\delta}|v|^2\, dt\, d\theta.\end{gathered} \tag{8.23}$$

Let $\mathscr{F}(\mathscr{D}) = C_0^\infty(0, \infty)$, $\mathscr{F}(\mathscr{N}) = C_0^\infty[0, \infty)$ and let W_j, $j = 0, 1, \dots$, be the space of spherical functions of order j on $\mathbb{S}^{d-1}$. The closures of the spaces $\mathscr{F}(\varepsilon)\otimes W_j$ diagonalize the quadratic forms in both parts of (8.23). Hence $A^{(\varepsilon)}$ splits into the orthogonal sum of the operators $A_j^{(\varepsilon)}$ acting on the corresponding "cells". On the jth cell, (8.23) splits into $k_j = \dim W_j$ inequalities for scalar functions

$$\int_0^\infty (|y_t'|^2 + \gamma_j|y|^2)\,dt < \alpha\int_0^\infty t^{-\delta}|y|^2\,dt, \qquad \gamma_j := \Lambda_j + \left(\frac{d-2}{2}\right)^2. \quad (8.24)$$

Maximal dimension of the subspaces in $\mathring{H}^1(\mathbb{R}_+)$ (the case $\varepsilon = \mathscr{D}$) or in $H^1(\mathbb{R}_+)$ (the case $\varepsilon = \mathscr{N}$) on which (8.24) holds equals $N(-\gamma_j;\check{A}^{(\varepsilon)}(\alpha))$, where $\check{A}^{(\varepsilon)}(\alpha)$ is the operator on $L_2(\mathbb{R}_+)$ generated by the quadratic form $\int_0^\infty(|y_t'|^2-\alpha t^{-\delta}|y|^2)\,dt$ under the corresponding boundary condition at $t=0$. By Remark 4 to Theorem 8.1,

$$N(-\gamma_j;\check{A}^{(\varepsilon)}(\alpha)) \sim \frac{\alpha^q}{2\sqrt{\pi}}\frac{\Gamma(q-1/2)}{\Gamma(q)}\gamma_j^{1/2-q}, \quad (8.25)$$

as $\alpha\to\infty$. Clearly

$$N(0, A^{(\varepsilon)}(\alpha)) = \sum_{j=0}^\infty k_j N(-\gamma_j;\check{A}^{(\varepsilon)}(\alpha)),$$

whence

$$\lim_{\alpha\to\infty}(\alpha^{-q}N(0;A^{(\varepsilon)}(\alpha))) = \lim_{\alpha\to\infty}\Big(\sum_{j=0}^\infty k_j\alpha^{-q}N(-\gamma_j;\check{A}^{(\varepsilon)}(\alpha))\Big).$$

The limit of each particular summand has been calculated in (8.25). The termwise passage to the limit can easily be justified and gives (8.21). □

We remark that the method employed cannot be applied if one multiplies the potential in (8.20) by a function $\Phi(x/|x|)\neq\text{const}$.

§9. Magnetic Schrödinger operator. Periodic operator

The two methods of deriving nonregular estimates, used in §§4–7, are nonspecific. All results have been derived from fundamental estimates guaranteeing a regular order of growth for $N(\cdot)$ and Hardy type inequalities, by means of a unified approach. These basic facts (and hence all intermediate results as well) can also be valid for operators differing from (0.1). Let us give two typical examples; in them $l=1$.

1. Schrödinger operator with magnetic field. On $L_2(\mathbb{R}^d)$ consider the operator

$$A(\vec{w}, \alpha V) = -(\nabla - i\vec{w})^2 - \alpha V, \quad (9.1)$$

where $\vec{w} = (w_1, \dots, w_d)$ is a vector field with real components of class $L_{2,\mathrm{loc}}(\mathbb{R}^d)$ and $V=\overline{V}\in L_{1,\mathrm{loc}}(\mathbb{R}^d)$. The main facts about this operator that we use below are taken from [AHS].

To give a consistent definition of the operator (9.1), one considers the quadratic form

$$a(\vec{w})[u] := \int |(\nabla - i\vec{w})u|^2\,dx$$
$$= \int \left(|\nabla u|^2 + 2\,\mathrm{Re}\left(i \sum_{1\leqslant j\leqslant d} \frac{\partial u}{\partial x_j} w_j \bar{u} \right) + |\vec{w}|^2 |u|^2 \right) dx. \quad (9.2)$$

This (nonnegative) form is originally defined on $C_0^\infty(\mathbb{R}^d)$ and then we take its closure in $L_2(\mathbb{R}^d)$. The selfadjoint operator $A(\vec{w})$ generated by this form can be chosen for a natural realization of the operator (9.1) with $V = 0$.

The following result has been essentially proved in [AHS] (cf. [AHS, Theorems 2.3, 2.5]).

PROPOSITION 9.1. *Let* $d \geqslant 3$. *Then*

(1) $\psi \in \mathscr{W}(d, 1)$ *implies*

$$\int \psi |u|^2\,dx \leqslant M_0(\psi) a(\vec{w})[u], \qquad u \in C_0^\infty(\mathbb{R}^d); \quad (9.3)$$

(2) $\psi \in W(d, 1)$ *implies*

$$\int \psi |u|^2\,dx \leqslant M(\psi)\big(a(\vec{w})[u] + \|u\|^2\big), \qquad u \in C_0^\infty(\mathbb{R}^d). \quad (9.4)$$

The constants M_0, M *in* (9.4) *are the same as in* (4.1), (4.2) *for* $l = 1$.

If (9.4) holds (*a fortiori* if (9.3) holds) then, for sufficiently small $\alpha > 0$ at least, the form $a(\vec{w})[u] - \alpha \int V|u|^2\,dx$ is semibounded and closed on $d[a(\vec{w})]$. The corresponding selfadjoint operator on $L_2(\mathbb{R}^d)$ is taken as the realization of the operator (9.1).

The next result (Lieb's estimate) was also discussed in [AHS].

PROPOSITION 9.2. *Let* $d \geqslant 3$ *and* $V \in L_{d/2}(\mathbb{R}^d)$. *Then*

$$N(0; A(\vec{w}, \alpha V)) \leqslant C\alpha^{d/2} \int V_+^{d/2}\,dx, \qquad C = C_{4.4}(d, 1). \quad (9.5)$$

In view of (9.5) and Proposition 9.1, all results of Subsection 4.7 concerning the operator $-\Delta - \alpha V$, $d \geqslant 3$, can be transferred (with the same constants in the estimates!) to the operator (9.1). In particular, the following theorem holds.

THEOREM 9.1. *Let* $d \geqslant 3$. *Then the results of Theorems* 4.2, 4.3, *Proposition* 7.1, *Theorem* 7.4, *and Corollary* 7.1 *with* $l = 1$ *remain true for the operator given by* (9.1).

See [BR] (this volume) for more information on the discrete spectrum of the operator given by (9.1).

We want to express our deep thanks to prof. Raĭkov with whom we consulted when we were writing this Subsection.

2. Operator with periodic coefficients. On $L_2(\mathbb{R}^d)$, $d \geqslant 1$, consider the elliptic operator $A = A(f, g)$ given by the expression

$$-\operatorname{div}(g(x)\nabla u) + f(x)u, \tag{9.6}$$

where the matrix-function g and the function f are measurable real-valued and satisfy

$$g_0 1 \leqslant g \leqslant g_1 1, \; 0 < g_0, \; g(x+n) = g(x), \; x \in \mathbb{R}^d, \; n \in \mathbb{Z}^d, \tag{9.7}$$

$$|f(x)| \leqslant f_0, \quad f(x+n) = f(n), \quad x \in \mathbb{R}^d, \quad n \in \mathbb{Z}^d. \tag{9.8}$$

A will be considered as a nonperturbed operator. A consistent definition of A can be given by using the quadratic form

$$a[u] = \int (\langle g\nabla u, \nabla u\rangle + f|u|^2)\, dx, \qquad u \in H^1(\mathbb{R}^d). \tag{9.9}$$

We assume that *the greatest lower bound of this form is* 0; this can always be ensured by adding a constant to f. Note also that the case of an arbitrary lattice of periods for (9.6) can easily be reduced to the case of the lattice $\mathbb{Z}^d$.

The constructions below are based on an application of a multiplicative transformation to the form (9.9) (compare with [B2]). Denote by R the selfadjoint operator on $L_2(Q)$ generated by the expression (9.6) under the periodic boundary conditions. The form corresponding to R is given by

$$r[u] = \int_Q (\langle g\nabla u, \nabla u\rangle + f|u|^2)\, dx, \qquad u \in \tilde{H}^1(Q), \tag{9.10}$$

where $\tilde{H}^1(Q)$ consists of restrictions to Q of periodic functions of class $H^1_{\mathrm{loc}}(\mathbb{R}^d)$. The spectrum of R is discrete. The following fact is basic.

PROPOSITION 9.3. *Suppose that* (9.7), (9.8) *are satisfied. Then*

(1) *the smallest eigenvalue of* R *equals* 0;

(2) *the zero eigenvalue of* R *is simple; the corresponding eigenfunction* ω *is continuous and does not change its sign* (*is positive*);

(3) *we have*

$$\omega_1 \leqslant C\omega_0, \qquad 0 < \omega_0 = \min \omega(x), \qquad \omega_1 = \max \omega(x), \tag{9.11}$$

$$C = (c(d))^{g_1 g_0^{-1} + f_0^{1/2} g_0^{-1/2}}. \tag{9.12}$$

The proposition stated above should be supplied with comments and references. This will be done in Subsection 4.

Now consider the operator $A(f, g, \alpha V) = A(f, g) - \alpha V$; V is subject to the same conditions as the operator $A_1(\alpha V) = -\Delta - \alpha V$. Under the substitution $u = \omega v$, the forms $\int V|u|^2\, dx$ and $a[u] + \gamma\|u\|^2$ will turn into

$$\int \omega^2 V|v|^2\, dx, \tag{9.13}$$

$$\int \omega^2 (\langle gv, v\rangle + \gamma|v|^2)\, dx, \qquad v \in H^1(\mathbb{R}^d), \tag{9.14}$$

respectively. It is important that the substitution $u \mapsto v = \omega^{-1}u$ maps $H^1(\mathbb{R}^d)$ onto itself. (This is evident if ω is smooth enough, but remains true under the general conditions (9.7), (9.8) as well.) Now the variational principle implies the following two-sided estimate.

PROPOSITION 9.4. *Under the conditions* (9.7), (9.8) *for* $\gamma \geqslant 0$ *we have*

$$N\left(-\frac{\gamma}{g_1}; A_1(\sigma_0\alpha V)\right) \leqslant N(-\gamma; A(f, g, \alpha V)) \leqslant N\left(-\frac{\gamma}{g_0}; A_1(\sigma_1\alpha V)\right),$$

$$\sigma_0 = \omega_0^2\omega_1^{-2}g_1^{-1}, \quad \sigma_1 = \omega_1^2\omega_0^{-2}g_0^{-1}. \tag{9.15}$$

This inequality allows one to extend automatically to $A(f, g, \alpha V)$ all estimates (both regular and nonregular) obtained for $A_1(\alpha V)$. In particular, (4.4) implies the following statement.

PROPOSITION 9.5. *Suppose that* (9.7), (9.8) *hold,* $V \in L_{1,\mathrm{loc}}(\mathbb{R}^d)$, $V_+ \in L_{d/2}(\mathbb{R}^d)$, $d \geqslant 3$. *Then*

$$N(-\gamma; A(f, g, \alpha V)) \leqslant C_{4.4}(\sigma_1\alpha)^{d/2}\int V_+^{d/2}\,dx. \tag{9.16}$$

REMARK 9.1. Unlike (9.5), the constant in (9.16) depends on the parameters involved. By (9.11), (9.12) this dependence is merely on the constants g_0, g_1, f_0 in (9.7), (9.8).

3. Starting with (9.16), it is possible to get an asymptotic formula for $N(-\gamma; A(f, g, \alpha V))$.

THEOREM 9.2. *Under the hypotheses of Proposition* 9.5, *let*

$$V \in L_{d/2}(\mathbb{R}^d).$$

Then

$$N(-\gamma; A(f, g, \alpha V)) \sim v_d(2\pi)^{-d}\alpha^{d/2}\int V_+^{d/2}(\det g)^{-1/2}\,dx, \quad \alpha \to \infty. \tag{9.17}$$

This formula can be derived from results of [R2]. We give, however, a short proof here. In view of (1.15), (1.16), the problem reduces to the investigation of the distribution function $n_+(\alpha^{-1})$ for the negative spectrum of the ratio of the forms given by (9.13), (9.14). By (9.16) it is sufficient to know the asymptotics for $V \in C_0^\infty(\mathbb{R}^d)$, assuming that f, g are fixed. Imposing Dirichlet or Neumann boundary conditions we get (coinciding!) lower and upper asymptotic estimates that lead to (9.17). The crucial point here is the spectral asymptotics for Dirichlet and Neumann problems in bounded domains obtained in [BS2]. It is essential that no smoothness conditions are imposed in [BS2] on the matrix of leading coefficients. Since the asymptotic coefficient depends only on the ratio of principal symbols, the factor ω^2 cancels and does not enter into (9.17).

REMARK 9.2. The formula (9.17) is true for $d = 1, 2$ as well, but this time under other assumptions on V. We shall not discuss this matter.

REMARK 9.3. If $\gamma > 0$, then one can obtain an estimate like (9.16), and hence also the formula (9.17) by elementary methods, avoiding the use of the multiplicative transformation. This method gives an unbounded constant in (9.6) as $\gamma \to 0$.

4. Let us give some hints to the proof of Proposition 9.3. As a matter of fact, it reduces to a series of references. We begin with statements (2), (3), assuming that 0 is the minimum of the form given by (9.10). Let $\tilde{\omega} \in \tilde{H}^1(Q)$ be a real-valued function on which this minimum is attained. Then (cf. [LU, Lemma 3.3 of Chapter II]) $\omega := |\tilde{\omega}| \in \tilde{H}^1(Q)$ and the form given by (9.10) evaluated at ω is also equal to 0. Thus ω is a nontrivial nonnegative periodic solution of the homogeneous equation $-\operatorname{div} g\nabla\omega + f\omega = 0$ in the entire space $\mathbb{R}^d$. More precisely, this means that

$$\int (\langle g\nabla\omega, \nabla\varphi\rangle + f\omega\overline{\varphi})\, dx = 0, \qquad \forall\varphi \in C_0^\infty(\mathbb{R}^d). \tag{9.18}$$

Hence (cf. [LU, Theorem 14.1 of Chapter III]) ω (and $\tilde{\omega}$) are Hölder continuous. Now (9.18) and $\omega(x) \geqslant 0$ imply (cf. [GT, Theorem 8.20]) that $\omega(x) > 0$, and (9.11), (9.12) hold. Thus $\tilde{\omega}$ does not change its sign and one can assume $\tilde{\omega} = \omega$. We have checked that every solution $\tilde{\omega} \in \tilde{H}^1(Q)$ is strictly positive (up to a constant factor), and consequently $\dim\ker R = 1$. It remains to prove statement (1). Transforming the quadratic form given by (9.9) to that given by (9.14) with $\gamma = 0$, we see that the greatest lower bound of (9.9) also equals 0. □

Proposition 9.3 can also be derived from results of Agmon's paper [A].

Further results on the discrete spectrum of the periodic operator of the form (9.6) perturbed by a decreasing potential can be found in [B3].

Added in proof. Recently Laptev [La] found an asymptotic formula for operator like (8.20), with the complementary factor $\Psi(x/|x|) > 0$, $\Psi \in L_{d/2}(\mathbb{S}^{d-1})$ included in the potential (cf. the end of §8). Namely,

$$N(0; A(\alpha)) \sim \pi^{-1} \sum_l \int_0^\infty \left(\Lambda_l(t) + \left(\frac{d-2}{2}\right)^2\right)_-^{1/2} dt,$$

where $\Lambda_l(t)$ are the eigenvalues of the operator $-\Delta_\theta - t^{-\alpha}\Psi(\theta)$ on $\mathbb{S}^{d-1}$ and $a_- = \max(-a, 0)$.

REFERENCES

[A] S. Agmon, *On positive solutions of elliptic equations with periodic coefficients in* $\mathbb{R}^n$, *spectral results and extensions to elliptic operators on Riemannian manifolds*, Differential Equations, North-Holland Math. Stud. 92, North-Holland, Amsterdam and New York, 1984, pp. 7–17.

[AHS] J. Avron, I. Herbst, and B. Simon, *Schrödinger operators with magnetic fields.* I. *General interactions*, Duke Math. J. **45** (1978), 847–883.

[BL] J. Bergh and J. Löfström, *Interpolation spaces. An introduction*, Springer-Verlag, Berlin, Heidelberg, New York, 1976.

[B1] M. Sh. Birman, *On the spectrum of the Schrödinger and Dirac operators*, Dokl. Akad. Nauk SSSR **129** (1959), 239–241. (Russian)

[B2] ——, *On the spectrum of singular boundary value problems*, Mat. Sb. **55** (1961), 125–174; English transl., Amer. Math. Soc. Transl. (2) **53** (1966), 23–80.

[B3] ——, *Discrete spectrum in the gaps of a continuous one for perturbations with large coupling constant*, (this volume).

[BB] M. Sh. Birman and V. V. Borzov, *On the asymptotics of the discrete spectrum for certain singular differential operators*, Probl. Mat. Fiz. **5** (1971), 24–38, English transl. in Topics in Math. Phys. **5** (1972), 19–30.

[BKS] M. Sh. Birman, G. E. Karadzhov, and M. Z. Solomyak, *Boundedness conditions and spectrum estimates for the operators* $b(X)a(D)$ *and their analogs*, (this volume).

[BR] M. Sh. Birman and G. D. Raĭkov, *Discrete spectrum in the gaps for perturbations of the magnetic Schrödinger operator*, (this volume).

[BS1] M. Sh. Birman and M. Z. Solomyak, *Piecewise-polynomial approximations of functions of the classes* W_p^α, Mat. Sb. **73 (115)** (1967), 331–355; English transl. in Math. USSR-Sb. **2** (1967).

[BS2] ——, *Spectral asymptotics of nonsmooth elliptic operators.* I, Trudy Moskov. Mat. Obshch. **27** (1972), 3–52; English transl. in Trans. Moscow Math. Soc. **27** (1972).

[BS3] ——, *Qualitative analysis in Sobolev imbedding theorems and applications to spectral theory*, Tenth Math. Summer School (Katsiveli/Nalchik, 1972), Izdanie Inst. Mat. Akad. Nauk Ukrain. SSR, Kiev, 1974, pp. 5–189; English transl., Amer. Math. Soc. Transl. (2) **114** (1980).

[BS4] ——, *Estimates for the singular numbers of integral operators*, Uspekhi Mat. Nauk **32** (1977), no. 1, 17–84; English transl. in Russian Math. Surveys **32** (1977), no. 1, 15–89.

[BS5] ——, *Spectral theory of selfadjoint operators in Hilbert space*, Izdat. Leningrad. Univ., Leningrad, 1980; English transl., Reidel, Dordrecht, 1987.

[BS6] ——, *Interpolation estimates for the number of negative eigenvalues of a Schrödinger operator*, Schrödinger Operators: Standard and Nonstandard, Dubna, USSR, Sept. 6–10, 1988, World Scientific, Singapore, 1989, pp. 3–18.

[BS7] ——, *The estimates for the number of negative bound states of the Schrödinger operator for large coupling constant*, Proc. Conf. Integral Equations and Inverse Problems, Varna, 1989, pp. 49–57.

[BS8] ——, *Negative discrete spectrum of the Schrödinger operator with large coupling constant: a qualitative discussion*, Oper. Theory: Adv. Appl. **46** (1990), 3–16.

[BS9] ——, *Discrete negative spectrum under nonregular perturbations* (*polyharmonic operator; Schrödinger operator with magnetic field; periodic operator*), Proc. Conf. Exact Methods in Quantum Dynamics, Prague, 1990 (to appear).

[BS10] ——, *Schrödinger operator. Estimates of the number of the bound states as a function-theoretic problem*, Lectures of the XIV School on Operator Theory on Functional Spaces, I, Novgorod, 1990; English transl., Amer. Math. Soc., Providence, RI (to appear).

[BSR] P. Blanchard, J. Stubbe, and J. Rezende, *New estimates on the number of bound states of Schrödinger operator*, Lett. Math. Phys. **14** (1987), 215–225.

[C] M. Cwikel, *Weak type estimates for singular values and the number of bound states of Schrödinger operators*, Ann. Math. **106** (1977), 93–100.

[EK1] Yu. V. Egorov and V. A. Kondrat′ev, *On an estimate the number of points of the negative spectrum of Schrödinger operator*, Mat. Sb. **134** (1987), 556–570; English transl. in Math. USSR-Sb **62** (1989), 551–566.

[EK2] ——, *On the negative spectrum of elliptic operator of any order*, Schrödinger Operators: Standard and Nonstandard, Dubna, USSR, Sept. 6–10, 1988, World Scientific, Singapore, 1989, pp. 21–27.

[EK3] ——, *Estimates of the negative spectrum for an elliptic operator*, Lectures of XIV School on Operator Theory on Functional Spaces, Novgorod, 1990; English transl., Amer. Math. Soc., Providence, RI (to appear).

[EK4] ———, *On the negative spectrum of an elliptic operator*, Mat. Sb. **181** (1990), 147–166; English transl. in Math. USSR-Sb. **69** (1991), 155–177.

[GT] D. Gilbarg and N. S. Trudinger, *Elliptic partial differential equations of second order*, Springer-Verlag, Berlin and New York, 1983.

[GK] I. C. Gokhberg and M. G. Kreĭn, *Introduction to the theory of linear nonselfadjoint operators in Hilbert space*, "Nauka", Moscow, 1965; English transl., Amer. Math. Soc., Providence, RI, 1969.

[LU] O. A. Ladyzhenskaya and N. N. Ural′tseva, *Linear and quasilinear elliptic equations*, "Nauka", Moscow, 1964; English transl., Academic Press, New York, 1968.

[La] A. Laptev, *Asymptotics of the negative discrete spectrum of a class of Schrödinger operators with large coupling constant*, Preprint, 1991.

[L1] E. Lieb, *The number of bound states of one-body Schrödinger operators and the Weyl problem*, Bull. Amer. Math. Soc. **82** (1976), 751–753.

[L2] ———, *The number of bound states of one-body Schrödinger operators and the Weyl problem*, Proc. Sympos. Pure Math. **36** (1980), 241–252.

[MS] V. G. Maz′ya and T. O. Shaposhnikova, *Theory of multipliers in spaces of differentiable functions*, Izdat. Leningrad. Univ., Leningrad, 1986; English transl. Pitman (Advanced Publishing Program), Boston, MA, and London, 1985.

[N] S. M. Nikol′skiĭ, *Approximation of functions of several variables and embedding theorems*, "Nauka", Moscow, 1969; English transl., Springer-Verlag, Berlin and New York, 1975.

[RS] M. Reed and B. Simon, *Methods of modern mathematical physics*. IV: *Analysis of operators*, Academic Press, New York, 1978.

[R1] G. V. Rozenblyum, *The distribution of the discrete spectrum for singular differential operators*, Dokl. Akad. Nauk SSSR **202** (1972), 1012–1015; English transl. in Soviet Math. Dokl. **13** (1972).

[R2] ———, *Distribution of the discrete spectrum of singular differential operators*, Izv. Vyssh. Uchebn. Zaved. Mat. no. 1, **1976**, 75–86; English transl. in Soviet Math. (Iz. VUZ.) **20** (1976).

[Sch] J. Schwinger, *On the bound states of a given potential*, Proc. Nat. Acad. USA **47** (1961), 122–129.

[S] B. Simon, *Weak trace ideals and the number of bound states of Schrödinger operators*, Trans. Amer. Math. Soc. **224** (1976), 367–380.

[SW] E. Stein and G. Weiss, *Introduction to Fourier analysis on Euclidean spaces*, Princeton Univ. Press, Princeton, NJ, 1971.

[T] H. Triebel, *Interpolation theory. Function spaces. Differential operators*, VEB Deutscher Verlag, Berlin, 1978.

[VS] I. L. Vulis and M. Z. Solomyak, *Spectral asymptotics of second-order degenerate elliptic operators*, Izv. Akad. Nauk SSSR Ser. Mat. **38** (1978), 1362–1392; English transl. in Math. USSR-Izv. **8** (1974).

Translated by S. KISLYAKOV

ADVANCES IN SOVIET MATHEMATICS
Volume 7, 1991

Discrete Spectrum in the Gaps of a Continuous One for Perturbations with Large Coupling Constant

M. SH. BIRMAN

ABSTRACT. Given two operators $A = A^*$, $V \geqslant 0$, we consider the quantity $N_\pm(\lambda, \alpha)$ (where $\alpha > 0$ and λ is a regular point for A) defined as the number of eigenvalues of $A \mp tV$ crossing λ as t increases from 0 to α. An abstract theorem on the stability of the asymptotics for $N_\pm$ (as $\alpha \to \infty$) under variations of A and λ is proved. Some applications to differential operators are given, in particular, to the Schrödinger operator.

Introduction

Let A be a semibounded selfadjoint operator, λ_0 be the lower bound of its spectrum, and λ be a point in a gap Λ of the spectrum of A. For a perturbation $V = W^*W$ and a parameter $\alpha > 0$ (the coupling constant), let $A_\pm(\alpha) = A \mp \alpha V$. Denote by $N_\pm(\lambda, A, W, \alpha)$ the number of eigenvalues of $A_\pm(t)$ that cross the point λ as t increases from 0 to a given α. If $\lambda < \lambda_0$ then N_+ coincides with the number of eigenvalues of $A_+(\alpha)$ lying to the left of λ. Estimates and asymptotic formulas (as $\alpha \to \infty$) for N_+ have been intensively studied for the Schrödinger operator and its generalizations in the case $\lambda < \lambda_0$. See [BS6] in this collection for more details. It is important that if $\lambda < \lambda_0$, then one can use the variational technique directly. This simplifies the investigation to a great extent. Here we propose an abstract theorem (Theorem 1.2) on the stability of power-type asymptotic formulas for $N_\pm$ under variations of A and λ. In particular, it turns out to be possible to transfer the calculations to the case of a semibounded gap. This often allows one to apply already known results directly. Of course, this approach works only when the asymptotics of $N_+(\lambda, \alpha)$ does not depend on λ for $\lambda > \lambda_0$. The perturbations V for which this is really the case are called *regular* in [BS6]. Theorem 1.2 gives interesting information only for such perturbations. In applications, this corresponds to a power-type asymptotics for $N_+(\lambda, \alpha)$,

1991 *Mathematics Subject Classification.* Primary 35J10, 47A75.

1051-8037/91 \$1.00 + \$.25 per page

$\lambda \in \Lambda$, where the exponent is the standard semiclassical exponent $\varkappa$ (see §2 below). For N_- this approach gives only the order $o(\alpha^\varkappa)$. For *nonregular* perturbations (in the terminology of [BS6]), Theorem 1.2 does not work.

We could give a generalization of Theorem 1.2 that would allow some non-power-type asymptotics. But no applications of such a generalization have been found so far.

There is a number of papers concerning the asymptotics of $N_\pm(\lambda, \alpha)$, $\lambda \in \Lambda$, as $\alpha \to \infty$. We mention, first and foremost, [H, ADH] and references therein. In particular, an asymptotic formula for N_+ has been obtained in [H] for the multidimensional Schrödinger operator in the regular case. This result will be considerably generalized and strengthened here, cf. §2, Subsection 3. There are also examples of a different nature in §2. We could extend substantially both the number of examples and the degree of their generality. The calculation of asymptotics for N_+ in the case when A is the Schrödinger operator with magnetic field deserves a special mention. This material is presented in a separate paper [BR] in this collection. It is important that there, in contrast with [B2], only mild restrictions of local character are imposed on the magnetic potential.

The approach proposed here enables one to single out relatively simple questions about the asymptotics of $N_+(\lambda, \alpha)$ for regular V and to give them a unified treatment. The following subjects should be classified as more difficult.

(1) Determining the sharp order of asymptotics for N_- and calculating the asymptotics itself. An interesting result on this matter for the Schrödinger operator is obtained in [ADH].
(2) The case when λ is an endpoint of the gap and V is not necessarily regular. See [B3] for the asymptotics of $N_\pm$ in the case of periodic operators of the second order.
(3) Studying the asymptotics of the difference $N_+(\lambda, \alpha) - N_+(\mu, \alpha)$, where $[\mu, \lambda] \subseteq \operatorname{clos} \Lambda$. See [So] on this matter for the one-dimensional periodic Schrödinger operator.
(4) The case of not necessarily positive V. For the Schrödinger operator, see [ADH] and references therein.

A short announcement of the results of this paper can be found in [B2]. In the present exposition, we enlarge the part containing applications. There is also a deviation in the abstract part. It was operator sums that were used in [B2] to define perturbations of A. Here we use form-sums instead. This is more convenient for the major part of the applications, but is closely associated with the semiboundedness of operators. The perturbation scheme recently developed by Yafaev [Y] for the purposes of scattering theory allows one to use form-sums in the case of nonsemibounded operators as well. We have not chosen that approach for the sake of simplicity.

§1. The main theorem

1. Notation. In what follows $\mathfrak{H}$ is the principal Hilbert space and $\mathfrak{G}$ is the "auxiliary" one. The scalar product $(\cdot, \cdot)$ and the norm $\|\cdot\|$ will not be supplied with indices referring to $\mathfrak{H}$ and $\mathfrak{G}$. This will not lead to any confusion. For a linear mapping M, we use the following notation: $\mathfrak{D}(M)$ is the domain of M, $\operatorname{Im} M$ is its image, $\rho(M)$ is its resolvent set, and M^* is the adjoint operator. The class of continuous operators will be denoted by $\mathfrak{R}$ and that of compact operators by $\mathfrak{S}_\infty$. We often use more detailed notation such as $\mathfrak{R}(\mathfrak{H}, \mathfrak{G})$, $\mathfrak{S}_\infty(\mathfrak{H})$, etc. Let $T \in \mathfrak{S}_\infty(\mathfrak{H}_1, \mathfrak{H}_2)$ and let $s_k(T)$, $k \in \mathbb{N}$, be the singular numbers of T (the eigenvalues of $(T^*T)^{1/2}$). Let us introduce the distribution function for singular numbers (s-numbers)

$$\nu(s, T) = \operatorname{card}\{k : s_k(T) > s\}, \qquad s > 0.$$

If $T = T^* \in \mathfrak{S}_\infty(\mathfrak{H})$, then $n_\pm(\cdot, T)$ denotes the distribution function for the positive (resp., negative) eigenvalues of T. We then have $\nu = n_+ + n_-$. For $0 < p < \infty$, the class $\Sigma_p \subset \mathfrak{S}_\infty$ is distinguished by the condition

$$|T|_p^p := \sup_{s>0} s^p \nu(s, T) < \infty.$$

The functional $|\cdot|_p$ is taken for the quasinorm on Σ_p. By Σ_p^0 we denote the separable subspace of Σ_p defined by

$$\Sigma_p^0 := \{T \in \Sigma_p : \nu(s, T) = o(s^{-p}),\ s \to 0\}.$$

For $T \in \Sigma_p$, we introduce the functionals

$$\Delta_p(T) := \limsup_{s\to 0} s^p \nu(s, T),$$
$$\delta_p(T) := \liminf_{s\to 0} s^p \nu(s, T).$$

If $T = T^*$, then the functionals $\Delta_p^{(\pm)}(T)$, $\delta_p^{(\pm)}(T)$ for the distribution functions $n_\pm(\cdot, T)$ are defined similarly. All the functionals Δ_p, δ_p, $\Delta_p^{(\pm)}$, and $\delta_p^{(\pm)}$ are continuous on Σ_p. They are defined, in fact, on the quotient space Σ_p/Σ_p^0 (i.e., adding a summand from Σ_p^0 to the argument does not change their values) and they are continuous with respect to the quasinorm $(\Delta_p(\cdot))^{1/p}$ on Σ_p/Σ_p^0. More details on the classes Σ_p can be found in [BS4].

By C, c (with or without indices) we denote various constants in the estimates.

2. Let $a[\cdot, \cdot]$ be a sesquilinear form on $\mathfrak{H}$ which is semibounded from below and closed. Its domain of definition $d[a]$ is assumed to be dense in $\mathfrak{H}$. The form a generates a uniquely determined selfadjoint operator A on $\mathfrak{H}$. Let us fix a number $\gamma \in \mathbb{R}$ such that $a_\gamma := a + \gamma \geqslant 1$, that is

$$a_\gamma[x, x] := a[x, x] + \gamma(x, x) \geqslant (x, x), \qquad x \in d[a].$$

Denote by $H_\gamma[a]$ the (complete) Hilbert space with metric form

$$a_\gamma[x, x] = \|(A+\gamma I)^{1/2}x\|^2, \qquad x \in d[a].$$

Let f be a Hermitian form on $d[a]$ satisfying

$$|f[x, x]| \leqslant \varepsilon a_\gamma[x, x] + c(\varepsilon)(x, x), \qquad x \in d[a], \qquad \forall\varepsilon > 0. \tag{1.1}$$

Consider the form $b := a + f$ on $d[a]$. The following elementary statement is certainly guaranteed by (1.1).

PROPOSITION 1.1. *If* (1.1) *is satisfied, then the form* $b = a + f$ *is semi-bounded from below and closed on* $d[a]$.

By an appropriate choice of γ, we can assume without loss of generality that $b_\gamma := b + \gamma \geqslant 1$. Then the norms in $H_\gamma[a]$ and $H_\gamma[b]$ are equivalent or, which is the same,

$$\mathfrak{D}((A+\gamma I)^{1/2}) = \mathfrak{D}((B+\gamma I)^{1/2}), \tag{1.2}$$

where $B = B^*$ is the operator corresponding to the form b. At the same time, (1.2) is equivalent to the following pair of relations:

$$(A+\gamma I)^{1/2}(B+\gamma I)^{-1/2} \in \mathfrak{R}(\mathfrak{H}), \tag{1.3}$$

$$(B+\gamma I)^{1/2}(A+\gamma I)^{-1/2} \in \mathfrak{R}(\mathfrak{H}). \tag{1.4}$$

It is not difficult to establish a version of the resolvent identity (which is now expressed in terms of forms) for the operators A, B. Set

$$M_\lambda := (A-\lambda I)^{-1} - (B-\lambda I)^{-1}, \qquad \lambda \in \rho(A) \cap \rho(B). \tag{1.5}$$

PROPOSITION 1.2. *Under the hypotheses of Proposition* 1.1, *we have*

$$\begin{aligned}(M_\lambda h, h) &= f[(B-\lambda I)^{-1}h, (A-\bar{\lambda} I)^{-1}h] \\ &= f[(A-\lambda I)^{-1}h, (B-\bar{\lambda} I)^{-1}h], \qquad h \in \mathfrak{H}.\end{aligned} \tag{1.6}$$

Note the following simple condition guaranteeing (1.1) (compare with [B1, BS2]).

PROPOSITION 1.3. *Let a Hermitian form* f *be compact on* $d[a]$. *Then* (1.1) *holds.*

We recall that a sesquilinear form v is said to be compact on $d[a]$ if v is continuous on $H_\gamma[a]$ and the corresponding operator $Q = Q(v)$ defined by

$$v[x, y] = a_\gamma[Qx, y], \qquad \forall x, y \in d[a], \tag{1.7}$$

is compact on $H_\gamma[a]$. If $v[x, x] \geqslant 0$, then v is compact on $d[a]$ if and only if the embedding of $H_\gamma[a]$ into the space with the seminorm $(v[x, x])^{1/2}$ is compact.

It is easy to derive the following statement from (1.6) (compare with [B1]).

PROPOSITION 1.4. *Under the hypotheses of Proposition* 1.3,

$$M_\lambda \in \mathfrak{S}_\infty(\mathfrak{H}, H_\gamma[a]) \subset \mathfrak{S}_\infty(\mathfrak{H}). \tag{1.8}$$

3. Let $W : \mathfrak{H} \to \mathfrak{G}$ be a closable linear operator, $\mathfrak{D}(W) \supset d[a]$ and

$$G(A) := W(A+\gamma I)^{-1/2} \in \mathfrak{R}(\mathfrak{H}, \mathfrak{G}).$$

If $\lambda = \overline{\lambda} \in \rho(A)$, then the operator $X_\lambda(A)$ given by

$$\begin{aligned} X_\lambda(A) &:= G(A)(A+\gamma I)(A-\lambda I)^{-1}G(A)^* \\ &= W|A-\lambda I|^{-1/2}\operatorname{sign}(A-\lambda I)(W|A-\lambda I|^{-1/2})^*, \end{aligned} \tag{1.9}$$

is continuous and selfadjoint on $\mathfrak{G}$. The operator $X_\lambda(A)$ can be equivalently defined by

$$X_\lambda(A)g = W(A-\lambda I)^{-1}W^*g, \qquad g \in \mathfrak{D}(W^*). \tag{1.10}$$

Note that $X_\lambda(A)$ extends analytically to all of $\rho(A)$.

If (1.1) holds then, by (1.3), (1.4), for $\mu = \overline{\mu} \in \rho(B)$ the operator

$$X_\mu(B) = G(B)(B+\gamma I)(B-\mu I)^{-1}G(B)^*$$

is well defined, continuous, and selfadjoint on $\mathfrak{G}$. Below we shall have to replace (1.1) by a somewhat stronger condition: the Hermitian form f has a nonnegative majorant, that is there exists a nonnegative sesquilinear form f_* with

$$|f[x,y]|^2 \leqslant f_*[x,x]f_*[y,y], \qquad x, y \in d[a], \tag{1.11}$$

and, moreover, this majorant satisfies a condition like (1.1):

$$f_*[x,x] \leqslant \varepsilon a_\gamma[x,x] + c(\varepsilon)\|x\|^2, \qquad \forall \varepsilon > 0. \tag{1.12}$$

Clearly, (1.11) is equivalent to

$$|f[x,x]| \leqslant f_*[x,x], \qquad x \in d[a]. \tag{1.13}$$

For a form f of definite sign, we always take $f_* = f$ or $f_* = -f$.

The following theorem plays an important role in the sequel.

THEOREM 1.1. *Let* $b = a + f$ *and suppose that* (1.13), (1.12) *hold. Let for some* $p > 0$

$$W(A+\gamma I)^{-1/2} \in \Sigma_{2p}(\mathfrak{H}, \mathfrak{G}), \tag{1.14}$$

$$W(A+\gamma I)^{-1} \in \Sigma^0_{2p}(\mathfrak{H}, \mathfrak{G}). \tag{1.15}$$

Then for $\lambda = \overline{\lambda} \in \rho(A)$, $\mu = \overline{\mu} \in \rho(B)$, *we have*

$$Z := X_\lambda(A) - X_\mu(B) \in \Sigma^0_p. \tag{1.16}$$

PROOF. By adding $(-\lambda)I$ to A and B we can assume that $\lambda = 0$. Moreover, we can then add the form of the operator $(-\mu I)$ to f (and simultaneously that of $|\mu|I$ to f_*), which enables us to assume that $\mu = 0$ as well.

Starting with (1.10) and a similar representation for $X_0(B)$, by using (1.3), (1.4), we get

$$(Zg, g) = f[A^{-1}W^*g, B^{-1}W^*g], \qquad g \in \mathfrak{D}(W^*).$$

Let us set $x = A^{-1}W^*g$, $y = B^{-1}W^*g$. For every $\varepsilon > 0$,

$$\begin{aligned}|(Zg, g)| &\leqslant (f_*[x, x]f_*[y, y])^{1/2} \\ &\leqslant \varepsilon^{-1/2}f_*[x, x] + \varepsilon^{1/2}f_*[y, y].\end{aligned}$$

To estimate $f_*[x, x]$, we use (1.12). Moreover, (1.12), (1.3) imply

$$f_*[y, y] \leqslant a_\gamma[y, y] + c(1)\|y\|^2 \leqslant Cb_\gamma[y, y].$$

Thus

$$|(Zg, g)| \leqslant \varepsilon^{1/2}(a_\gamma[x, x] + Cb_\gamma[y, y] + \varepsilon^{-1/2}c(\varepsilon)\|x\|^2). \tag{1.17}$$

Let us estimate the forms on the right. Since $0 \in \rho(A) \cap \rho(B)$, we obtain

$$\begin{aligned}a_\gamma[x, x] &= \|(A + \gamma I)^{1/2}A^{-1}W^*g\|^2 \\ &\leqslant C\|(A + \gamma I)^{-1/2}W^*g\|^2 = C\|G(A)^*g\|^2,\end{aligned} \tag{1.18}$$

and, taking into account (1.3),

$$\begin{aligned}b_\gamma[y, y] &\leqslant C\|(B + \gamma I)^{-1/2}W^*g\|^2 \\ &\leqslant C\|(A + \gamma I)^{1/2}(B + \gamma I)^{-1/2}\|^2\|(A + \gamma I)^{-1/2}W^*g\|^2 \\ &= C_1\|G(A)^*g\|^2.\end{aligned} \tag{1.19}$$

At the same time

$$\|x\|^2 = \|A^{-1}W^*g\|^2 = \|(WA^{-1})^*g\|^2. \tag{1.20}$$

Considering $a_\gamma[x, x]$, $b_\gamma[y, y]$ as quadratic forms in g, we see, by (1.14), (1.18), (1.19), that they correspond to operators of class Σ_p. By virtue of (1.15), (1.20), the form $\|x\|^2$, when treated similarly, corresponds to an operator of class Σ_p^0. Since the factor $\varepsilon^{1/2}$ in (1.17) can be arbitrarily small, Z satisfies (1.16). □

4. Let us discuss in more detail the case when the form of the perturbation is *of fixed sign and compact on* $d[a]$. We shall introduce a numerical parameter that will absorb the sign of the form. So the form of the perturbation will now be supposed nonnegative and denoted by v. Let W be a closable operator with $\mathfrak{D}(W) \supset d[a]$ and let

$$G(A) = W(A + \gamma I)^{-1/2} \in \mathfrak{S}_\infty(\mathfrak{H}, \mathfrak{G}). \tag{1.21}$$

Then the form

$$v[x, y] = (Wx, Wy) \tag{1.22}$$

is clearly compact on $d[a]$.

REMARK 1.1. A compact nonnegative form v on $d[a]$ always has a representation of type (1.22). Indeed, let Q be the operator from (1.7). Then setting $\mathfrak{G} = H_\gamma[a]$, $Wx = Q^{1/2}x$ for $x \in d[a] = \mathfrak{D}(W)$, we see that (1.21), (1.22) hold.

We shall use the representation (1.22) for v. Since v satisfies an estimate of type (1.1), it is possible to introduce a family of semibounded forms closed on $d[a]$ by setting

$$a_\pm(\alpha) = a \mp \alpha v, \qquad \alpha > 0. \tag{1.23}$$

The corresponding selfadjoint operators on $\mathfrak{H}$ will be denoted by $A_\pm(\alpha)$. In view of Proposition 1.4, the difference of the resolvents of A and $A_\pm(\alpha)$ is compact. Therefore, in the gaps of the spectrum of A, the spectrum of $A_\pm(\alpha)$ is discrete. The description of the discrete spectrum can be reduced to the investigation of the spectrum of a certain compact operator on $\mathfrak{G}$. It is standard to proceed in this way for additive operator perturbations. Here we present a version suitable for perturbations of forms.

PROPOSITION 1.5. *Let $\lambda = \bar{\lambda} \in \rho(A)$. The following two statements are equivalent*

(1) *The point λ is an eigenvalue of multiplicity τ for $A_\pm(\alpha)$.*

(2) *The point $\pm\alpha^{-1}$ is an eigenvalue of multiplicity τ for $X_\lambda(A)$, the selfadjoint compact operator in $\mathfrak{G}$ defined in* (1.9), (1.10).

PROOF. Adding $(-\lambda)I$, we reduce the matter to the case $\lambda = 0$. Denote $\Omega := \operatorname{sign} A$. To be definite, we consider the case of "+"-sign.

(1) $\Rightarrow$ (2). $0 \neq x \in \operatorname{Ker} A_+(\alpha) \iff a[x, \varphi] = \alpha(Wx, W\varphi)$, $\forall \varphi \in d[a]$ $\iff (\Omega|A|^{1/2}x, |A|^{1/2}\varphi) = \alpha((W|A|^{-1/2})^* Wx, |A|^{1/2}\varphi) \iff \alpha^{-1}\Omega|A|^{1/2}x = (W|A|^{-1/2})^* Wx \iff \alpha^{-1}x = |A|^{-1/2}\Omega(W|A|^{-1/2})^* Wx$

$$\Longrightarrow \alpha^{-1}z = W|A|^{-1/2}\Omega(W|A|^{-1/2})^* z, \tag{1.24}$$

where $z := Wx \neq 0$. The equality in (1.24) will precisely mean that $X_0(A)z = \alpha^{-1}z$.

(2) $\Rightarrow$ (1). Let $0 \neq z \in \mathfrak{G}$, (1.24) be satisfied, and

$$x := \alpha|A|^{-1/2}\Omega(W|A|^{-1/2})^* z.$$

Then $0 \neq x \in d[a]$, $z = Wx$, and $\alpha^{-1}x = |A|^{-1/2}\Omega(W|A|^{-1/2})^* Wx$. It remains to use the arrows "$\Longleftarrow$" from the first part of the proof. □

Let the interval $\Lambda = (\lambda_-, \lambda_+) \subset \mathbb{R}$ be a gap in the spectrum of A. Let $\lambda_0 \in \Lambda$ be an eigenvalue of $A_+(\alpha_0)$ of multiplicity τ; then α_0^{-1} is an eigenvalue (of the same multiplicity) for $X_{\lambda_0}(A)$. The operator $X_\lambda(A)$ is an increasing function of the variable $\lambda \in \Lambda$. Therefore, there exist *decreasing* (analytic) functions $\alpha_j(\lambda)$, $j = 1, \dots, \tau$, defined near λ_0, such that $(\alpha_j(\lambda))^{-1}$ are eigenvalues of $X_\lambda(A)$. For every $j = 1, \dots, \tau$ the inverse function of

$\alpha_j(\lambda)$, denoted by $\lambda_j(\alpha)$, is also *decreasing*. The functions $\lambda_j(\alpha)$ give the eigenvalues of $A_+(\alpha)$ near α_0. Thus *the eigenvalues of* $A_+(\alpha)$ *inside the gap* Λ *move monotonically to the left as* α *increases.* Moreover, it is immediate from Proposition 1.5 that, for a given $\lambda \in \Lambda$, we have $\lambda \in \rho(A_+(\alpha))$ for all sufficiently small $\alpha > 0$. Finally, it is clear from what has been said above that λ_- is not an accumulation point for the spectrum of $A_+(\alpha)$ in Λ. The properties of the spectrum of $A_-(\alpha)$ in Λ arc similar. The eigenvalues of $A_-(\alpha)$ move monotonically to the right as α increases, and λ_+ is not an accumulation point for the spectrum of $A_-(\alpha)$ in Λ.

Denote by $N_\pm(\lambda, A, W, \alpha)$ the number of eigenvalues of $A_\pm(t)$ crossing $\lambda = \overline{\lambda} \in \rho(A)$ as the parameter t increases from 0 to $\alpha > 0$. If λ belongs to a semi-infinite gap, then $N_-(\lambda, A, W, \alpha) = 0$, and $N_+(\lambda, A, W, \alpha)$ is the number of eigenvalues lying to the left of λ.

Let us combine monotonicity considerations with Proposition 1.5. We then obtain

PROPOSITION 1.6. *Let* $a_\pm(\alpha)$ *be the family of forms given by* (1.23), *the form* v *be defined in* (1.22), *and suppose that* (1.21) *is satisfied. Then*

$$N_\pm(\lambda, A, W, \alpha) = n_\pm(\alpha^{-1}, X_\lambda(A)), \qquad \lambda = \overline{\lambda} \in \rho(A). \tag{1.25}$$

We remark that for λ in a semi-infinite gap, (1.25) corresponds to (1.16) in [BS6].

5. In this subsection we present the main result on stability of asymptotic characteristics of $N_\pm(\lambda, A, W, \alpha)$, as $\alpha \to \infty$, under variations of A and λ.

Let $b = a + f$ be the same form as in Theorem 1.1. By virtue of (1.3) and (1.4), condition (1.21) is equivalent to the condition $G(B) \in \mathfrak{S}_\infty(\mathfrak{H}, \mathfrak{G})$. Therefore, all that was said in Subsection 4 applies to the family of forms

$$b_\pm(\alpha) = b \mp \alpha v, \qquad \alpha > 0. \tag{1.26}$$

In particular, along with (1.25), we have

$$N_\pm(\mu, B, W, \alpha) = n_\pm(\alpha^{-1}, X_\mu(B)), \qquad \mu \in \rho(B). \tag{1.27}$$

Suppose now that (1.14) holds. Then $X_\lambda(A) \in \Sigma_p$ for $\lambda \in \rho(A)$ and, by (1.3), $X_\mu(B) \in \Sigma_p$ for $\mu \in \rho(B)$. Now (1.25) implies that the following quantities are finite:

$$\Delta_p^{(\pm)}(\lambda, A, W) := \limsup_{\alpha\to\infty} \alpha^{-p} N_\pm(\lambda, B, W, \alpha) = \Delta_p^{(\pm)}(X_\lambda(A)), \tag{1.28}$$

$$\delta_p^{(\pm)}(\lambda, A, W) := \liminf_{\alpha\to\infty} \alpha^{-p} N_\pm(\lambda, B, W, \alpha) = \delta_p^{(\pm)}(X_\lambda(A)). \tag{1.29}$$

The same is true for the operator B at any point $\mu \in \rho(B)$.

THEOREM 1.2. *Suppose that the assumptions of Theorem* 1.1 *are satisfied and let the form* v *in* (1.23), (1.26) *be defined by* (1.22). *Then for every* $\lambda \in \rho(A)$, $\mu \in \rho(B)$ *we have*

$$\Delta_p^{(+)}(\lambda, A, W) = \Delta_p^{(+)}(\mu, B, W), \tag{1.30}$$

$$\delta_p^{(+)}(\lambda, A, W) = \delta_p^{(+)}(\mu, B, W), \tag{1.31}$$

$$\Delta_p^{(-)}(\lambda, A, W) = \Delta_p^{(-)}(\mu, B, W) = 0. \tag{1.32}$$

PROOF. We apply Theorem 1.1. By virtue of (1.16), $X_\lambda(A)$ can be replaced by $X_\mu(B)$ on the right-hand sides of (1.28), (1.29). This together with (1.27) immediately leads to (1.30), (1.31). To obtain (1.32), it should be also noted that $N_-(\lambda, A, W, \alpha) = 0$ for $\lambda > -\gamma$. □

REMARK 1.2. Theorem 1.2 remains meaningful and interesting if $f = 0$. In this case $A = B$ and (1.30)–(1.32) imply that the functionals defined by (1.28), (1.29) do not depend on $\lambda \in \rho(A)$.

6. We return here to the material of Subsection 4 and discuss the possibility of transfering the identities (1.25) to the endpoints of the gap $\Lambda = (\lambda_-, \lambda_+)$. The result corresponding to this case will not be used here but constitutes the background for the author's investigations [B3] on perturbations of the periodic Schrödinger operator.

To be definite, we consider the point λ_+. Suppose that (1.21) holds and

$$\sup_{\lambda\in\Lambda}(X_\lambda(A)g, g) \leqslant C\|g\|^2, \qquad g \in \mathfrak{G}. \tag{1.33}$$

Since the operator-valued function $X_\lambda(A)$ increases, the strong limit

$$X(\lambda_+, A) := \operatorname*{s\text{-}lim}_{\lambda\to\lambda_+} X_\lambda(A) \tag{1.34}$$

exists. (To calculate $X(\lambda_+, A)$, it is sufficient to evaluate the weak limit of $X_\lambda(A)$ on a pair of dense subsets of $\mathfrak{G}$.) If it turns out that

$$X(\lambda_+, A) \in \mathfrak{S}_\infty(\mathfrak{G}), \tag{1.35}$$

then, as easily seen, it will be possible to replace the strong convergence in (1.34) by convergence in norm. This in turn will imply that the limits $\lim_{\lambda\to\lambda_+} n_\pm(\beta, X_\lambda(A))$, $\beta = \alpha^{-1}$ exist, coincide with $n_\pm(\beta, X(\lambda_+, A))$, and according to (1.25),

$$\lim_{\lambda\to\lambda_+} N_\pm(\lambda, A, W, \alpha) = n_\pm(\alpha^{-1}, X(\lambda_+, A)). \tag{1.36}$$

It follows from (1.36) that for λ sufficiently close to λ_+ the functions $N_\pm(\lambda, A, W, \alpha)$ have constant values. It is natural to take these values for $N_\pm(\lambda_+, A, W, \alpha)$. Then (1.36) turns into

$$N_\pm(\lambda_+, A, W, \alpha) = n_\pm(\alpha^{-1}, X(\lambda_+, A)). \tag{1.37}$$

The quantity $N_+(\lambda_+, A, W, \alpha)$ should be interpreted as the number of eigenvalues of $A_+(t)$ "born" at the right endpoint λ_+ of the gap Λ as t was growing from 0 to $\alpha > 0$. The quantity $N_-(\lambda_+, A, W, \alpha)$ should be interpreted as the number of eigenvalues of $A_-(t)$ that "disappeared" at λ_+ as t was growing from 0 to $\alpha > 0$. Thus (1.37) *holds if, along with* (1.21), *conditions* (1.33) *and* (1.35) *are satisfied.* A similar argument leads to the relation

$$N_\pm(\lambda_-, A, W, \alpha) = n_\pm(\alpha^{-1}, X(\lambda_-, A)).$$

For this it should be assumed that the operator-valued function $X_\lambda(A)$ be *bounded from below* for $\lambda \in \Lambda$ and the limit operator $X(\lambda_-, A)$ be *compact.*

§2. Applications

1. We present here some applications of Theorem 1.2 to the investigation of differential operators. All cases discussed admit some immediate generalizations. We did not aim at the greatest generality, intending only to demonstrate typical ways to verify the hypotheses of Theorem 1.2. The more "difficult" example of the Schrödinger operator with magnetic field is treated in a separate paper [BR] of the present collection. That paper is closely linked with the subject-matter of this section. All the analytic facts that we require (estimates and asymptotic formulas) are taken from various preceding papers. We refer, when possible, to [BS6], where further literature can be found.

In what follows the integration is supposed to be extended to the entire space $\mathbb{R}^d$ if the domain is not indicated; $\omega_d = \operatorname{vol}\{x \in \mathbb{R}^d : |x| < 1\}$; D_j, D^β, ∇ are the ordinary notations for differentiations on $\mathbb{R}^d$; Φ is the Fourier transformation on $\mathbb{R}^d$; $\hat{u} := \Phi u$. The operator of multiplication by a function is usually denoted by the same letter as the function itself. By $H^s(\Omega)$, $\Omega \subset \mathbb{R}^d$, $s \in \mathbb{R}_+$, we denote the Sobolev function classes. Below, $Q = [0, 1)^d$ is the unit cube in $\mathbb{R}^d$; the cubes $Q_n = Q + n$, $n \in \mathbb{Z}^d$, constitute a partition of $\mathbb{R}^d$. Starting with this partition, one defines the function classes $l_p(\mathbb{Z}^d; L_q(Q))$, $0 < p, q \leqslant \infty$. The (quasi)norm in such class is the l_p (quasi)norm of the sequence $\{\|\cdot\|_{L_q(Q_n)}\}$.

The standard scalar product in $\mathbb{C}^d$ is denoted by $\langle \cdot, \cdot \rangle$.

2. In our first example, $\mathfrak{H} = \mathfrak{G} = L_2(\mathbb{R}^d)$, $d[a] = H^s(\mathbb{R}^d)$, $s \in \mathbb{R}_+$,

$$a[u, u] = \int |\xi|^{2s} |\hat{u}(\xi)|^2 \, d\xi, \qquad \hat{u} = \Phi u. \tag{2.1}$$

If s is an integer, then A is the polyharmonic operator, $A = (-\Delta)^s$. For an arbitrary $s > 0$, A can be expressed in terms of Fourier transforms. For f one can take any Hermitian form on $d[a]$ for which (1.13) is satisfied, with

f_* defined by

$$f_*[u, u] = \int (|\xi|^{2\sigma} + 1)|\hat{u}(\xi)|^2 \, d\xi, \qquad 0 \leqslant \sigma < s. \tag{2.2}$$

For example, (1.13) holds for

$$\left.\begin{aligned} &f[u, u] = \sum_{|\beta| \leqslant m} \int f_\beta(x) |D^\beta u|^2 \, dx, \\ &f_\beta \in L_\infty(\mathbb{R}^d), \qquad |\beta| \leqslant m \leqslant \sigma. \end{aligned}\right\} \tag{2.3}$$

It is clear that the forms given by (2.1), (2.2) satisfy (1.12). Now let V be a measurable function,

$$v[u, u] = \int V(x)|u|^2 \, dx, \qquad V(x) \geqslant 0, \quad W(x) = V^{1/2}. \tag{2.4}$$

The function V should be subject to certain conditions that guarantee (1.14), (1.15). In the situation considered, only the value $p = \varkappa$ is possible for the parameter $p > 0$, where $\varkappa$ is the standard semiclassical exponent,

$$\varkappa = d/2s. \tag{2.5}$$

Let us take $\gamma = 1$. Suppose that

$$\left.\begin{aligned} &V \in L_\varkappa(\mathbb{R}^d) && \text{if } \varkappa > 1, \\ &V \in l_1(\mathbb{Z}^d; L_q(Q)), \quad q > 1 && \text{if } \varkappa = 1, \\ &V \in l_\varkappa(\mathbb{Z}^d; L_1(Q)) && \text{if } \varkappa < 1, \end{aligned}\right\} \tag{2.6}$$

and let $[V]_\varkappa$ be the (quasi)norm in any of the classes appearing in (2.6). Then (1.14) holds for the exponent given by (2.5) and, moreover,

$$|W(A+I)^{-1/2}|_{2\varkappa} \leqslant C[V]_\varkappa^{1/2}. \tag{2.7}$$

Note that (2.7) reduces, depending on $\varkappa$, to estimates (4.5), (5.17) or (5.5) from [BS6]. Let us pass to condition (1.15). It clearly follows from (2.7) that

$$|W(A+I)^{-1}|_{2\varkappa} \leqslant C[V]_\varkappa^{1/2}. \tag{2.8}$$

In view of (2.8), it is sufficient to check (1.15) on a certain set of functions V dense with respect to $[\cdot]_\varkappa$. We shall assume that

$$V \in C_0^\infty(\mathbb{R}^d). \tag{2.9}$$

Then, applying an estimate of type (2.7) with doubled s, we find that under condition (2.9)

$$W(A+I)^{-1} \in \Sigma_\varkappa \subset \Sigma_{2\varkappa}^0. \tag{2.10}$$

Thus (1.15) is verified completely and we can apply Theorem 1.2.

THEOREM 2.1. *Let* a, v *be the forms defined in* (2.1), (2.4) *and suppose* (2.6) *holds. Let* f *be a Hermitian form satisfying* (1.13) *with respect to the majorant given by* (2.2), $b = a + f$, *and* $\varkappa$ *be the exponent defined by* (2.5). *Then for every* $\mu \in \rho(B)$ *the asymptotic behavior of the functions* $N_{\pm}(\mu, B, W, \alpha)$ *as* $\alpha \to \infty$ *is determined by the equations*

$$\Delta_{\varkappa}^{(-)}(\mu, B, W) = 0, \tag{2.11}$$

$$\Delta_{\varkappa}^{(+)}(\mu, B, W) = \delta_{\varkappa}^{(+)}(\mu, B, W) = (2\pi)^{-d}\omega_d \int V^{\varkappa}\,dx. \tag{2.12}$$

PROOF. The relation (2.11) coincides with (1.32). By virtue of (1.30) and (1.31), it is sufficient to verify that A has asymptotics of the form (2.12) for $\lambda = -1$. This is the "regular" version of the situation discussed in [BS6]. If s is an integer, then the desired asymptotic behavior is well known (e.g., see the introduction to [BS6]). For an arbitrary $s > 0$, the asymptotics can easily be recovered from the results of [BS3], especially if one keeps in mind that, by (2.7), it is sufficient to obtain the asymptotic formula under the additional assumption (2.9). □

3. Let us discuss some consequences of Theorem 2.1. Let f be a form of type (2.3) with $m = 0$,

$$f[u, u] = \int f_0(x)|u|^2\,dx, \tag{2.13}$$

i.e., $B = A + f_0$ (in the sense of form-sums). We impose on f_0 the condition

$$\left.\begin{array}{ll} f_0 \in l_\infty(\mathbb{Z}^d; L_r(Q)), & \\ r > \varkappa & \text{if } \varkappa \geqslant 1, \\ r = 1 & \text{if } \varkappa < 1, \end{array}\right\} \tag{2.14}$$

which is much less restrictive than the condition $f_0 \in L_\infty(\mathbb{R}^d)$ in (2.3). It is easily verified that (1.13) will then hold for the form given by (2.13) with the majorant defined in (2.2), provided that $s - \sigma$ is sufficiently small. Thus Theorem 2.1 implies the following statement.

THEOREM 2.2. *Let* f *be given by* (2.13) *with* f_0 *satisfying* (2.14), *and suppose that in all other respects the assumptions of Theorem* 2.1 *are valid. Then the asymptotic formulas* (2.11), (2.12) *hold.*

We remark that for $s = 1$ the asymptotic formula (2.12) was obtained earlier in [H] under the conditions $f_0 \in L_\infty(\mathbb{R}^d)$, $|V(X)| \leqslant C(1+x)^{-\eta}$, $\eta > 2$. These conditions are far more restrictive than (2.14), (2.6). If $\varkappa > 1$, then (2.6) is necessary for (2.12). The question about sharp asymptotics for N_- is more difficult. This problem was considered in [ADH]. It was assumed there that the operator $B = -\Delta + f_0$ with $f_0 \in L_\infty(\mathbb{R}^d)$ possesses the so-called density of states (this is true, say, for periodic f_0) and that

$V(x) \sim c|x|^{-\beta}$ as $|x| \to \infty$, $c > 0$, $\beta > 0$. Under these requirements the limit $\Delta_p^{(-)} = \delta_p^{(-)} > 0$, $p\beta = d$, has been calculated. The "poorer" relation (2.11) nevertheless contains some independent information because of different conditions on f_0, V.

Now consider another choice of f under the assumptions of Theorem 2.1. Let us restrict ourselves to the case $s = 1$. Let Γ be a plane in $\mathbb{R}^d$, $d > 1$, $\operatorname{codim}\Gamma = 1$, and let

$$f[u, u] = \int_\Gamma \varphi|u|^2\, d\Gamma, \qquad \varphi \in L_\infty(\Gamma).$$

Then (1.13) holds with respect to the form given by (2.2), provided that $2\sigma > 1$, and thus (2.11), (2.12) hold under the condition (2.6) with $2\varkappa = d$. In this example $B = -\Delta$ in $\mathbb{R}^d \setminus \Gamma = \mathbb{R}^d_+ \cup \mathbb{R}^d_-$, and on Γ the following boundary condition holds (in the sense of distributions if φ is not smooth):

$$\left.\frac{\partial u}{\partial \nu_+} + \frac{\partial u}{\partial \nu_-} + \varphi u\right|_\Gamma = 0.$$

Here $\partial/\partial\nu_\pm$ is differentiation on Γ in the direction of the outer normal relative to the subspace $\mathbb{R}^d_\pm$.

4. In Theorem 2.1, W is the operator of multiplication by a function. This option is not compulsory. Moreover, in applications of Theorem 1.2 one can avoid using explicit expressions for W by working directly with the form v. Let us illustrate this possibility. Let the forms a, f be the same as in Theorem 2.1, $s > 1$ be an integer,

$$v[u, u] = \int \langle V(x)\nabla u, \nabla u\rangle\, dx, \tag{2.15}$$

$V = \{V_{jk}\}_1^d$ be a real nonnegative matrix. In contrast to (2.5), the exponent $\varkappa$ is defined in this subsection by

$$2\varkappa(s-1) = d. \tag{2.16}$$

Only in order to facilitate the exposition and the choice of references, we suppose further that $\varkappa > 1$; for the same purposes we have restricted ourselves to a form of the first order in (2.15). It is assumed (compare with (2.6)) that $V \in L_\varkappa(\mathbb{R}^d)$.

Let $\mathfrak{G} = H^s(\mathbb{R}^d)$ $(= d[a])$ and the operator W (acting from $\mathfrak{H}$ to $\mathfrak{G}$) be defined with the help of the operator $Q(v) \in \mathfrak{S}_\infty(\mathfrak{G})$ as in Remark 1.1. Then

$$|W(A+I)^{-1/2}|_\varkappa^2 = |Q(v)|_\varkappa.$$

Furthermore, the spectrum of $Q(v)$ coincides with the spectrum of the form-quotient

$$v[u, u]/\|(A+I)^{1/2}u\|^2, \qquad u \in H^s(\mathbb{R}^d), \tag{2.17}$$

whose estimate is known (cf., for example, [BS2, Theorem 5.12]). The result is that we get (2.7) again, but $\varkappa$ is now defined by (2.16), V being the matrix from (2.15). So we have checked (1.14). The verification of (1.15) follows the outline of Subsection 2. Estimate (2.8) allows us to restrict ourselves to matrices V satisfying (2.9). The squares of the s-numbers of $W(A+I)^{-1}$ coincide with the eigenvalues of the form-quotient

$$v[u, u]/\|(A+I)u\|^2, \qquad u \in \mathfrak{D}(A) = H^{2s}(\mathbb{R}^d). \tag{2.18}$$

The denominator in (2.18) is equivalent to $\|u\|^2_{H^{2s}}$. By virtue of (2.9), spectrum estimates for (2.18) are dominated by those for the ratio

$$v[u, u]/\|u\|^2_{H^{2s}(K)}, \qquad u \in H^{2s}(K), \tag{2.19}$$

in a ball $K \supset \operatorname{supp} V$. The required spectrum estimate for the problem (2.19) is contained, for example, in [BS1, Theorem 3.1]. It follows that (2.10) holds if (2.9) does. This finishes the verification of (1.15), and we can apply Theorem 1.2. For $\mu \in \rho(B)$, we obtain (2.11) and the asymptotic formula

$$\Delta^{(+)}_{\varkappa}(\mu, B, W) = \delta^{(+)}_{\varkappa}(\mu, B, W) = \Theta, \tag{2.20}$$

where $\varkappa$ is defined by (2.16), $\varkappa > 1$,

$$\Theta = (2\pi)^{-d} \operatorname{vol}\{(x, \xi) \in \mathbb{R}^{2d} : \langle V(x)\xi, \xi\rangle > |\xi|^{2s}\}.$$

Indeed, it is sufficient to establish (2.20) for the operator A and $\mu = -1$. This is the same as determining the spectral asymptotics for the quotient (2.17). The required result can be found in [Ro].

5. Finally, we consider a periodic elliptic operator A of the second order. Its coefficients are supposed to be only measurable. This will cause certain difficulties for the verification of condition (1.15) that will be postponed until Subsection 6.

Let g be a matrix-function with measurable real-valued entries,

$$c_0|\xi|^2 \leqslant \langle g(x)\xi, \xi\rangle \leqslant c_1|\xi|^2, \qquad c_0 > 0, \quad \xi \in \mathbb{R}^d, \tag{2.21}$$

$$g(x+n) = g(x), \qquad x \in \mathbb{R}^d, \quad n \in \mathbb{Z}^d, \tag{2.22}$$

and let a be a form of the first order defined by

$$a[u, u] = \int \langle g(x)\nabla u, \nabla u\rangle\, dx, \qquad u \in H^1(\mathbb{R}^d) = d[a]. \tag{2.23}$$

If the matrix g is smooth, then, by (2.21), (2.22), we have $\mathfrak{D}(A) = H^2(\mathbb{R}^d)$, $Au = -\operatorname{div}(g\nabla u)$. In the general case, this differential expression should be understood in the sense of distributions.

As before, the forms f, v are defined by (2.13), (2.4) and they are subject to conditions (2.14), (2.6) with $\varkappa = d/2$. Note that the function f_0 is not assumed to be periodic. In the situation discussed, (2.7) evidently still holds true, and therefore (1.14) is satisfied. Assuming that (1.15) has already been checked as well, we obtain the following theorem.

THEOREM 2.3. *Suppose that the forms* a, f, v *are defined in* (2.23), (2.13), (2.4), *conditions* (2.21), (2.22), (2.14), (2.6) *are satisfied, and* $b = a+f$. *Then for* $\mu \in \rho(B)$, $\varkappa = d/2$, (2.11) *holds and we have the asymptotic formula*

$$\Delta_{\varkappa}^{(+)}(\mu, B, W) = \delta_{\varkappa}^{(+)}(\mu, B, W) = (2\pi)^{-d}\omega_d \int V^{\varkappa}(\det g)^{-1/2}\,dx. \quad (2.24)$$

PROOF. It suffices to check (2.24) for the operator A provided that $\mu = -1$. This is equivalent to determining the spectral asymptotics for the quotient

$$\left(\int V|u|^2\,dx\right) : \left(\int (\langle g\nabla u, \nabla u\rangle + |u|^2)\,dx\right), \qquad u \in H^1(\mathbb{R}^d). \quad (2.25)$$

By (2.7), it is possible to impose condition (2.9) on V. Let $\operatorname{supp} V \subset K$, where K is a ball. Consider a quotient of the type (2.25) for $u \in H^1(K)$ (Neumann problem) and for $u \in H^1(K)$ with $u|_{\partial K} = 0$ (Dirichlet problem). We use spectral asymptotics for bounded domains obtained in [BS1]. This will lead to coinciding upper and lower asymptotic estimates of the distribution functions for the spectrum of the quotient (2.25), that give (2.24). □

Note that in [BS6, §9] the asymptotic formula (2.24) is also established for $\mu = \mu_+$, the endpoint of a semi-infinite spectral gap for the operator B. The function f_0 is assumed to be bounded and periodic there.

6. We are going to verify (1.15) under the assumptions made in Subsection 5. In view of (2.8), it can be assumed that (2.9) holds. Let a matrix $\tilde{g} \in C^{\infty}(\mathbb{R}^d)$ satisfy (2.22) and (2.21) *with the same constants* c_0, c_1. Let $\tilde{a}$ be a form of type (2.23) for $\tilde{g}$, and $\widetilde{A}$ be the corresponding operator. Then $\mathfrak{D}(\widetilde{A}) = H^2(\mathbb{R}^d)$ and therefore (compare with (2.10))

$$W\widetilde{A}_1^{-1} \in \Sigma_{\varkappa}, \qquad \widetilde{A}_1 = \widetilde{A} + I, \qquad 2\varkappa = d. \quad (2.26)$$

Our purpose is to obtain the relation

$$WA_1^{-1} \in \Sigma_d^0, \qquad A_1 = A + I \quad (2.27)$$

in the general case, approximating A by "smooth" operators $\widetilde{A}$. Set $\tilde{g} - g = z = \{\zeta_{jk}\}$. To estimate the operator $\widetilde{\Gamma} := W(A_1^{-1} - \widetilde{A}_1^{-1})$ in Σ_d or, which is more convenient, the conjugate operator

$$\widetilde{\Gamma}^* = (A_1^{-1} - \widetilde{A}_1^{-1})W,$$

we use the representation (cf. [BS2, Lemma 5.2])

$$A_1^{-1} - \widetilde{A}_1^{-1} = \sum_{j,k} A_1^{-1/2}(D_k A_1^{-1/2})^* L_{jk}, \qquad L_{jk} = \zeta_{jk} D_j \widetilde{A}_1^{-1}. \quad (2.28)$$

The operators $A_1^{-1/2}(D_k A_1^{-1/2})^*$ are continuous and do not depend on $\tilde{g}$. Therefore, it is sufficient to estimate the operators $L_{jk}W$; the indices j, k

will be omitted in the sequel. We begin by transforming the expression for LW and extracting summands of class Σ_d^0. *In what follows such summands will be denoted by dots.* We need the following lemma.

LEMMA. *We have*

$$\begin{aligned} LW &= \zeta D\widetilde{A}_1^{-1}W \\ &= \zeta DW\widetilde{A}_1^{-1} + \cdots \\ &= \zeta WD\widetilde{A}_1^{-1} + \cdots \\ &= \zeta W\widetilde{A}_1^{-1}D + \cdots . \end{aligned} \tag{2.29}$$

PROOF. When making the first transposition, one should take into account the relations

$$\widetilde{A}_1^{-1}W - W\widetilde{A}_1^{-1} = \widetilde{A}_1^{-1}(W\widetilde{A} - \widetilde{A}W)\widetilde{A}_1^{-1} = \widetilde{A}_1^{-1}S\widetilde{A}_1^{-1},$$

where the first order operator S can be expressed in the form

$$S = \sum_1^d D_m s_m(x) + s_0(x), \qquad s_m \in C_0^\infty(\mathbb{R}^d), \qquad m = 0, \ldots, d.$$

The operators $\zeta D\widetilde{A}_1^{-1}$, $\zeta D\widetilde{A}_1^{-1}D = \zeta D\widetilde{A}_1^{-1/2}(D\widetilde{A}_1^{-1/2})^*$ are continuous, the operators $s_m\widetilde{A}_1^{-1}$ lie in $\Sigma_\varkappa$ (compare with (2.26)). This justifies the first transposition. Under the second transposition, the summand $\zeta W_0\widetilde{A}_1^{-1}$ (of class $\Sigma_\varkappa$) is extracted, where $W_0 = DW \in C_0^\infty(\mathbb{R}^d)$. Finally, the third transposition gives rise to the summand $\zeta W\widetilde{A}_1^{-1}(\widetilde{A}D - D\widetilde{A})\widetilde{A}_1^{-1}$ of class $\Sigma_\varkappa$. Indeed, $R = \widetilde{A}D - D\widetilde{A}$ is a second order differential operator with smooth periodic coefficients. Consequently, $R\widetilde{A}_1^{-1} \in \mathfrak{R}$. It remains to take into account (2.26). □

Now write (2.29) in the form

$$LW = \zeta WA_1^{-1/2}(A_1^{1/2}\widetilde{A}_1^{-1/2})(D\widetilde{A}_1^{-1/2})^* + \cdots .$$

The two last factors here are continuous operators; their norms can be estimated in terms of c_0, c_1 from (2.21) only. Now $|\zeta WA_1^{-1/2}|_d$ can be estimated, in view of (2.6), (2.7), in terms of the L_r-norm ($r > \max(2, d)$) of the function ζW on the ball $K \supset \operatorname{supp} V$. This L_r-norm can be made arbitrarily small by taking $\tilde{g}$ sufficiently close to g in $L_r(Q)$. All this together means that $\widetilde{\Gamma}^* \in \Sigma_d^0$, and hence $\widetilde{\Gamma} \in \Sigma_d^0$. Combined with (2.26), this leads to (2.27), as desired.

REFERENCES

[ADH] S. Alama, P. A. Dieft, and R. Hempel, *Eigenvalue brunches of the Schrödinger operator $H - \lambda W$ in a gap of $\sigma(H)$*, Comm. Math. Phys. **121** (1989), 291–321.

[B1] M. Sh. Birman, *On the spectrum of singular boundary value problems*, Mat. Sb. **55** (1961), 125–174; English transl., Amer. Math. Soc. Transl. (2) **53** (1966), 23–80.

[B2] ______, *Discrete spectrum in the gaps of the continuous one in the large-coupling-constant limit*, Oper. Theory: Adv. Appl. **46** (1990), 17–25.

[B3] ______, *The discrete spectrum in a gap of the perturbed periodic operator at large coupling constants*, Proc. Conf. Rigorous Results in Quantum Dynamics, Liblice Castle, 1990 (to appear).

[BR] M. Sh. Birman and G. D. Raĭkov, *Discrete spectrum in the gaps for perturbations of the magnetic Schrödinger operator*, (this volume).

[BS1] M. Sh. Birman and M. Z. Solomyak, *Spectral asymptotics of nonsmooth elliptic operators*. I, Trudy Moskov. Mat. Obshch. **27** (1972), 3–52; English transl. in Trans. Moscow Math. Soc. **27** (1972).

[BS2] ______, *Qualitative analysis in Sobolev imbedding theorems and applications to spectral theory*, Tenth Math. Summer School (Katsiveli/Nalchik, 1972), Izdanie Inst. Mat. Akad. Nauk Ukrain. SSR, Kiev, 1974, pp. 5–189; English transl., Amer. Math. Soc. Transl. (2) **114** (1980).

[BS3] ______, *Asymptotics of the spectrum of pseudodifferential operators with anisotropic-homogeneous symbols*, Vestnik Leningrad. Univ. Mat. Mekh. Astronom. **13** (1977), 13–21; English transl. in Vestnik Leningrad Univ. Math. **10** (1982), 237–247.

[BS4] ______, *Spectral theory of self-adjoint operators in Hilbert space*, Izdat. Leningrad. Univ., Leningrad, 1980; English transl., Reidel, Dordrecht, 1987.

[BS5] ______, *Schrödinger operator. Estimates of the number of the bound states as a function-theoretic problem*, Lectures of the XIV School on Operator Theory on Functional Spaces, I, Novgorod, 1990; English transl., Amer. Math. Soc., Providence, RI (to appear).

[BS6] ______, *Estimates on the number of the negative eigenvalues of the Schrödinger operator and its generalizations*, (this volume).

[H] R. Hempel, *On the asymptotic distribution of the eigenvalue branches of a Schrödinger operator* $H - \lambda W$ *in a spectral gap of* H, J. Reine Angew. Math. **399** (1989), 38–59.

[Ro] G. V. Rozenblyum, *Distribution of the discrete spectrum of singular differential operators*, Izv. Vyssh. Uchebn. Zaved. Mat. **1976**, no. 1, 75–86; English transl. in Soviet Math. (Iz. VUZ.) **20** (1976).

[So] A. V. Sobolev, *Weyl asymptotics for the discrete spectrum of the perturbed Hill operator*, (this volume).

[Y] D. R. Yafaev, *Mathematical scattering theory*, vol. I. *General theory*, Amer. Math. Soc., Providence, RI (to appear).

Translated by S. KISLYAKOV

ADVANCES IN SOVIET MATHEMATICS
Volume 7, 1991

Discrete Spectrum in the Gaps for Perturbations of the Magnetic Schrödinger Operator

M. SH. BIRMAN AND G. D. RAĬKOV

ABSTRACT. We consider the operator $B(\alpha) = A+F-\alpha V$ on $L_2(\mathbb{R}^d)$, where $A = (-i\nabla - \mathfrak{A})^2$, $\mathfrak{A}$ is the vector potential, F, V are electric potentials, $V(x) \geqslant 0$, $\alpha > 0$. We obtain a Weyl type asymptotic formula, as $\alpha \to \infty$, for the number of eigenvalues of $B(\alpha)$ that have crossed a fixed point in a gap of the spectrum of $A + F$.

This paper adjoins [B2], but in fact can be read independently. The main result is an example of the application of the abstract Theorem 1.2 in [B2]. For earlier results on the discrete spectrum asymptotics for the perturbed magnetic Schrödinger operator, see the authors' papers [B1, R].

§1. Formulation of the problem. Statement of the main result

1. Let $\mathfrak{H}$ be a Hilbert space, $\|\cdot\|$ and $(\cdot, \cdot)$ be the norm and the scalar product in $\mathfrak{H}$. If T is a linear operator on $\mathfrak{H}$, then $\mathfrak{D}(T)$ stands for its domain and $\rho(T)$ for its resolvent set. If T is compact and $s_k(T)$ are its singular numbers, then

$$\nu(s, T) := \operatorname{card}\{ k : s_k(T) > s \}, \qquad s > 0,$$

is the corresponding distribution function. If, moreover, T is selfadjoint, then $n_\pm(\cdot, T)$ are the distribution functions for its positive and negative eigenvalues; clearly $\nu = n_+ + n_-$. The class Σ_p, $p > 0$, is distinguished by the condition

$$|T|_p^p := \sup_{s>0} s^p \nu(s, T) < \infty.$$

We denote by Σ_p^0 the separable subclass of Σ_p:

$$\Sigma_p^0 = \{ T \in \Sigma_p : \nu(s, T) = o(s^{-p}),\ s \to 0 \}.$$

In the sequel the role of the principal Hilbert space $\mathfrak{H}$ will be played by $L_2(\mathbb{R}^d)$. *Throughout* $d \geqslant 3$. For $\Omega \subset \mathbb{R}^d$, by $H^1(\Omega)$ we denote the Sobolev

---

1991 *Mathematics Subject Classification.* Primary 35J10, 35P20.

1051-8037/91 $1.00 + $.25 per page

class with the metric form $\int_\Omega(|\nabla u|^2 + |u|^2)\,dx$. The class $\overset{\circ}{H}{}^1(\Omega)$ is the closure of $C_0^\infty(\Omega)$ in $H^1(\Omega)$. Below

$$D := -i\nabla;$$

$$\omega_d := \operatorname{vol}\{ x \in \mathbb{R}^d : |x| < 1 \};$$

integration without indication of the domain extends over all of $\mathbb{R}^d$. By C, c (possibly with subscripts) we denote various constants in the estimates. For the operator of multiplication by a function, we use the same notation as for the function itself.

2. Let $\mathfrak{A}(X) = \{\mathfrak{A}_1, \dots, \mathfrak{A}_d\}$ be the magnetic potential (a real vector-valued function). Let

$$\mathfrak{A} \in L_{d,\mathrm{loc}}(\mathbb{R}^d; \mathbb{R}^d). \tag{1.1}$$

The magnetic Schrödinger operator $A = A(\mathfrak{A})$ is introduced with the help of the sesquilinear form

$$a[u_1, u_2] = \int (D - \mathfrak{A})u_1 \overline{(D - \mathfrak{A})u_2}\,dx$$

whose domain $d[a]$ is given by

$$d[a] = \{ u \in L_2(\mathbb{R}^d) : u \in H^1_{\mathrm{loc}}(\mathbb{R}^d),\ a[u, u] < \infty \}.$$

The set $d[a]$ is a complete Hilbert space under the metric form $a[u, u] + \|u\|^2$. The class $C_0^\infty(\mathbb{R}^d)$ is dense in $d[a]$. According to general theory,

$$a[u, v] = (A^{1/2}u, A^{1/2}v), \qquad d[a] = \mathfrak{D}(A^{1/2}).$$

The free particle energy operator $A(0)$ corresponds to the potential $\mathfrak{A} \equiv 0$. The associated quadratic form is

$$a(0)[u, u] = \int |\nabla u|^2\,dx, \qquad d[a(0)] = H^1(\mathbb{R}^d).$$

The operator $A(0)$ turns out to be extremal in many respects. In particular, the following estimate is true (cf. [AHS, Theorems 2.3, 2.5]).

Proposition 1.1. *Let φ be multiplication by a measurable function. Then*

$$\|\varphi(A + \gamma I)^{-1/2}\| \leqslant \|\varphi(A(0) + \gamma I)^{-1/2}\|, \qquad \gamma > 0. \tag{1.2}$$

The following proposition can easily be derived from (1.2) (compare with Proposition 9.1 from [BS3]).

Proposition 1.2. *If for some $\gamma \geqslant 0$*

$$\|\varphi u\|^2 \leqslant C \int (|\nabla u|^2 + \gamma|u|^2)\,dx, \qquad u \in C_0^\infty(\mathbb{R}^d),$$

then with the same C

$$\|\varphi u\|^2 \leqslant C(a[u, u] + \gamma\|u\|^2), \qquad u \in C_0^\infty(\mathbb{R}^d).$$

We emphasize that $\gamma = 0$ is not forbidden in Proposition 1.2. This proposition implies, in particular, the "magnetic" version of the Hardy inequality

$$\int \frac{|u|^2}{|x|^2}\,dx \leqslant \frac{4}{(d-2)^2} a[u, u]. \tag{1.3}$$

Now let Ω be a ball. Then

$$\int_\Omega (|\nabla u|^2 + |u|^2)\,dx \leqslant C(\mathfrak{A}, \Omega) a[u, u], \qquad u \in d[a]. \tag{1.4}$$

Indeed, under condition (1.1), the following inequality can easily be derived from the Sobolev embedding theorem of limit order:

$$\left| \int_\Omega |(\mathfrak{D} - \mathfrak{A})u|^2\,dx - \int_\Omega |u|^2\,dx \right| \leqslant \varepsilon \int_\Omega |\nabla u|^2\,dx + c(\varepsilon) \int_\Omega |u|^2\,dx, \qquad \forall \varepsilon > 0.$$

This, together with (1.3), implies (1.4).

3. Now let us introduce the form

$$v[u, u] = \int V(x)|u|^2\,dx, \tag{1.5}$$

$$V = \overline{V} \in L_{d/2}(\mathbb{R}^d). \tag{1.6}$$

By $2V_\pm$ we denote the function $|V| \pm V$.

Let $n_\pm(\gamma, s)$ be the distribution function for the positive eigenvalues (i.e., the consecutive maxima) of the quotient of quadratic forms

$$\pm v[u, u]/(a[u, u] + \gamma\|u\|^2), \qquad \gamma \geqslant 0. \tag{1.7$_\pm$}$$

In other words, $n_\pm(\gamma, s) = n_\pm(s, T)$, where T is the operator generated by the form (1.5) on the space with the metric form $a[u, u] + \gamma\|u\|^2$. For $\gamma > 0$ the quotient (1.7) is considered for $u \in d[a]$. For $\gamma = 0$ one should consider (1.7) on the set $\tilde{d}[a]$, the completion of $C_0^\infty(\mathbb{R}^d)$ in the metric corresponding to the form $a[u, u]$. It should be emphasized that $\tilde{d}[a]$ is a function space by (1.3).

The following *Lieb estimate* is crucial for what follows.

PROPOSITION 1.3. *Under conditions* (1.1), (1.6), *the following estimate for* $n_\pm(\gamma, s)$ *holds*:

$$n_\pm(\gamma, s) \leqslant C(d) s^{-d/2} \int V_\pm^{d/2}\,dx. \tag{1.8$_\pm$}$$

The constant $C(d)$ *in* (1.8) *does not depend on* $\mathfrak{A}$ *and on* $\gamma \geqslant 0$.

We make some necessary explanations. Under condition (1.6) the form $a_\pm(\alpha) := a \mp \alpha v$ is bounded from below and closed on $d[a]$, whatever be $\alpha > 0$. Let $A_\pm(\alpha)$ be the operator generated by this form on $L_2(\mathbb{R}^d)$.

Denote by $N_+(\gamma, \mathfrak{A}, V; \alpha)$ the number of eigenvalues of $A_+(\alpha)$ lying to the left of the point $\lambda = -\gamma \leqslant 0$. Then

$$N_+(\gamma, \mathfrak{A}, V; \alpha) = n_+(\gamma, \alpha^{-1}), \tag{1.9}$$

and (1.8_+) turns into an estimate for N_+. If $\mathfrak{A} = 0$, this estimate becomes the well-known inequality of Rozenblyum, Lieb, and Cwikel (cf., e.g., [RS] or [BS2]). In [AHS] it was indicated (Theorem 2.15) that Lieb's method of deriving estimates for $N_+(\gamma, 0, V; \alpha)$ extends directly to the "magnetic" case; moreover, the constant in the estimate does not become larger.

REMARK 1.1. Of course, (1.8_-) follows from (1.8_+) by substituting $-V$ for V.

4. In what follows we suppose that V in (1.5), (1.6) is nonnegative (the only exception is Subsection 5 of §2). In accordance to that, (1.5) can be rewritten as

$$v[u, u] = \|Wu\|^2, \qquad W(x) \geqslant 0, \qquad W \in L_d(\mathbb{R}^d). \tag{1.10}$$

In the case considered, (1.8_+) with $\gamma > 0$ is equivalent to the following statement.

PROPOSITION 1.4. *Let* (1.1) *be satisfied and* $W \in L_d(\mathbb{R}^d)$. *Then for* $\gamma > 0$ *we have*

$$W(A + \gamma I)^{-1/2} \in \Sigma_d \tag{1.11}$$

and

$$|W(A + \gamma I)^{-1/2}|_d \leqslant (C(d))^{1/d} \|W\|_{L_d(\mathbb{R}^d)}, \qquad \gamma > 0. \tag{1.12}$$

5. To pass to the operator $A_\pm(\alpha)$ means to incorporate the electric potential $\mp\alpha W^2$ into the magnetic operator $A(\mathfrak{A})$. Here $\alpha > 0$ is the coupling constant that we shall suppose large in the sequel. Besides that, we incorporate into $A(\mathfrak{A})$ an electric potential F that does not contain α. Let us impose the condition

$$\int |F(x)||u|^2\,dx \leqslant \varepsilon \int |\nabla u|^2\,dx + c(\varepsilon) \int |u|^2\,dx, \qquad \forall \varepsilon > 0, \quad u \in C_0^\infty(\mathbb{R}^d). \tag{1.13}$$

PROPOSITION 1.5. *Let the function* F *satisfy* (1.13). *Then, with the same constant* $c(\varepsilon)$, *we have*

$$\int |F(x)||u|^2\,dx \leqslant \varepsilon a[u, u] + c(\varepsilon)\|u\|^2, \quad \forall \varepsilon > 0, \quad u \in C_0^\infty(\mathbb{R}^d). \tag{1.14}$$

The proof reduces to a reference to (1.2). It should only be taken into account that (1.14) is equivalent to the inequality

$$\|\varphi(A + \gamma I)^{-1/2}\| \leqslant \varepsilon^{1/2}, \qquad \varphi = |F|^{1/2}, \qquad \gamma = \varepsilon^{-1} c(\varepsilon).$$

REMARK 1.2. A sufficient condition for (1.13) (and then also for (1.14)) is

$$\sup_{n\in\mathbb{Z}^d}\int_{Q_n}|F|^r\,dx<\infty,\qquad Q_n=[0,1)^d+n,\qquad 2r>d.$$

This condition is a particular case of condition (2.14) in [B2].

6. Let $F=\overline{F}$ and (1.14) hold. Then the form b,

$$b:=a+f,\qquad f[u_1,u_2]:=(Fu_1,u_2),\tag{1.15}$$

is bounded from below and closed on $d[a]$. The same is true for the form

$$b_\pm(\alpha):=b\mp\alpha v=a+f\mp\alpha v,\qquad \alpha>0,\tag{1.16}$$

where v is the form given by (1.10). Let us denote by B, $B_\pm(\alpha)$ the selfadjoint operators on $L_2(\mathbb{R}^d)$ generated by the forms b, $B_\pm(\alpha)$. Furthermore, let $\lambda=\overline{\lambda}\in\rho(B)$. *We denote by* $N_\pm(\lambda,B,W;\alpha)$ *the number of eigenvalues of the operator* $B_\pm(t)$ *that cross the point* λ *as* t *increases from* 0 *to* $\alpha>0$.

The following theorem is the main result of this paper.

THEOREM 1.1. *Suppose that conditions* (1.1), (1.13) *are satisfied and let the form* v *be defined by* (1.10). *Let the operators* B, $B_\pm(\alpha)$ *be generated by the forms* b, $b_\pm(\alpha)$ *defined in* (1.15), (1.16). *Then for every* $\lambda=\overline{\lambda}\in\rho(B)$

$$\lim_{\alpha\to\infty}\alpha^{d/2}N_+(\lambda,B,W;\alpha)=(2\pi)^{-d}\omega_d\int W^d\,dx,\tag{1.17}$$

$$\lim_{\alpha\to\infty}\alpha^{d/2}N_-(\lambda,B,W;\alpha)=0.\tag{1.18}$$

REMARK 1.2. For $\lambda<0$ the function $N_+(\lambda,A,W;\cdot)$ coincides with the number of eigenvalues of A lying to the left of λ. In this case considerations of variational nature can be applied to obtain asymptotic formulas (see [R], and also Theorem 2.1 below). The general case can be reduced to the case $B=A$, $\lambda<0$ with the help of the abstract Theorem 1.2 from [B2]. This plan of the proof will be implemented in §2.

REMARK 1.3. Gaps inside the spectrum of B can appear, for example, in the following cases. 1) The potentials $\mathfrak{A}$, F are periodic; 2) $B=A$ and the spectrum of $A=A(\mathfrak{A})$ is discrete ("magnetic bottles"; cf., say, [AHS, CV]); 3) the case of constant magnetic field of maximal rank (this is possible only if d is even).

REMARK 1.4. The asymptotic formula (1.17) was obtained in [B1] under much more restrictive assumptions on $\mathfrak{A}$. In particular, the boundedness of $\mathfrak{A}$ and $\operatorname{div}\mathfrak{A}$ was required.

§2. Proof of Theorem 1.1

1. Under the assumptions of Theorem 1.1, the relations (1.11), (1.14) hold. Suppose also that the following relation has already been established under these assumptions:

$$W(A+I)^{-1}\in\Sigma_d^0.\tag{2.1}$$

Then, by virtue of Theorem 1.2 from [B2], (1.18) holds and it is sufficient to obtain the asymptotic formula (1.17) for $B = A$ (that is, for $F = 0$) and $\lambda < 0$. The last result is contained in Theorem 1.1 of [R]. However, here we shall give an independent argument, differing from that in [R] (see Theorem 2.1 below). Our main objective is to prove inclusion (2.1).

2. We begin with statements of technical nature pertaining to the case of a bounded domain. Let Ω be a ball in $\mathbb{R}^d$; by $\|\cdot\|_\Omega$ we denote the norm in $L_2(\Omega)$. The following lemma can easily be derived from the Sobolev embedding theorem of limit order.

LEMMA 2.1. *Let* $h \in L_d(\Omega)$. *Then the form* $\|hu\|^2_\Omega$, $u \in H^1(\Omega)$, *is compact on* $H^1(\Omega)$.

Let us introduce the form a_Ω, closed on $L_2(\Omega)$ and defined on $d[a_\Omega] = H^1(\Omega)$ by the formula:

$$a_\Omega[u, u] := \int_\Omega |(D - \mathfrak{A})u|^2 \, dx$$
$$= x - 2\,\mathrm{Re} \int_\Omega u\mathfrak{A}\overline{Du}\, dx, \quad u \in H^1(\Omega). \tag{2.2}$$

LEMMA 2.2. *Let* (1.1) *be satisfied. Then the form* $a_\Omega[u, u] - \|\nabla u\|^2_\Omega$ *is compact on* $H^1(\Omega)$.

Indeed, by Lemma 2.1, the second (and hence also the third) summand on the right-hand side of (2.2) is a compact form on $H^1(\Omega)$. □

Below we denote by $\overset{0}{a}_\Omega$ the restriction of a_Ω to $d[\overset{0}{a}_\Omega] := \overset{\circ}{H}{}^1(\Omega)$; let A_Ω and $\overset{0}{A}_\Omega$ be the selfadjoint operators on $L_2(\Omega)$ generated by the forms a_Ω and $\overset{0}{a}_\Omega$ respectively.

LEMMA 2.3. *Let* $\varphi = \overline{\varphi} \in L_{d/2}(\Omega)$ *and*

$$n_\pm(\gamma, \Omega, \mathfrak{A}, \varphi; \cdot), \qquad \overset{0}{n}_\pm(\gamma, \Omega, \mathfrak{A}, \varphi; \cdot)$$

be the distribution functions for the positive spectrum of the quotient of quadratic forms

$$\pm \int_\Omega \varphi |u|^2 \, dx \, (a_\Omega[u, u] + \gamma \|u\|^2_\Omega)^{-1}, \qquad \gamma > 0, \quad u \in H^1(\Omega), \tag{2.3$_\pm$}$$

$$\pm \int_\Omega \varphi |u|^2 \, dx \, (a_\Omega[u, u] + \gamma \|u\|^2_\Omega)^{-1}, \qquad \gamma \geqslant 0, \quad u \in \overset{\circ}{H}{}^1(\Omega). \tag{2.4$_\pm$}$$

Under condition (1.1), *we have the following asymptotic formula*:

$$\lim_{s \to 0} s^{d/2} n_\pm(\gamma, \Omega, \mathfrak{A}, \varphi; s) = \lim_{s \to 0} s^{d/2} \overset{0}{n}_\pm(\gamma, \Omega, \mathfrak{A}, \varphi; s)$$
$$= (2\pi)^{-d} \omega_d \int_\Omega \varphi_\pm^{d/2} \, dx. \tag{2.5$_\pm$}$$

PROOF. A power type spectral asymptotic formula for the quotient of two quadratic forms will not change if a compact perturbation is incorporated into the denominator (i.e., into the metric form; cf. [BS1, Lemma 1.3]). Consequently, Lemma 2.2 allows one to replace the form $a[u, u]$ in (2.3), (2.4) by the form $\|\nabla u\|_{\Omega}^{2}$, that is, to reduce the verification of $(2.5_{\pm})$ to the case $\mathfrak{A} \equiv 0$. In this case the desired result is well known (see, e.g., [BS2] or [Rs]). □

REMARK 2.1. If $\varphi \equiv 1$, then (2.5_{+}) gives a spectral asymptotic formula for compact operators $(A_{\Omega} + \gamma I)^{-1}$, $\gamma > 0$, and $(A_{\Omega}^{0} + \gamma I)^{-1}$, $\gamma \geqslant 0$. In particular, these operators belong to $\Sigma_{d/2}$.

3. Let us return to the operator $A = A(\mathfrak{A})$ on $L_2(\mathbb{R}^d)$.

LEMMA 2.4. *Let* $\zeta \in C_0^{\infty}(\mathbb{R}^d)$, $u \in \mathfrak{D}(A)$. *Then* $\zeta u \in \mathfrak{D}(A)$ *and*

$$A(\zeta u) - \zeta A u = -2i(\nabla\zeta)(D - \mathfrak{A})u - (\Delta\zeta)u =: \tilde{h}. \tag{2.6}$$

PROOF. The relation $u \in \mathfrak{D}(A)$ means that $u \in H^1_{\mathrm{loc}}(\mathbb{R}^d)$, $(D - \mathfrak{A})u \in L_2(\mathbb{R}^d)$ and

$$a[u, \psi] = (h, \psi), \qquad h \in L_2(\mathbb{R}^d), \qquad \forall\psi \in C_0^{\infty}(\mathbb{R}^d). \tag{2.7}$$

Moreover, $h = Au$. Substituting $\zeta\psi$ for ψ in (2.7), we obtain after elementary calculations

$$a[\zeta u, \psi] = (\zeta h + \tilde{h}, \psi), \qquad \forall\psi \in C_0^{\infty}(\mathbb{R}^d). \tag{2.8}$$

Since evidently $\zeta u \in d[a]$, we get the desired result from (2.8). □

LEMMA 2.5. *Let* $\zeta \in C_0^{\infty}(\mathbb{R}^d)$. *Then*

$$\|(A + I)(\zeta u)\| \leqslant C(\zeta)\|(A + I)u\|, \qquad u \in \mathfrak{D}(A). \tag{2.9}$$

PROOF. By (2.6) we have

$$\begin{aligned}\|(A + I)(\zeta u)\| &\leqslant \|\zeta(A + I)u\| + \|\tilde{h}\| \\ &\leqslant c(\|(A + I)u\| + \|(D - \mathfrak{A})u\| + \|u\|).\end{aligned}$$

It remains to take into account the inequalities

$$\|(D - \mathfrak{A})u\| = \|A^{1/2}u\| \leqslant \|(A + I)u\|, \qquad \|u\| \leqslant \|(A + I)u\|. \ \square$$

Observe also the following version of Lemma 2.4 involving the operator A_{Ω}^{0} from Subsection 2.

LEMMA 2.6. *Suppose that under the assumptions of Lemma* 2.4 *we have* $\operatorname{supp}\zeta \subset \Omega$. *Then*

$$u_0 := (\zeta u)|_{\Omega} \in \mathfrak{D}(A_{\Omega}^{0}), \qquad A_{\Omega}^{0}u_0 = \zeta h + \tilde{h}|_{\Omega} = A(\zeta u)|_{\Omega}. \tag{2.10}$$

PROOF. Since $u_0 \in \mathring{H}^1(\Omega)$, the desired result follows from (2.8) with $\psi \in C_0^{\infty}(\Omega)$. □

4. Now we proceed to the proof of (2.1). Since clearly (1.12) implies the estimate

$$|W(A+I)^{-1}|_d \leqslant C\|W\|_{L_d(\mathbb{R}^d)},$$

it is sufficient to obtain (2.1) for

$$W \in C_0^\infty(\mathbb{R}^d) \tag{2.11}$$

(compare with [B, §2]). In this case a sharper estimate holds.

LEMMA 2.7. *Let* (1.1), (2.11) *be satisfied. Then for* $A(\mathfrak{A})$ *we have the inclusion*

$$W(A+I)^{-1} \in \Sigma_{d/2}. \tag{2.12}$$

PROOF. The squares of the singular numbers of $W(A+I)^{-1}$ coincide with the consecutive maxima (i.e. the spectrum) of the quotient of quadratic forms

$$\|Wu\|^2/\|(A+I)u\|^2, \qquad u \in \mathfrak{D}(A). \tag{2.13}$$

Variational considerations allow one to obtain upper estimates for the spectrum of the quotient (2.13). Let Ω be a ball, $\operatorname{supp} W \subset \Omega$, and the function $\zeta \in C_0^\infty(\Omega)$ satisfy $\zeta(x)W(x) = W(x)$. Let us use (2.9), (2.10). We obtain

$$\|Wu\|^2 = \|Wu_0\|_\Omega^2, \qquad \|(A+I)u\|^2 \geqslant c\|(\overset{0}{A}_\Omega + I)u_0\|_\Omega^2 \quad (c>0),$$

and so the quotient (2.13) is estimated from above (we omit the constant factor) by the quotient

$$\|Wu_0\|_\Omega^2/\|(\overset{0}{A}_\Omega + I)u_0\|_\Omega^2, \qquad u_0 \in \mathfrak{D}(\overset{0}{A}_\Omega). \tag{2.14}$$

Moreover, an arbitrary $u_0 \in \mathfrak{D}(\overset{0}{A}_\Omega)$ instead of the u_0 from (2.10) can be allowed in (2.14), and that can only increase the consecutive maxima. Finally, it is possible to estimate W in the numerator in (2.14) by a constant. Again omitting a constant factor, we arrive at the spectrum of the quotient

$$\|u_0\|_\Omega^2/\|(\overset{0}{A}_\Omega + I)u_0\|_\Omega^2, \qquad u_0 \in \mathfrak{D}(\overset{0}{A}_\Omega).$$

The latter coincides with the spectrum of the operator $(\overset{0}{A}_\Omega + I)^{-2}$. In view of Remark 2.1, we have $(\overset{0}{A}_\Omega + I)^{-2} \in \Sigma_{d/4}$. Thus the eigenvalues (the consecutive maxima) σ_n of the quotient (2.13) also satisfy the estimate $\sigma_n = O(n^{-d/4})$. This is equivalent to (2.12).

Along with Lemma 2.7, we have established estimate (2.1) under the assumptions of Theorem 1.1.

5. It remains to establish (1.17) for $A = B$, $\lambda = -\gamma < 0$. Let us prove a somewhat more general statement. As in Subsection 3 of §1, we assume that the form v is defined by (1.5), (1.6) rather than by (1.10). The operator $A_+(\alpha)$ is generated by the form $a - \alpha v$, $\alpha > 0$; $N_+(\gamma, \mathfrak{A}, V; \alpha)$ is the number of eigenvalues of $A_+(\alpha)$ lying to the left of $\lambda = -\gamma \leqslant 0$.

THEOREM 2.1. *Let* (1.1), (1.6) *be satisfied. Then*

$$\lim_{\alpha\to\infty} \alpha^{-d/2} N_+(\gamma, \mathfrak{A}, V; \alpha) = (2\pi)^{-d}\omega_d \int V_+^{d/2}\, dx, \qquad \gamma \geqslant 0. \tag{2.15}$$

PROOF. By virtue of (1.9), the problem is equivalent to that of determining the asymptotics for $n_+(\gamma, s)$ (we recall that this is the spectrum distribution function for the quotient (1.7_+)). Estimate (1.8) enables us to restrict ourselves to calculating the asymptotics under the condition $V \in C_0^\infty(\mathbb{R}^d)$. Let Ω be a ball, $\operatorname{supp} V \subset \Omega$. The lower estimate for $n_+(\gamma, s)$ follows if we pass from (1.7_+) to the quotient (2.4_+) with $\varphi = V$. Similarly for $\gamma > 0$, the upper estimate follows by passing to the quotient (2.3_+) with $\varphi = V$. If $\gamma = 0$, one should first add the form $\|u\|_\Omega^2$ to the denominator of the quotient (1.7_+). By (1.4), this form is compact with respect to the form a, and hence adding this form does not change the asymptotics of $n_+(0, s)$. After that the upper estimate follows by passing to the quotient (2.3_+) with $\varphi = V$, $\gamma = 1$. These two-sided estimates together with (2.5_+), (1.9) lead to (2.15). □

If the form v is defined by (1.10) and $\gamma \geqslant 0$, then evidently

$$N_+(\gamma, \mathfrak{A}, V; \alpha) = N_+(-\gamma, A, W; \alpha).$$

Therefore, once the asymptotic formula (2.15) established, the proof of Theorem 1.1 is complete.

Finally we note that the second-named author has established (cf. [R, Theorem 1.1]) that the asymptotic formula (2.15) holds for every $\gamma = -\lambda$, $\lambda < \lambda_e$; here λ_e is the lower bound of the essential spectrum of $A(\mathfrak{A})$. If $\lambda_e = 0$, then formally (2.15) with $\gamma = 0$ is not implied by this result, but can be obtained by the method of [R]. The proof of Theorem 2.1 presented here is slightly simpler than the argument in [R].

REFERENCES

[AHS] J. Avron, I. Herbst, and B. Simon, *Schrödinger operators with magnetic fields*. I. *General interactions*, Duke Math. J. **45** (1978), 847–883.

[B1] M. Sh. Birman, *Discrete spectrum in the gaps of the continuous one in the large-coupling-constant limit*, Oper. Theory: Adv. Appl. **46** (1990), 17–25.

[B2] ______, *Discrete spectrum in the gaps of a continuous one for perturbations with large coupling constant*, (this volume).

[BS1] M. Sh. Birman and M. Z. Solomyak, *Spectral asymptotics of nonsmooth elliptic operators*. I, Trudy Moskov. Mat. Obshch. **27** (1972), 3–52; English transl. in Trans. Moscow Math. Soc. **27** (1972).

[BS2] ______, *Qualitative analysis in Sobolev imbedding theorems and applications to spectral theory*, Tenth Math. Summer School (Katsiveli/Nalchik, 1972), Izdanie Inst. Mat. Akad. Nauk Ukrain. SSR, Kiev, 1974, pp. 5–189; English transl., Amer. Math. Soc. Transl. (2) **114** (1980).

[BS3] ______, *Estimates for the number of negative eigenvalues of the Schrödinger operator and its generalizations*, (this volume).

[CV] Y. Colin de Verdière, *L'asymptotique de Weyl pour les bouteilles magnétiques*, Comm. Math. Phys. **105** (1986), 327–335.

[R] G. D. Raĭkov, *Strong electric field eigenvalue asymptotics for the Schrödinger operator with electromagnetic potential*, Lett. Math. Phys. **21** (1991), 41–49.

[RS] M. Reed and B. Simon, *Methods of modern mathematical physics*. IV: *Analysis of operators*, Academic Press, New York, 1978.

Translated by S. KISLYAKOV

ADVANCES IN SOVIET MATHEMATICS
Volume 7, 1991

Boundedness Conditions and Spectrum Estimates for the Operators $b(X)a(D)$ and Their Analogs

M. SH. BIRMAN, G. E. KARADZHOV, AND M. Z. SOLOMYAK

ABSTRACT. Some conditions on functions a, b on $\mathbb{R}^d$ are found which guarantee the boundedness of the operator $b(X)a(D)$ on $L_p(\mathbb{R}^d)$. Connections with general Hardy-type inequalities are discussed. For $p = 2$, some estimates of the s-numbers of $b(X)a(D)$ are considered as well. In fact, the main results of the paper are related to a certain class of operators containing $b(X)a(D)$ as a particular case.

CONTENTS

§0. Introduction

This paper is devoted to boundedness and compactness conditions and singular numbers estimates for a certain class of integral operators on L_2. The simplest (and the most important) example of operators of this class on $L_2(\mathbb{R}^d)$ is given by

$$(E_{ab}f)(x) = b(x)\int e^{ix\xi}a(\xi)f(\xi)\,d\xi = (2\pi)^{d/2}b(x)\Phi^+_{\xi\to x}(af), \qquad (0.1)$$

Φ^+ being the formal adjoint of the unitary Fourier operator Φ. This type

1991 *Mathematics Subject Classification.* Primary 47G30, 35S30.

1051-8037/91 $1.00 + $.25 per page

of operators arises naturally when one studies diverse analytic and operator-theoretic problems. We can mention here such subjects as spectral properties of convolution operators on weighted L_2-spaces [BS1]; pseudodifferential operators of negative order [BS2]; Schrödinger operators [S1, BS4, BS5]; scattering theory [B1]. The operators E_{ab} were studied in this connection in [BS1, C, S2]. Here we obtain some supplementary results about them. We also give (in §5) a list of singular numbers estimates for E_{ab} known at the present time.

Let X be the family of operators of multiplication by coordinates $x_1, \dots, x_d$ in $\mathbb{R}^d$ and D be the family of differentiation operators: $D = \{D_1, \dots, D_d\}$, $D_j = -i\partial/\partial x_j$. According to the general definition of a function of a commuting family of selfadjoint operators, we have:

$$E_{ab}\Phi = (2\pi)^{d/2} b(X)a(D). \tag{0.2}$$

Therefore, all results true for E_{ab} remain valid for $b(X)a(D)$ as well. In particular, E_{ab} is bounded on $L_2(\mathbb{R}^d)$ if and only if the following inequality holds:

$$\|b(X)u\| \leqslant c\|(a(D))^{-1}u\|, \qquad u \in \mathscr{S}(\mathbb{R}^d). \tag{0.3}$$

It is natural to treat estimates of this kind and their analogs for L_p-norms as "general Hardy inequalities". One of our main results consists in proving that (0.3) is valid for weights a, b belonging to the "weak L_δ-class" $L_{\delta,w}(\mathbb{R}^d)$ with $2 < \delta < \infty$.

In most cases only some elementary properties of Φ are used in the proofs, namely that Φ maps continuously L_1 into L_∞ and L_2 into L_2. We introduce the "class $\mathscr{E}$" of operators T satisfying such properties. Then we show that some fundamental results about the operators given by (0.1) hold true for operators $T_{ab}\colon \mapsto bTaf$ with $T \in \mathscr{E}$ as well. The important Cwikel estimate [C] is among these results.

This paper was written in connection with investigations of the negative spectrum of the Schrödinger operator and its analogs. Properties of the operators E_{ab} play an important role there (see [BS5] and especially §4 there). General operators T_{ab} appear when one studies spectral problems for some other classes of differential operators (see [B2, B3]).

A few remarks about notation. C (or c) denotes different constants. We write $C_{m.n}$ if it is necessary to indicate that C is the constant from formula $(m.n)$. If A_i, $i = 1, 2$, are some (quasi)normed spaces with (quasi)norms $|\cdot|_i$, then $T\colon A_1 \to A_2$ means that T maps A_1 into A_2 continuously, and we write

$$\|T\|_{A_1 \to A_2} = \sup_{0 \neq x \in A_1} (|Tx|_2 / |x|_1).$$

If $r \in [1, \infty]$, then $r' := (1 - r^{-1})^{-1}$.

To conclude the Introduction, we remark that the material of §§4, 5 is due to the first and the third named authors alone.

§1. Lorentz classes and weak L_p-classes

1.1. Let $(\mathscr{X}, \rho)$ be a measure space. Note that we consider only separable measure spaces with σ-finite measure. Recall the notion of Lorentz classes $L_{p,s} = L_{p,s}(\mathscr{X}, \rho)$, $0 < p < \infty$, $0 < s \leqslant \infty$ (see e.g., [BL]). For a measurable a.e. finite complex-valued function f on $\mathscr{X}$, let $\mu_f(t)$ be the distribution function of $|f|$:

$$\mu_f(t) = \rho\{x \in \mathscr{X} : |f(x)| > t\}, \qquad t > 0.$$

By definition,

$$f \in L_{p,s} \iff t\mu_f^{1/p}(t) \in L_s(\mathbb{R}_+, t^{-1}dt).$$

Taking the norm (quasinorm for $s < 1$) of $t\mu_f^{1/p}(t)$ in $L_s(\mathbb{R}_+, t^{-1}dt)$, we define a functional on $L_{p,s}$ denoted by $|f|_{p,s} = |f|_{L_{p,s}(\mathscr{X},\mu)}$. Thus,

$$|f|_{p,s} = \left(\int_0^\infty t^{s-1}\mu_f^{s/p}(t)\,dt\right)^{1/s}, \qquad 0 < s < \infty;$$

$$|f|_p := |f|_{p,\infty} = \sup_{t>0}(t\mu_f^{1/p}(t)).$$

This is, by definition, the quasinorm on $L_{p,s}$. If $p > 1$, $1 \leqslant s \leqslant \infty$, or $p = s = 1$, then there exists a norm on $L_{p,s}$ equivalent to this quasinorm. For other p, s, the spaces $L_{p,s}$ are nonnormable. It is clear that $L_{p,p} = L_p$ and, moreover, $p^{1/p}|f|_{p,p} = \|f\|_p := \|f\|_{L_p}$. It is convenient to put $L_{\infty,s} := L_\infty$, $\forall s$.

The classes $L_{p,s}$ are separable for $p, s < \infty$. It is customary to denote nonseparable spaces $L_{p,\infty}$, $p > \infty$, by $L_{p,w}$ ("weak L_p-classes"). $L_{p,w}$ has a distinguished separable subspace

$$L^0_{p,w} = \{f \in L_{p,w} : \mu_f(t) = o(t^{-p}),\ t \to 0,\ t \to \infty\}.$$

In particular, let $\mathscr{X}$ be a countable set and ρ be the "counting measure": $\rho(e) = \operatorname{card} e$, $\forall e \subset \mathscr{X}$. The corresponding classes $L_{p,s}$ are denoted by $l_{p,s}(\mathscr{X})$. We write briefly $l_{p,s}$ if $\mathscr{X} = \mathbb{N}$. So, $l_{p,w} = l_{p,\infty}$ consists of sequences $v = \{v_n\}$ such that

$$\nu_v(t) := \operatorname{card}\{n : |v_n| > t\} = O(t^{-p}).$$

An equivalent description: $\{v_n\} \in l_{p,w}$ if and only if for the nonincreasing rearrangement $\{v_n^*\}$ of $\{|v_n|\}$ we have $v_n^* = O(n^{-1/p})$. To obtain a description of $l^0_{p,w}$, one should replace the symbol O with o in any of the above two definitions.

The following simple assertion (see [O'N]) is important for the sequel. For this reason we present it with a proof.

1.2. LEMMA. *Let* $a \in L_{r,w}(\mathscr{X}, \rho)$, $u \in L_{p,s}(\mathscr{X}, \rho)$, $0 < r \leqslant \infty$, $0 < p < \infty$, $0 < s \leqslant \infty$. *Then*

$$|au|_{q,s} \leqslant C|a|_r |u|_{p,s}, \qquad q^{-1} = r^{-1} + p^{-1}, \quad C = C(r, p, s). \tag{1.1}$$

PROOF. The case $r = \infty$ is evident. If $a \in L_r$, $0 < r < \infty$, and $u \in L_\beta$, then, by the Hölder inequality,

$$\|au\|_\gamma \leqslant \|a\|_r \|u\|_\beta, \qquad 0 < \beta < \infty, \quad \gamma^{-1} = r^{-1} + \beta^{-1}. \tag{1.2}$$

Denote by Q_f the multiplication operator $v \mapsto fv$. For $u \in L_\beta$ fixed, (1.2) means that $Q_u\colon L_r \to L_\gamma$ and its norm equals $\|u\|_\beta$. In particular, for $0 < \varepsilon < r^{-1}$ we have $Q_u\colon L_{r_i} \to L_{\gamma_i}$, $i = 0, 1$, where $r_0^{-1} = r^{-1} + \varepsilon$, $r_1^{-1} = r^{-1} - \varepsilon$, $\gamma_i^{-1} = r_i^{-1} + \beta^{-1}$. Using the interpolation functor $K_{1/2,\infty}$ (see [BL]), we obtain that $Q_u\colon L_{r,w} \to L_{\gamma,w}$. In other words,

$$|au|_\gamma \leqslant c|a|_r \|u\|_\beta, \qquad 0 < \beta < \infty, \quad c = c(r, \beta). \tag{1.3}$$

Now fix $a \in L_{r,w}$. The inequality (1.3) means that $Q_a\colon L_\beta \to L_{\gamma,w}$. Choosing $\varepsilon \in (0, p^{-1})$, we see that $Q_a\colon L_{\beta_i} \to L_{\gamma_i,w}$, $i = 0, 1$, where $\beta_0^{-1} = p^{-1} + \varepsilon$, $\beta_1^{-1} = p^{-1} - \varepsilon$, $\gamma_i^{-1} = r^{-1} + \beta_i^{-1}$. Now we apply the functor $K_{1/2,s}$. Then we get $Q_a\colon L_{p,s} \to L_{q,s}$, $q^{-1} = r^{-1} + p^{-1}$. This is equivalent to (1.1). □

§2. Operators of class $\mathscr{E}$

2.1. Let $(\mathscr{X}, \rho)$, $(\mathscr{Y}, \sigma)$ be measure spaces and let T be a linear mapping acting from $L_1(\mathscr{X}, \rho) + L_2(\mathscr{X}, \rho)$ to some space of measurable functions on $(\mathscr{Y}, \sigma)$. We say that T belongs to the class $\mathscr{E} = \mathscr{E}(\mathscr{X}, \rho, \mathscr{Y}, \sigma)$ if the following conditions hold:

1°) $T\colon L_1(\mathscr{X}, \rho) \to L_\infty(\mathscr{Y}, \sigma)$, with the norm $M_0 = M_0(T)$;

2°) $T\colon L_2(\mathscr{X}, \rho) \to L_2(\mathscr{Y}, \sigma)$, with the norm $M_1 = M_1(T)$.

Sometimes we shall use a shortened notation, namely,

$$\mathscr{E}(\mathscr{X}, \rho) := \mathscr{E}(\mathscr{X}, \rho, \mathscr{X}, \rho),$$

and

$$\mathscr{E}(\mathscr{X}) := \mathscr{E}(\mathscr{X}, \rho),$$

if $\mathscr{X} \subset \mathbb{R}^d$ and ρ is the Lebesgue measure on $\mathbb{R}^d$.

Condition 1° is equivalent to the assumption that T is an integral operator with L_∞-kernel, $M_0(T)$ coinciding with the L_∞-norm of this kernel (see [KA], § XI.1).

If $T \in \mathscr{E}$, then, by interpolation,

$$T\colon L_{r,s}(\mathscr{X}, \rho) \longrightarrow L_{r',s}(\mathscr{Y}, \sigma), \qquad 1 < r < 2, \quad 0 < s \leqslant \infty; \tag{2.1}$$

$$T\colon L^0_{r,w}(\mathscr{X}, \rho) \longrightarrow L^0_{r',w}(\mathscr{Y}, \sigma), \qquad 1 < r < 2; \tag{2.2}$$

$$\|T\|_{L_{r,s}\to L_{r',s}} \leqslant cM_0^{1-\theta}M_1^{\theta}, \qquad r'\theta = 2, \quad c = c(r, s); \quad (2.3)$$

$$\|T\|_{L^0_{r,w}\to L^0_{r',w}} \leqslant c_{2.3}(r, \infty)M_0^{1-\theta}M_1^{\theta}, \quad r'\theta = 2.$$

Evidently, $\Phi \in \mathscr{E}(\mathbb{R}^d)$ and $M_0(\Phi) = (2\pi)^{-d/2}$, $M_1(\Phi) = 1$. A few more examples:

(1) Suppose $\mathscr{X} = (1, \infty)$ and T is the integral operator with kernel $(x+y)^{-1}$. Then $T \in \mathscr{E}(\mathscr{X})$ and $M_0 = 1/2$, $M_1 = \pi$.

(2) Suppose $\mathscr{X} = \mathbb{R}^d$ and T is the operator of convolution with the function $\exp(-i|x|^2/2h)$, $h > 0$. Here we have $T \in \mathscr{E}(\mathbb{R}^d)$, $M_0 = 1$, $M_1 = (2\pi h)^{d/2}$.

(3) The same remains true (including the values of M_0, M_1) for the operator of convolution with the function $\exp(-|x|^2/2h)$.

The last two kernels represent (after an appropriate normalization) the fundamental solutions for the Schrödinger and heat equations.

The operators investigated in this paper are compositions of mappings $T \in \mathscr{E}$ and multiplication operators Q_f (we shall write f instead of Q_f hereafter). It is convenient to single out the following auxiliary result.

2.2. LEMMA. *Let $T \in \mathscr{E}(\mathscr{X}, \rho, \mathscr{Y}, \sigma)$ and $a \in L_{r,w}(\mathscr{Y}, \sigma)$, $0 < r < \infty$. Then*

$$aT\colon L_{p,s}(\mathscr{X}, \rho) \longrightarrow L_{q,s}(\mathscr{Y}, \sigma), \qquad q^{-1} = 1 + r^{-1} - p^{-1}, \qquad (2.4)$$

where either $1 < p < 2$, $0 < s \leqslant \infty$ or $p = s = 2$. The following inequality is valid:

$$\|aT\|_{L_{p,s}\longrightarrow L_{q,s}} \leqslant cM_0^{1-\theta}M_1^{\theta}, \qquad p'\theta = 2, \quad c = c(p, r, s).$$

The result for $s = \infty$ remains true if we replace $L_{p,\infty} = L_{p,w}$ by $L^0_{p,w}$ and $L_{q,\infty}$ by $L^0_{q,w}$.

PROOF. The validity of (2.4) follows from (1.1) and (2.1), since aT can be represented as the composition mapping:

$$L_{p,s} \xrightarrow{T} L_{p',s} \xrightarrow{a} L_{q,s}.$$

For $p = 2$, we use the property 2° from the definition of $\mathscr{E}$ instead of (2.1). The result for $L^0_{p,w}$ follows immediately from (2.4), because $L_{p,s}$, $s < \infty$, is dense in $L^0_{p,w}$. □

The following statement collects the main inequalities exploited in this work.

2.3. THEOREM. *Let $(\mathscr{X}, \rho)$, $(\mathscr{Y}, \sigma)$, and $(\mathscr{Z}, \tau)$ be measure spaces and $T_1 \in \mathscr{E}(\mathscr{X}, \rho, \mathscr{Y}, \sigma)$, $T_2 \in \mathscr{E}(\mathscr{Y}, \sigma, \mathscr{Z}, \tau)$.*

(i) *Let* $a \in L_{r,w}(\mathscr{Y}, \sigma)$, $b \in L_{r,w}(\mathscr{Z}, \tau)$, $1 < r < \infty$, *and let* $1 < p \leqslant 2$, $p < r$, $q' := (p^{-1} - r^{-1})^{-1} \in (2, \infty)$. *Then*

$$bT_2aT_1 : L_p(\mathscr{X}, \rho) \longrightarrow L_p(\mathscr{Z}, \tau)$$

and

$$\|bT_2aT_1\|_{L_p \to L_p} \leqslant C|a|_r|b|_r M_0^{1-\theta_1}(T_1)M_1^{\theta_1}(T_1)M_0^{1-\theta_2}(T_2)M_1^{\theta_2}(T_2),$$
$$p'\theta_1 = q'\theta_2 = 2, \quad C = C(p, r). \tag{2.5}$$

(ii) *Let* $a \in L_{r,w}(\mathscr{Y}, \sigma)$, $b_1 \in L_{r,w}(\mathscr{X}, \rho)$, $b_2 \in L_{r_2}(\mathscr{Z}, \tau)$, *where* $1 < r < \infty$, $r \leqslant r_1, r_2 \leqslant \infty$, $r^{-1} = r_1^{-1} + r_2^{-1}$, *and let* $1 < p < \infty$,

$$1/2 < p^{-1} + r_1^{-1} < 1, \qquad 1/2 < (p')^{-1} + r_2^{-1} < 1. \tag{2.6}$$

Then

$$b_2T_2aT_1b_1 : L_p(\mathscr{X}, \rho) \longrightarrow L_p(\mathscr{Z}, \tau)$$

and

$$\|b_2T_2aT_1b_1\|_{L_p \to L_p}$$
$$\leqslant C|a|_r|b_1|_{r_1}|b_2|_{r_2} M_0^{1-\theta_1}(T_1)M_1^{\theta_1}(T_1)M_0^{1-\theta_2}(T_2)M_1^{\theta_2}(T_2),$$

$$\begin{aligned} &p_1'\theta_1 = q'\theta_2 = 2, \quad p_1 := (r_1^{-1} + p^{-1})^{-1}, \\ &q' := (p^{-1} - r_2^{-1})^{-1}, \quad C = C(r_1, r_2, p) \end{aligned} \tag{2.7}$$

PROOF. (i) According to Lemma 2.2, $aT_1 : L_p \longrightarrow L_{q,p}$. The restrictions on p, r imply $1 < q < 2$. Thus we can use Lemma 2.2 once more. This gives us the desired result.

(ii) For $f \in L_p$, by Lemma 1.2 we have $b_1 f \in L_{p_1,p}$, $p_1^{-1} = p^{-1} + r^{-1}$. Then we apply Lemma 2.1 twice (as in the proof of (i)). The restrictions on p, r_1, r_2 allow us to do so. □

Note that (ii) includes (i) (take $b_1 \equiv 1$, $b_2 = b$).

2.4. In particular, let $\mathscr{X} = \mathscr{Y} = \mathscr{Z} = \mathbb{R}^d$, $\rho = \sigma = \tau$ being the Lebesgue measure, and $T_1 = \Phi$, $T_2 = \Phi^*$. On $L_2(\mathbb{R}^d)$ the following equalities are true (cf. (0.2)):

$$b\Phi^* a\Phi = b(X)a(D),$$
$$b_2\Phi^* a\Phi b_1 = b_2(X)a(D)b_1(X).$$

We shall use these equalities for the spaces $L_p(\mathbb{R}^d)$, $p \neq 2$, as well. In this case the right-hand sides should be considered simply as notation for the operators on the left.

2.5. COROLLARY. (i) *Let* p, r *be as in Theorem* 2.3(i). *Then* $b(X)a(D)$ *is bounded on* $L_p(\mathbb{R}^d)$ *and*

$$\|b(X)a(D)\|_{L_p\to L_p} \leqslant C|a|_r|b|_r, \qquad C = C_{2.5}(2\pi)^{-d/r}. \tag{2.8}$$

(ii) *Let* p, r_1, r_2 *be as in Theorem* 2.3(ii). *Then* $b_2(X)a(D)b_1(X)$ *is bounded on* $L_p(\mathbb{R}^d)$ *and*

$$\|b_2(X)a(D)b_1(X)\|_{L_p\to L_p} \leqslant C|a|_r|b_1|_{r_1}|b_2|_{r_2}, \qquad C = C_{2.7}(2\pi)^{-d/r}.$$

2.6. Any operator of the form $a(D)$ coincides with the convolution operator $f \mapsto (2\pi)^{-d/2}\check{a}*f$, where $\check{a} := \Phi^* a$. Compare Corollary 2.5 with the "exact form" of the Young convolution Theorem ([O'N]; see also [T, Subsection 1.18.9]):

$$\begin{gathered}|h*f|_{q,s} \leqslant c|h|_{r'}|f|_{p,s}, \\ q^{-1} = p^{-1} - r^{-1}, \quad 1 < p < r < \infty, \quad 0 < s \leqslant \infty.\end{gathered} \tag{2.9}$$

In view of (2.9) and Lemma 1.2, one has:

$$\|b(h*f)\|_p \leqslant c|h|_{r'}|b|_r\|f\|_p, \qquad 1 < p < r < \infty; \tag{2.10}$$

$$\begin{gathered}\|b_2(h*(b_1 f))\|_p \leqslant c|h|_{r'}|b_1|_{r_1}|b_2|_{r_2}\|f\|_p, \qquad 1 < r < \infty, \\ r \leqslant r_1, r_2 \leqslant \infty, \quad r^{-1} = r_1^{-1} + r_2^{-1}, \quad 1 < p < r_2.\end{gathered} \tag{2.11}$$

Because of the exact Hausdorff-Young inequality (the special case of (2.1) for $T = \Phi^*$), the inclusion $a \in L_{r,w}$, $1 < r < 2$, implies $\check{a} \in L_{r',w}$. Therefore, both assertions of Corollary 2.5 for $1 < r < 2$ follow from the Young theorem. On the contrary, the case $r \geqslant 2$ does not depend on this Theorem. However, for some functions $a \in L_{r,w}$, $r \geqslant 2$, it may occur that $\check{a} \in L_{r',w}$. If this is the case, then Young's theorem gives a better result than Corollary 2.5. Indeed, the conditions on r in (2.10) and on r, r_1, r_2 in (2.11) are less restrictive than those in Theorem 2.3.

2.7. Now let us discuss the boundedness conditions for operators bTa on L_2-spaces. The analog of Theorem 2.3(i) holds for these operators. In fact, a more general "individual" result is valid, in which we do not suppose $T \in \mathscr{E}$.

2.8. PROPOSITION. *Let* $(\mathscr{X}, \rho)$, $(\mathscr{Y}, \sigma)$ *be measure spaces,*

$$T: L_{p,2}(\mathscr{X}, \rho) \longrightarrow L_{p',2}(\mathscr{Y}, \sigma)$$

for some $p \in [1, 2)$, *and let*

$$a \in L_{r,w}(\mathscr{X}, \rho), \qquad b \in L_{r,w}(\mathscr{Y}, \sigma),$$

where $r^{-1} = p^{-1} - 1/2$. *Then*

$$bTa: L_2(\mathscr{X}, \rho) \to L_2(\mathscr{Y}, \sigma)$$

and

$$\|bTa\|_{L_2\to L_2} \leqslant C\,|a|_r\,|b|_r\|T\|_{L_{p,2}\to L_{p',2}}, \qquad C = C(p). \tag{2.12}$$

PROOF. One has, by virtue of (1.1):

$$L_2 \xrightarrow{a} L_{p,2} \xrightarrow{T} L_{p',2} \xrightarrow{b} L_2. \quad \square$$

2.9. Here we consider an example of more general nature. At first we introduce some additional notation. For $u \in \mathscr{S}(\mathbb{R}^d)$, let

$$\int_{\mathbb{R}^d} |\nabla_\alpha u|^2\,dx := \int |\xi|^{2\alpha}|\hat{u}(\xi)|^2\,d\xi, \qquad \hat{u} := \Phi u. \tag{2.13}$$

The quadratic functional $\int(|\nabla_\alpha u|^2 + |u|^2)\,dx$ defines the standard norm on the Sobolev class $H^\alpha(\mathbb{R}^d)$, and $\mathscr{S}(\mathbb{R}^d)$ is dense in $H^\alpha(\mathbb{R}^d)$, $\forall\alpha > 0$. On the other hand, completing $\mathscr{S}(\mathbb{R}^d)$ with respect to the metric form given by (2.13), we obtain a space of functions only for $2\alpha < d$. This space is called "homogeneous Sobolev class" and is denoted by $\mathscr{H}^\alpha(\mathbb{R}^d)$. Recall the "embedding theorem with limit exponent":

$$\mathscr{H}^\alpha(\mathbb{R}^d) \subset L_{p',2}(\mathbb{R}^d), \qquad (p')^{-1} = 1/2 - \alpha d^{-1}. \tag{2.14}$$

Now, let $\mathfrak{A}[\cdot\,,\cdot]$ be a sesquilinear form on $\mathscr{S}(\mathbb{R}^d)$ satisfying the following inequality:

$$\mathfrak{A}[u, u] \geqslant c\int |\nabla_\alpha u|^2\,dx, \qquad u \in \mathscr{S}(\mathbb{R}^d), \quad 2\alpha < d. \tag{2.15}$$

$\mathfrak{A}$ can be extended by continuity to some Hilbert space $\mathfrak{D}$ densely embedded into $\mathscr{H}^\alpha(\mathbb{R}^d)$. Let $f \in L_{p,2}$, $p^{-1} = 1/2 + \alpha d^{-1}$. Consider the "equation in variations" with the unknown function $u \in \mathfrak{D}$:

$$\mathfrak{A}[u, v] = \int f\bar{v}\,dx, \qquad \forall v \in \mathscr{S}(\mathbb{R}^d).$$

According to (2.14), (2.15), it has a unique solution which belongs to $L_{p',2}$. Thus we have constructed an operator $T_{\mathfrak{A}}\colon L_{p,2} \to L_{p',2}$, $u = T_{\mathfrak{A}}f$. In particular, if the form $\mathfrak{A}$ is associated with some elliptic differential expression, then $T_{\mathfrak{A}}$ is an integral operator. Its kernel is the corresponding Green function. Using Proposition 2.8, we see that for $a, b \in L_{d/\alpha,w}$ the operator $bT_{\mathfrak{A}}a$ is continuous on $L_2(\mathbb{R}^d)$ and

$$\|bT_{\mathfrak{A}}a\|_{L_2\to L_2} \leqslant c_{2.15}C(d, \alpha)\,|a|_r\,|b|_r, \qquad 2\alpha < d, \quad r = d\alpha^{-1}. \tag{2.16}$$

It is easy to see that (2.16) is equivalent to some Hardy-type inequality (cf. (3.9) in the next Section).

Suppose now that, instead of (2.15), $\mathfrak{A}$ satisfies the following conditon:

$$\mathfrak{A}[u, u] \geqslant c\int (|\nabla_\alpha u|^2 + |u|^2)\,dx, \qquad u \in \mathscr{S}(\mathbb{R}^d), \quad 2\alpha < d. \tag{2.17}$$

Then $T_{\mathfrak{A}}: L_{p,2} \longrightarrow L_{p',2}$ as before. Besides, we now have $T_{\mathfrak{A}}: L_2 \longrightarrow L_2$ and hence, by interpolation,

$$T_{\mathfrak{A}}: L_{q,2} \longrightarrow L_{q',2}, \qquad q \in [p, 2].$$

Using Proposition 2.8 once more, we see that now

$$\|bT_{\mathfrak{A}}a\|_{L_2 \to L_2} \leqslant c|a|_r |b|_r, \qquad c = c(d, \alpha, r), \quad d/\alpha \leqslant r \leqslant \infty \tag{2.18}$$

(cf. (3.6)).

2.10. All these facts can be carried over automatically to the case of domains $\Omega \subset \mathbb{R}^d$. For differential operators the condition $u \in C_0^\infty(\Omega)$ corresponds to the first boundary condition. Now let an inequality of type (2.17) (resp., (2.15)) be valid on the entire space $H^\alpha(\Omega)$ (resp., $\mathscr{H}^\alpha(\Omega)$). Then for the approach described above to be applicable, it is sufficient to assume that Ω admits a continuous extension operator $H^\alpha(\Omega) \longrightarrow H^\alpha(\mathbb{R}^d)$ (resp., $\mathscr{H}^\alpha(\Omega) \longrightarrow \mathscr{H}^\alpha(\mathbb{R}^d)$; see [P] in this connection).

§3. General Hardy inequalities

3.1. Let the measurable and a.e. finite functions a, b on $\mathbb{R}^d$ be such that the operator $b(X)a(D)$ is bounded on $L_p(\mathbb{R}^d)$ for some $p \in [1, \infty]$. Suppose besides that

$$g := 1/a \in L_{1,\text{loc}}(\mathbb{R}^d). \tag{3.1}$$

Then $g\hat{u} \in L_{1,\text{loc}} \subset \mathscr{D}'$ for $u \in \mathscr{S}$. Assume that $f := \Phi^*(g\hat{u}) \in L_p$. In this case evidently

$$\int |bu|^p \, dx \leqslant c(a, b) \int |g(D)u|^p \, dx, \quad g = 1/a, \tag{3.2}$$
$$c(a, b) = \|b(X)a(D)\|^p_{L_p \to L_p}.$$

If $f \notin L_p$ (this means that either the distribution f is not a function, or f is a function not in L_p), then it is natural to agree that the integral on the right-hand side of (3.2) equals ∞. Under this convention (3.2) is valid on the entire set $\mathscr{S}$.

We call inequalities like the one in (3.2) "general Hardy inequalities". Now we can use Corollary 2.5(i) to get concrete conditions on a, b guaranteeing the validity of (3.2).

3.2. THEOREM. *Let* $1 \leqslant p \leqslant 2$, $p < r < \infty$, $0 < p^{-1} - r^{-1} < 1/2$. *Let* $a, b \in L_{r,w}(\mathbb{R}^d)$ *and* (3.1) *be true. Then for* $u \in \mathscr{S}$ (3.2) *is fulfilled, with*

$$c(a, b) \leqslant ((2\pi)^{-d/r} C_{2.5} |a|_r |b|_r)^p.$$

It is sometimes reasonable to impose conditions on a in terms of its inverse Fourier transform $\check{a}$. Indeed, in this case conditions on p, r become less restrictive (see Subsection 2.6). For convenience in further applications, we give a precise formulation here.

3.3. PROPOSITION. *Let* $1 < p < r < \infty$, $\check{a} \in L_{r',w}(\mathbb{R}^d)$, $b \in L_{r,w}(\mathbb{R}^d)$ *and* (3.1) *be fulfilled. Then for* $u \in \mathscr{S}$ (3.2) *holds with*

$$c(a, b) \leqslant (C_{2.10} | \check{a} |_{r'} | b |_r)^p.$$

3.4. EXAMPLE. Let $a(\xi) = |\xi|^{-\alpha}$, $0 < \alpha < d$. Then $a \in L_{r,w}$ and $\check{a} = c|x|^{-(d-\alpha)} \in L_{r',w}$, where $r = d/\alpha$. By virtue of Proposition 3.3, for $b \in L_{r,w}$ the following inequality holds:

$$\begin{gathered} \int |bu|^p \, dx \leqslant c | b |^p_{d/\alpha} \int |g(D)u|^p \, dx, \\ g(\xi) = |\xi|^\alpha, \quad u \in \mathscr{S}, \quad 1 < p < d/\alpha. \end{gathered} \tag{3.3}$$

We discuss (3.3) first when $p = 2$. According to (2.13), in this case we can rewrite (3.3) in the equivalent form:

$$\int |bu|^2 \, dx \leqslant c(b) \int |\nabla_\alpha u|^2 \, dx, \qquad 2\alpha < d, \quad u \in \mathscr{S}, \tag{3.4}$$

with $c(b) = c | b |^2_{d/\alpha}$. For $\alpha = 1$, $d \geqslant 3$, $b(x) = |x|^{-1}$ relation (3.4) becomes the classical Hardy inequality (but the constant is not sharp).

Maz′ya and Shaposhnikova (see [M, MS]) have found a necessary and sufficient condition on $b(x)$ for (3.4) to be true. This condition is formulated in terms of an appropriate capacity. Only the case of $\alpha \in \mathbb{N}$ was considered in [MS], but the result remains true for any $\alpha > 0$ as well (private communication of the authors of [MS]). The inclusion $b \in L_{d/\alpha,w}$ is only a sufficient condition for (3.4). However, the latter condition is formulated in terms which are much easier to verify. Theorem 3.2 (for $p = 2$) shows that this last condition is "nonspecific" in the sense that it does not involve the gradient.

Now consider an inequality resembling (3.4), but containing a term of lower order on the right-hand side:

$$\begin{aligned} \int |bu|^2 \, dx &\leqslant c(b, d, \alpha) \int (|\nabla_\alpha u|^2 + |u|^2) \, dx \\ &= c(b, d, \alpha) \int (|\xi|^{2\alpha} + 1) |\hat{u}(\xi)|^2 \, d\xi. \end{aligned} \tag{3.5}$$

Here $a(\xi) = (|\xi|^{2\alpha} + 1)^{-1/2} \in L_{d/\alpha,w} \cap L_\infty$ and therefore (3.5) holds for $b \in L_{d/\alpha,w} + L_\infty$. In particular, the weights b in $L_{r,w}(\mathbb{R}^d)$, $d/\alpha \leqslant r < \infty$, are suitable. A criterion on b for (3.5) to be fulfilled was found in [MS].

Now let $p \neq 2$. By definition, the functions $u \in L_{p,\mathrm{loc}}(\mathbb{R}^d)$ with the property $v(x) := \Phi^*_{\xi \to x}(|\xi|^\alpha u(\xi)) \in L_p$ constitute the "homogeneous Liouville class" $\mathscr{L}^\alpha_p(\mathbb{R}^d)$, with norm $\|u\|_{\mathscr{L}^\alpha_p} := \|v\|_p$. For $\alpha \in \mathbb{N}$, $p\alpha < d$, this class coincides with the homogeneous Sobolev class $\mathscr{W}^\alpha_p(\mathbb{R}^d)$; the latter is defined as the completion of $\mathscr{S}(\mathbb{R}^d)$ with respect to the norm $(\int |\nabla_\alpha u|^p \, dx)^{1/p}$. For arbitrary $\alpha > 0$ and $p < 2$, $p\alpha < d$, we have the continuous embedding $\mathscr{B}^{\alpha}_{p,p} \subset \mathscr{L}^{\alpha}_{p}$, $\mathscr{B}^{\alpha}_{p,p}$ being the homogeneous Besov space.

Thus, by Proposition 3.3, the following Hardy inequality is valid: for $b \in L_{d/\alpha,w}$

$$\int |bu|^p\,dx \leqslant c(b, d, \alpha, p)\|u\|^p_{\mathscr{L}^{\alpha}_p}, \qquad 1 < p < d/\alpha, \quad u \in \mathscr{S}(\mathbb{R}^d), \quad (3.6)$$

where $c(b, d, \alpha, p) = c(d, \alpha, p)|b|^p_{d/\alpha}$. Estimate (3.6) follows from Theorem 3.2 too, but under some unnecessary restrictions on p, r. One can replace the factor $\|u\|_{\mathscr{L}^{\alpha}_p}$ on the right-hand side of (3.6) by $\|u\|_{\mathscr{B}^{\alpha}_{p,p}}$ (case $p < 2$) and by $\|u\|_{\mathscr{W}^{\alpha}_p}$ (case $\alpha \in \mathbb{N}$, $1 < p < \infty$).

In [MS] a criterion on $b(x)$ for (3.6) to be true is given (for $\alpha \in \mathbb{N}$). This criterion is formulated in terms of capacity.

3.5. Example. Let $\vec{\alpha} = (\alpha_1, \dots, \alpha_d)$ be a set of positive numbers. For $\varkappa > 0$, put

$$a_{\varkappa,\vec{\alpha}}(\xi) = \Big(\sum_{1\leqslant j\leqslant d} |\xi_j|^{\varkappa\alpha_j}\Big)^{-1/\varkappa}, \qquad g_{\varkappa,\vec{\alpha}} = 1/a_{\varkappa,\vec{\alpha}}. \quad (3.7)$$

Then $a_{\varkappa,\vec{\alpha}} \in L_{r,w}(\mathbb{R}^d)$, $\forall \varkappa > 0$, where $r = r(\vec{\alpha}) := \sum_j \alpha_j^{-1}$. According to the Fourier multiplier theorem ([N, Theorem 1.5.4]) the following is true. Suppose $u \in \mathscr{S}$ and $g_{\varkappa,\vec{\alpha}}(D)u \in L_p$ for some $\varkappa > 0$. Then the same inclusion holds for any $\varkappa > 0$ and the norms $u \mapsto \|g_{\varkappa,\vec{\alpha}}(D)u\|_p$ are pairwise equivalent on $\mathscr{S}$ for various $\varkappa > 0$. For $\varkappa$ large, $a_{\varkappa,\vec{\alpha}}$ is smooth enough on $\mathbb{R}^d \setminus \{0\}$. Then, integrating by parts, one can check that $\check{a}_{\varkappa,\vec{\alpha}}(x)$ is continuous for $x \neq 0$. Besides, $\check{a}_{\varkappa,\vec{\alpha}}$ is "anisotropically homogeneous":

$$\check{a}_{\varkappa,\vec{\alpha}}(t^{1/\alpha_1}x_1, \dots, t^{1/\alpha_d}x_d) = t^{1-r(\vec{\alpha})}\check{a}_{\varkappa,\vec{\alpha}}(x), \qquad \forall t > 0.$$

These properties of $\check{a}_{\varkappa,\vec{\alpha}}$ entail $\check{a}_{\varkappa,\vec{\alpha}} \in L_{r'(\vec{\alpha}),w}(\mathbb{R}^d)$. Therefore for large $\varkappa$ one can use Proposition 3.3 and, for $r = r(\vec{\alpha}) > p > 1$ and for $b \in L_{r,w}$, obtain the following inequality:

$$\|bu\|_p \leqslant c|b|_r\|g_{\varkappa,\vec{\alpha}}(D)u\|_p, \qquad u \in \mathscr{S}. \quad (3.8)$$

This inequality is carried over to arbitrary $\varkappa > 0$.

For $p = \varkappa = 2$ (3.8) reads

$$\int |bu|^2\,dx \leqslant c^2|b|^2_r \int \sum_j |\xi_j|^{2\alpha_j}|\hat{u}(\xi)|^2\,d\xi, \qquad r(\vec{\alpha}) > 2, \quad u \in \mathscr{S}. \quad (3.9)$$

One does not need Proposition 3.3 here because (3.9) follows directly from Theorem 3.2. By definition, the integral on the right-hand side of (3.9) gives the norm on the homogeneous anisotropic Sobolev class $\mathscr{H}^{\vec{\alpha}}(\mathbb{R}^d)$. For $\alpha \in \mathbb{N}$, this integral equals $\int \sum_j |D^{\alpha_j}u|^2\,dx$. For $b = a_{2,\vec{\alpha}}$, inequality (3.9) was obtained in [A1].

For $\varkappa = p$ and $\alpha_j \in \mathbb{N}$, (3.8) turns into

$$\int |bu|^p \, dx \leqslant c^p \, \| b \|_r^p \int \sum_{1\leqslant j\leqslant d} |D_j^{\alpha_j} u|^p \, dx, \qquad r(\vec{\alpha}) > p > 1, \quad u \in \mathcal{S}.$$

For $b = a_{p,\vec{\alpha}}$ this inequality and some of its generalizations were obtained in [A2].

3.6. Let us discuss inequality (3.2) for $p = 2$ in more detail. We shall show that in this case the boundedness of $b(X)a(D)$ is *equivalent* to the validity of (3.2). For this be true, we have to strengthen our assumption on $a(x)$ slightly.

Let us agree on some notation. Let V be a measurable function on $\mathbb{R}^d$, and $u \mapsto \int Vu\,dx$ be continuous as a functional on $\mathcal{S}(\mathbb{R}^d)$. Then we shall write $V \in \mathcal{S}'(\mathbb{R}^d)$. By the L. Schwartz theorem, for $V \geqslant 0$ we have $V \in \mathcal{S}'(\mathbb{R}^d)$ if and only if

$$\int (1 + |\xi|)^{-m} V(\xi)\, d\xi < \infty \tag{3.10}$$

for some $m \geqslant 0$.

3.7. PROPOSITION. *Let a, b be measurable a.e. finite functions on $\mathbb{R}^d$ and let $|a|^{-2} \in \mathcal{S}'(\mathbb{R}^d)$. Then the operator $b(X)a(D)$ is bounded on $L_2(\mathbb{R}^d)$, with norm $\leqslant M$, if and only if*

$$\int |bu|^2 \, dx \leqslant M^2 \int |a|^{-2} |\hat{u}|^2 \, d\xi, \qquad u \in \mathcal{S}(\mathbb{R}^d). \tag{3.11}$$

PROOF. Let $b(X)a(D)$ be bounded. Take $u \in \mathcal{S}$. Our assumption on a implies that $\hat{f} := a^{-1} u \in L_2$. Therefore

$$\int |bu|^2 \, dx \leqslant \|b(X)a(D)f\|^2 \leqslant M^2 \|f\|^2 = M^2 \int |a|^{-2} |\hat{u}|^2 \, d\xi,$$

i.e., (3.11) holds.

For the converse, we check first that the functions $a^{-1}\hat{u}$, $u \in \mathcal{S}$, form a dense set in L_2. Indeed, let $h \in L_2$ and

$$\int h a^{-1} \hat{u} \, d\xi = 0, \qquad \forall u \in \mathcal{S}. \tag{3.12}$$

Using (3.10) for $V = |a|^{-2}$, we get

$$\int |h(\xi)| |a(\xi)|^{-1} (1 + |\xi|)^{-m/2} \, d\xi < \infty.$$

Hence $h/a \in \mathcal{S}'$. Now (3.12) means that $h/a = 0$ and consequently $h = 0$.

Thus, because of (3.11), $b(X)a(D)$ is bounded on a dense set and therefore on the entire space L_2. $\square$

3.8. To conclude this section, note that Corollary 2.5(ii) allows one to get some "Hardy inequalities with two weights" of the type

$$\|b_2 u\|_p \leqslant c\|b_1^{-1} g(D)u\|_p, \qquad g = 1/a, \quad u \in \mathscr{S}(\mathbb{R}^d).$$

§4. Estimates of singular numbers

4.1. Return to the case of arbitrary measure spaces $(\mathscr{X}, \rho)$, $(\mathscr{Y}, \sigma)$ and $p = 2$. We shall consider the operators (cf. Subsection 2.7)

$$T_{ab} := bTa, \tag{4.1}$$

where $T \in \mathscr{E} = \mathscr{E}(\mathscr{X}, \rho, \mathscr{Y}, \sigma)$.

4.2. PROPOSITION. (i) *Let* $a \in L_{r,w}(\mathscr{X}, \rho)$, $b \in L_{r,w}(\mathscr{Y}, \sigma)$, $r > 2$. *Then*

$$T_{ab}\colon L_2(\mathscr{X}, \rho) \longrightarrow L_2(\mathscr{Y}, \sigma)$$

and

$$\|T_{ab}\| \leqslant C\,|a|_r\,|b|_r M_0^{1-\theta} M_1^{\theta}, \qquad r\theta = 2, \quad C = C(r). \tag{4.2}$$

(ii) *If at least one of the functions* a, b *belongs to* $L^0_{r,w}$, *then the operator given by* (4.1) *is compact.*

Assertion (i) is merely a particular case of Proposition 2.8. It is more suitable for us to postpone the proof of (ii) until Subsection 4.10.

4.3. Our aim here is to investigate the rate of decrease of the singular numbers (s-numbers) of T_{ab} in terms of a, b. This rate will be measured by indicating p, q such that the operator in question belongs to $\mathfrak{S}_{p,q}$. Recall the definition of the classes $\mathfrak{S}_{p,q}$ (see [BS1]).

Let $\mathscr{H}_1$, $\mathscr{H}_2$ be Hilbert spaces and

$$\mathfrak{R} = \mathfrak{R}(\mathscr{H}_1, \mathscr{H}_2), \qquad \mathfrak{S}_\infty = \mathfrak{S}_\infty(\mathscr{H}_1, \mathscr{H}_2)$$

denote the linear spaces of all bounded and all compact operators from $\mathscr{H}_1$ into $\mathscr{H}_2$. For an operator $R \in \mathfrak{S}_\infty$, let $\{s_n(R)\}$ denote the sequence of its singular numbers. By definition,

$$\begin{gathered}\mathfrak{S}_{p,q} = \mathfrak{S}_{p,q}(\mathscr{H}_1, \mathscr{H}_2) = \{ R \in \mathfrak{S}_\infty(\mathscr{H}_1, \mathscr{H}_2) : \{s_n(R)\} \in l_{p,q} \}, \\ |R|_{p,q} = |\{s_n(R)\}|_{l_{p,q}}, \qquad 0 < p < \infty, \quad 0 < q \leqslant \infty.\end{gathered} \tag{4.3}$$

The functional given by (4.3) defines a quasinorm on $\mathfrak{S}_{p,q}$. Evidently $\mathfrak{S}_{p,p} = \mathfrak{S}_p$ (von Neumann-Schatten classes). The classes $\mathfrak{S}_{p,q}$ for $q < \infty$ are separable. The nonseparable class $\mathfrak{S}_{p,\infty}$ (other notation $\mathfrak{S}_{p,w}$, Σ_p) can be characterized by the following condition:

$$\mathfrak{S}_{p,\infty} = \Sigma_p = \{ R \in \mathfrak{S}_\infty : s_n(R) = O(n^{-1/p}) \}.$$

Σ_p has a distinguished separable subspace

$$\Sigma_p^0 := \{R \in \mathfrak{S}_\infty : s_n(R) = o(n^{-1/p})\} = \{ R \in \mathfrak{S}_\infty : \{s_n(R)\} \in l^0_{p,w} \}.$$

We denote the quasinorm on Σ_p by $|\cdot|_p$ in the sequel. Besides, $\|\cdot\|_p$ is the standard norm (quasinorm for $p < 1$) on $\mathfrak{S}_p$.

4.4. The main result of this section is a generalization of the well-known Simon-Cwikel estimate (see [S1], [C]). The proof we give here is based on an interpolation theorem obtained in [BS1]. This theorem deals with bilinear mappings acting from function spaces to operator spaces and satisfying a certain complementary "orthogonality condition". As compared to [BS1], the theorem we present here is widened to the case of vector-valued function spaces.

Let $\mathscr{F}$, $\mathscr{G}$ be normed spaces and $(\mathscr{X}, \rho)$, $(\mathscr{Y}, \sigma)$ be measure spaces. Let A (resp., B) be a (quasi)normed space whose elements are ρ-measurable $\mathscr{F}$-valued functions on $\mathscr{X}$ (resp., σ-measurable $\mathscr{G}$-valued functions on $\mathscr{Y}$). Let Γ be a bilinear mapping, $\Gamma : A \times B \longrightarrow \mathfrak{R}(\mathscr{H}_1, \mathscr{H}_2)$. We call this mapping "orthogonal", if it possesses the following properties.

1°) For each pair $a_1, a_2 \in A$ with disjoint supports and for each pair $b_1, b_2 \in B$ one has
$$(\Gamma(a_1, b_1))^*\Gamma(a_2, b_2) = 0.$$

2°) Similarly, for each pair $b_1, b_2 \in B$ with disjoint supports and for each pair $a_1, a_2 \in A$ one has
$$\Gamma(a_1, b_1)(\Gamma(a_2, b_2))^* = 0.$$

Now, let us agree to use a shortened notation of the type $L_{p,*}$, $\mathfrak{S}_{\delta,*}$ in the formulation of the following theorem. We use it whenever the values of the missing parameters are inessential. Besides, it is convenient to regard $\mathfrak{S}_{\infty,*}$ as $\mathfrak{R}$.

As usual, for $r_0, r_1 \in (0, \infty]$ and $\theta \in (0, 1)$, we put
$$r_\theta := ((1-\theta)r_0^{-1} + \theta r_1^{-1})^{-1}.$$

4.5. THEOREM. *Let $(\mathscr{X}, \rho)$, $(\mathscr{Y}, \sigma)$ be measure spaces, $\mathscr{F}$, $\mathscr{G}$ be complete normed spaces, $\mathscr{H}_1$, $\mathscr{H}_2$ be Hilbert spaces, and p_i, $\bar{p}_i$, δ_i, $i = 0, 1$, be certain indices. Moreover, let $\delta_0 \neq \delta_1$ and the triple $(p_1, \bar{p}_1, \delta)$ be a scalar multiple of the triple $(p_0, \bar{p}_0, \delta_0)$. Consider bilinear continuous orthogonal mappings Γ_i,*
$$\Gamma_i: L_{p_i,*}(\mathscr{X}, \rho, \mathscr{F}) \times L_{\bar{p}_i,*}(\mathscr{Y}, \sigma, \mathscr{G}) \longrightarrow \mathfrak{S}_{\delta_i,*}(\mathscr{H}_1, \mathscr{H}_2)$$
with the quasinorms $\|\Gamma_i\|_i$ and suppose that they coincide on
$$(L_{p_0,*} \cap L_{p_1,*})(\mathscr{X}, \rho, \mathscr{F}) \cap (L_{\bar{p}_0,*} \cap L_{\bar{p}_1,*})(\mathscr{Y}, \sigma, \mathscr{G}).$$
(In what follows we denote them by Γ, without indices.)

Let $\theta \in (0, 1)$, $\delta := \delta_\theta$,
$$s_1, s_2 \in [\delta, \infty], \qquad s^{-1} := s_1^{-1} + s_2^{-1} - \delta^{-1} \geqslant 0. \tag{4.4}$$
Then Γ acts continuously as follows:
$$\Gamma: L_{p_\theta, s_1}(\mathscr{X}, \rho, \mathscr{F}) \times L_{\bar{p}_\theta, s_2}(\mathscr{Y}, \sigma, \mathscr{G}) \longrightarrow \mathfrak{S}_{\delta, s}. \tag{4.5}$$

For the corresponding quasinorm, we have the inequality

$$\|\Gamma\|_{\theta, s_1, s_2} \leqslant c\|\Gamma\|_0^{1-\theta}\|\Gamma\|_1^{\theta},$$

the constant c *being independent of* Γ_i.

If in (4.4) $s = \infty$, $s_1, s_2 < \infty$, *then the image of the mapping given by* (4.5) *is contained in* Σ_δ^0. *Besides,*

$$\Gamma\colon L^0_{p_\theta, w}(\mathscr{X}, \rho, \mathscr{F}) \times L_{\bar{p}_\theta, \delta}(\mathscr{Y}, \sigma, \mathscr{G}) \longrightarrow \Sigma_\delta^0.$$

4.6. Some comments about the meaning of (4.5) are in order. Due to (4.4), at least one of the numbers s_1, s_2 (say, s_2) is finite. By linearity, Γ can be extended to the class

$$(L_{p_0,*} + L_{p_1,*})(\mathscr{X}, \rho, \mathscr{F}) \times (L_{\bar{p}_0,*} \cap L_{\bar{p}_1,*})(\mathscr{Y}, \sigma, \mathscr{G}).$$

In particular, Γ is well defined on

$$L_{p_\theta, s_1}(\mathscr{X}, \rho, \mathscr{F}) \times (L_{\bar{p}_0,*} \cap L_{\bar{p}_1,*})(\mathscr{Y}, \sigma, \mathscr{G}).$$

Theorem 4.5 states that Γ is continuous as a mapping from this product to $\mathfrak{S}_{\delta,s}$, provided that the second factor is equipped with the topology of $L_{\bar{p}_\theta, s_2}(\mathscr{Y}, \sigma, \mathscr{G})$. We mean in (4.5) the extension of this mapping by continuity. Recall that $(L_{\bar{p}_0,*} \cap L_{\bar{p}_1,*})(\mathscr{Y}, \sigma, \mathscr{G})$ is dense in $L_{\bar{p}_\theta, s_2}(\mathscr{Y}, \sigma, \mathscr{G})$ for $s_2 < \infty$ (see [T], Subsection 1.18.4).

4.7. The relation (4.5) was proved in [BS1, Theorem 3.5, c)], for $\mathscr{F} = \mathscr{G} = \mathbb{C}$. In fact, the proof given there remains valid for the general case as well. Thus, one must check the assertions concerning the class Σ_δ^0 only. They follow directly from (4.5). Indeed, let $s = \infty$, $s_1, s_2 < \infty$ and $\tau_1 < s_1$, $\tau_2 < s_2$. Then $\tau := (\tau_1^{-1} + \tau_2^{-1} - \delta)^{-1} < \infty$ and by (4.5)

$$\Gamma\colon L_{p_\theta, \tau_1}(\mathscr{X}, \rho, \mathscr{F}) \times L_{\bar{p}_\theta, \tau_2}(\mathscr{Y}, \sigma, \mathscr{G}) \longrightarrow \mathfrak{S}_{\delta,\tau} \subset \Sigma_\delta^0.$$

It remains to take into account the fact that the classes $L_{p,\tau_j}(\cdot)$ are dense in $L_{p,s_j}(\cdot)$ for $\tau_j < s_j < \infty$.

The proof of the second assertion about Σ_δ^0 is quite similar.

It is worth comparing Theorem 4.5 with the well-known general statement on interpolation of bilinear mappings (see [T, Theorem 1.19.5] and [BS1, Theorems 3.3, 3.4]). This comparison shows that using the orthogonality property of Γ, we have obtained a fairly delicate result.

Now we return to the operators given by (4.1). The following theorem provides an important supplement to Proposition 4.2. It shows that a slight additional assumption on one of the weights a, b leads to a noticeable sharpening of the statements about T_{ab}.

4.8. Theorem. *Let $T \in \mathscr{E}(\mathscr{X}, \rho, \mathscr{Y}, \sigma)$ and $\delta > 2$.*

(i) *Let $a \in L_{\delta, s_1}(\mathscr{X}, \rho)$, $b \in L_{\delta, s_2}(\mathscr{Y}, \sigma)$, where*

$$s_1, s_2 \in [\delta, \infty), \qquad s^{-1} := s_1^{-1} + s_2^{-1} - \delta^{-1} \geqslant 0. \tag{4.6}$$

Then $T_{ab} \in \mathfrak{S}_{\delta, s}$ and

$$|T_{ab}|_{\delta, s} \leqslant C|a|_{\delta, s_1}|b|_{\delta, s_2} M_0^{1-\theta}(T) M_1^{\theta}(T), \\ \delta\theta = 2, \qquad C = C(\delta, s_1, s_2). \tag{4.7}$$

If $s = \infty$ in (4.6), *then $T_{ab} \in \Sigma_\delta^0$.*

(ii) *Let $a \in L_{\delta, w}(\mathscr{X}, \rho)$, $b \in L_\delta(\mathscr{Y}, \sigma)$. Then $T_{ab} \in \Sigma_\delta$ and*

$$|T_{ab}|_\delta \leqslant C|a|_\delta \|b\|_\delta M_0^{1-\theta}(T) M_1^{\theta}(T), \qquad \delta\theta = 2, \qquad C = C(\delta). \tag{4.8}$$

If, besides, $\mu_a(t) = o(t^{-\delta})$ as $t \to 0$, then $T_{ab} \in \Sigma_\delta^0$.

Proof. Consider the bilinear mapping

$$\Gamma_T \colon (a, b) \longmapsto T_{ab}.$$

Evidently Γ_T is orthogonal. Recall that T is an integral operator with L_∞-kernel (see Subsection 1 of §2). Therefore $\Gamma_T(a, b) \in \mathfrak{S}_2$ for $a, b \in L_2$ and

$$\|\Gamma_T(a, b)\|_2 \leqslant M_0(T)\|a\|_2\|b\|_2.$$

Further,

$$\|\Gamma_T(a, b)\| \leqslant M_1(T)\|a\|_\infty\|b\|_\infty$$

because of the boundedness of T on L_2. Thus, the assumptions of Theorem 4.5 are fulfilled for Γ_T, with $p_0 = \bar{p}_0 = \delta_0 = \infty$, $p_1 = \bar{p}_1 = \delta_1 = 2$. Use this Theorem for $\theta = 2\delta^{-1}$. Then $p_\theta = \bar{p}_\theta = \delta_\theta = \delta$ and we get (4.7), (4.8), and also the assertion on Σ_δ^0 from (i).

It remains to prove the assertion on Σ_δ^0 from (ii). For this purpose, let us consider the truncated functions a_ε, $\varepsilon > 0$. Namely, let $a_\varepsilon(x) = a(x)$ for $|a(x)| > \varepsilon$ and $a_\varepsilon(x) = 0$ otherwise. Then $a_\varepsilon \in L_2$ and it follows from the assumptions on a that $a_\varepsilon \to a$ in the $L_{\delta, w}$-metric as $\varepsilon \to 0$. One can approximate b by means of functions $b_\varepsilon \in L_2$ as well. Then $T_{a_\varepsilon b_\varepsilon} \in \mathfrak{S}_2 \subset \Sigma_\delta^0$ and therefore $T_{ab} \in \Sigma_\delta^0$ by continuity of Γ_T. □

4.9. Let $\mathscr{X} = \mathscr{Y} = \mathbb{R}^d$, $\rho = \sigma$, be Lebesgue measure and $T = \Phi$. Then $T_{ab} = E_{ab}$, E_{ab} being the operator given by (0.1). For this case, (4.8) was first obtained by Cwikel [C]. The proof given above is not independent of [C]. In fact, the analysis of Cwikel's proof led the authors of [BS1] to Theorem 4.5; see the commentary to Theorem 6.5 in [BS1].

The possibility of extending the Cwikel theorem to arbitrary operators $T \in \mathscr{E}$ was indicated in [BS3]. The supplement concerning the classes Σ_δ^0

appeared in the same work. Note that the assumption $\mu_a(t) = o(t^{-\delta})$, $t \to 0$, in Theorem 4.8(ii) is less restrictive than $a \in L^0_{\delta,w}$. The two conditions coincide only if $\rho(\mathscr{X}) < \infty$. In the lectures [BS4], there is a proof of Theorem 4.8(ii) based of the construction from [C].

4.10. It is not difficult to complete the proof of Proposition 4.2 on the basis of Theorem 4.8. Indeed, $T_{ab} \in \mathfrak{S}_\infty$ for $a \in L_{\delta,w}$, $b \in L_\delta$, because of Theorem 4.8(ii). The class L_δ is dense in $L^0_{\delta,w}$. Therefore, the inclusion $T_{ab} \in \mathfrak{S}_\infty$ remains true for $b \in L^0_{\delta,w}$ by continuity.

4.11. Here we discuss estimates on s-numbers of T_{ab} intermediate between (4.2) and (4.8). Straightforward interpolation is inconvenient here since the function classes $L_{\delta,w}$ and L_δ are "too close" to each other. However, one can obtain some satisfactory results by introducing a "functional parameter" in the formulations.

Let χ be a measurable function on $\mathscr{Y}$. Then the symbol $L_{q,s}(\chi) := L_{q,s}(\mathscr{Y}, \sigma_\chi)$ will stand for the class $L_{q,s}$ with respect to the measure $\sigma_\chi(e) = \int_e \chi\, d\sigma$.

4.12. THEOREM. *Let* $T \in \mathscr{E}(\mathscr{X}, \rho, \mathscr{Y}, \sigma)$, $\delta > 2$, $a \in L_{\delta,w}(\mathscr{X}, \rho)$ *and* $\psi \in L_{\delta,w}(\mathscr{Y}, \sigma)$, $\psi > 0$, $|\psi|_\delta \leqslant 1$. *Then for* $b/\psi \in L_{q,s}(\psi^\delta)$, $\delta < q < \infty$, $0 < s \leqslant \infty$, *the inclusion* $T_{ab} \in \mathfrak{S}_{q,s}$ *is valid and*

$$|T_{ab}|_{q,s} \leqslant C|a|_\delta |b/\psi|_{L_{q,s}(\psi^\delta)} M_0^{1-\theta} M_1^\theta, \qquad \delta\theta = 2, \quad C = C(\delta, q, s). \tag{4.9}$$

If $b/\psi \in L^0_{q,w}(\psi^\delta)$, $\delta < q < \infty$, *then* $T_{ab} \in \Sigma^0_q$.

PROOF. It follows from (4.2), in view of the equality $T_{ab} = (b/\psi)T_{a\psi}$, that

$$\|T_{ab}\| \leqslant C_{4.2}|a|_\delta \|b/\psi\|_\infty M_0^{1-\theta} M_1^\theta, \qquad \delta\theta = 2. \tag{4.10}$$

Further, the following estimate in fact coincides with (4.8):

$$|T_{ab}|_\delta \leqslant C_{4.8}|a|_\delta \|b/\psi\|_{L_\delta(\psi^\delta)} M_0^{1-\theta} M_1^\theta, \qquad \delta\theta = 2. \tag{4.11}$$

Note that the norms in $L_\infty(\mathscr{Y}, \sigma)$ and in $L_\infty(\psi^\delta)$ coincide.

Now let the operator $T \in \mathscr{E}$ and the functions a, ψ be fixed. Consider the linear mapping $\Pi\colon b/\psi \mapsto T_{ab}$. Inequalities (4.10), (4.11) show that

$$\Pi\colon L_\infty(\psi^\delta) \longrightarrow \mathfrak{R}(L_2(\mathscr{X}, \rho), L_2(\mathscr{Y}, \sigma)),$$
$$\Pi\colon L_\delta(\psi^\delta) \longrightarrow \Sigma_\delta(L_2(\mathscr{X}, \rho), L_2(\mathscr{Y}, \sigma)),$$

with the obvious estimates for the norms. Applying the interpolation functor $K_{\eta,s}$, $\eta = \delta/q$, $0 < s \leqslant \infty$, one gets

$$\Pi\colon L_{q,s}(\psi^\delta) \longrightarrow \mathfrak{S}_{q,s}$$

with the standard estimate for $\|\Pi\|$. This is equivalent to (4.9).

The assertion on Σ_q^0 follows from (4.9), since L_q is dense in $L_{q,w}^0$. □

One can regard Theorem 4.12 as a generalization of Theorems 4.1, 4.4 from [BS5].

§5. Estimates of singular numbers of the operators E_{ab}

5.1. Here we present a list of s-number estimates for the operators E_{ab} defined in (0.1). Recall (see (0.2)) that the same estimates, up to the factor $(2\pi)^{-d/2}$, remain true for $b(X)a(D)$. As a rule, the results presented below are known. Some of them were given in the survey paper [BS1]. Some others are contained in §4 of the present paper as special cases of general estimates of the operators T_{ab}. To make our description complete, we include quite trivial cases as well.

5.2. Boundedness and compactness.

1°) If $a, b \in L_\infty$, then $E_{ab} \in \mathfrak{R}$ and

$$\|E_{ab}\| \leqslant (2\pi)^{d/2}\|a\|_\infty\|b\|_\infty. \tag{5.1}$$

If in addition $|a(x)|, |b(x)| \to 0$ as $|x| \to \infty$, then $E_{ab} \in \mathfrak{S}_\infty$.

2°) If $a, b \in L_{\delta,w}$ for $\delta > 2$, then $E_{ab} \in \mathfrak{R}$ and

$$\|E_{ab}\| \leqslant (2\pi)^{d/\delta} C_{4.2}(\delta) |a|_\delta |b|_\delta.$$

If in addition at least one of the functions a, b is in $L_{\delta,w}^0$, then $E_{ab} \in \mathfrak{S}_\infty$.

Statement 1° is obvious, 2° is a particular case of Proposition 4.2.

5.3. Conditions for belonging to Σ_δ, Σ_δ^0 for $\delta > 2$.

1°) If $a \in L_{\delta,w}$, $b \in L_\delta$ for $\delta > 2$, then $E_{ab} \in \Sigma_\delta$ and

$$|E_{ab}|_\delta \leqslant (2\pi)^{d/\delta} C_{4.8}(\delta) |a|_\delta \|b\|_\delta. \tag{5.2}$$

If in addition $\mu_a(t) \to 0$ as $t \to 0$, then $E_{ab} \in \Sigma_\delta^0$.

2°) If $a \in L_{\delta,s_1}$, $b \in L_{\delta,s_2}$ for $\delta > 2$, $s_1, s_2 > \infty$, $s_1^{-1} + s_2^{-1} = \delta^{-1}$, then $E_{ab} \in \Sigma_\delta^0$ and

$$|E_{ab}|_\delta \leqslant (2\pi)^{d/\delta} C_{4.7}(\delta, s_1, s_2) |a|_{\delta,s_1} |b|_{\delta,s_2}.$$

3°) If $a, \psi \in L_{r,w}$ for $2 < r < \delta$, $|\psi|_r \leqslant 1$, $\psi > 0$, and $b/\psi \in L_{\delta,w}(\psi^r)$, then $E_{ab} \in \Sigma_\delta$ and

$$|E_{ab}|_\delta \leqslant (2\pi)^{d/\delta} C_{4.9}(r, \delta, \infty) |a|_r |b/\psi|_{L_{\delta,w}(\psi^r)}.$$

If in addition $b/\psi \in L_{\delta,w}^0(\psi^r)$, then $E_{ab} \in \Sigma_\delta^0$.

All three statements are contained in Theorems 4.8 and 4.12.

5.4. Conditions for belonging to $\mathfrak{S}_{\delta,s}$ for $\delta > 2$, $s < \infty$.

1°) If $a \in L_{\delta,s_1}$, $b \in L_{\delta,s_2}$ for $s_1, s_2 \geqslant \delta > 2$, $s^{-1} := s_1^{-1} + s_2^{-1} - \delta^{-1} > 0$, then $E_{ab} \in \mathfrak{S}_{\delta,s}$ and

$$|E_{ab}|_{\delta,s} \leqslant (2\pi)^{d/\delta} C_{4.7}(\delta, s_1, s_2) |a|_{\delta,s_1} |b|_{\delta,s_2}. \tag{5.3}$$

If $a, b \in L_\delta$, $\delta > 2$, then

$$\|E_{ab}\|_\delta \leqslant (2\pi)^{\frac{d}{2}(1-2/\delta)} \|a\|_\delta \|b\|_\delta. \tag{5.4}$$

2°) If $a, \psi \in L_{r,w}$ for $2 < r < \delta$, $|\psi|_r \leqslant 1$, $\psi > 0$, and $b/\psi \in L_{\delta,s}(\psi^r)$, $0 < s < \infty$, then $E_{ab} \in \mathfrak{S}_{\delta,s}$ and

$$|E_{ab}|_{\delta,s} \leqslant (2\pi)^{d/\delta} C_{4.9}(r, \delta, s) |a|_r |b/\psi|_{L_{\delta,s}(\psi^r)}.$$

In particular,

$$\|E_{ab}\|_\delta \leqslant (2\pi)^{d/\delta} C_{4.9}(r, \delta, \delta) |a|_r \left(\int |b|^\delta \psi^{r-\delta}\, dx \right)^{1/\delta}.$$

Estimate (5.3) is contained in Theorem 4.8(i). Estimate (5.4) can be deduced from (5.1) and from the identity (5.5) by means of complex interpolation (see [S2, Theorem 4.4]). An inequality like (5.4) but with a bigger constant factor can be regarded as a particular case of (5.3) also.

Statement 2° is contained in Theorem 4.12.

5.5. The inclusion $E_{ab} \in \mathfrak{S}_2$ is evidently equivalent to the condition $a, b \in L_2$ and

$$\|E_{ab}\|_2 = \|a\|_2 \|b\|_2. \tag{5.5}$$

5.6. Here we introduce some function classes which will appear in estimates below. Their description involves the standard partition of $\mathbb{R}^d$ into the lattice of unit cubes.

Let $\Delta = [0, 1)^d$ and $\Delta_n = \Delta + n$, $n \in \mathbb{Z}^d$. All the functions we shall come across in the sequel are assumed to belong to $L_{2,\mathrm{loc}}(\mathbb{R}^d)$. With $f \in L_{2,\mathrm{loc}}$ we associate the numerical sequence $v(f) = \{v_n(f)\}$, $n \in \mathbb{Z}^d$, where $v_n(f) = (\int_{\Delta_n} |f|^2\, dx)^{1/2}$. The set $\{f : v(f) \in l_p(\mathbb{Z}^d)\}$ can be naturally identified with the space $l_p(\mathbb{Z}^d, L_2(\Delta))$. The function spaces $l_{p,s}(\mathbb{Z}^d, L_2(\Delta))$, and so on are introduced in a similar way.

The proofs of the assertions of Subsections 5.7, 5.8 that were not known before are postponed until Subsection 5.9.

5.7. Conditions for belonging to Σ_δ, Σ_δ^0 for $\delta < 2$.

1°) If $v(a) \in l_{\delta,w}$, $v(b) \in l_\delta$ for $0 < \delta < 2$, then $E_{ab} \in \Sigma_\delta$ and

$$|E_{ab}|_\delta \leqslant C |v(a)|_\delta \|v(b)\|_\delta, \qquad C = C(d, \delta). \tag{5.6}$$

If in addition $v(a) \in l_{\delta,w}^0$, then $E_{ab} \in \Sigma_\delta^0$.

2°) If $v(a) \in l_{\delta, s_1}$, $v(b) \in l_{\delta, s_2}$ with $s_1, s_2 < \infty$, $s_1^{-1} + s_2^{-1} = \delta^{-1}$, then $E_{ab} \in \Sigma_\delta^0$ and

$$|E_{ab}|_\delta \leqslant C |v(a)|_{\delta, s_1} |v(b)|_{\delta, s_2}, \qquad C = C(d, s_1, s_2). \tag{5.7}$$

5.8. Conditions for belonging to $\mathfrak{S}_{\delta, s}$ for $\delta < 2$, $s < \infty$. If $v(a) \in l_{\delta, s_1}$, $v(b) \in l_{\delta, s_2}$, where $s_1, s_2 \geqslant \delta$, $s^{-1} := s_1^{-1} + s_2^{-1} - \delta^{-1} > 0$, $0 < \delta < 2$, then $E_{ab} \in \mathfrak{S}_{\delta, s}$ and

$$|E_{ab}|_{\delta, s} \leqslant C |v(a)|_{\delta, s_1} |v(b)|_{\delta, s_2}, \qquad C = C(d, \delta, s_1, s_2). \tag{5.8}$$

In particular, if $v(a), v(b) \in l_\delta$ for $\delta < 2$, then

$$\|E_{ab}\|_\delta \leqslant C \|v(a)\|_\delta \|v(b)\|_\delta, \qquad C = C(d, \delta). \tag{5.9}$$

5.9. The estimate (5.9) was proved in [BS1, Theorem 11.1]. The remaining assertions of Subsections 5.7, 5.8 can be deduced from (5.9) in the following way.

Fix a number $\delta_0 \in (0, \delta)$ and consider the bilinear orthogonal mapping $\Gamma: (a, b) \mapsto E_{ab}$. According to (5.5) and (5.9),

$$\Gamma: l_2(\mathbb{Z}^d, L_2(\Delta)) \times l_2(\mathbb{Z}^d, L_2(\Delta)) (= L_2(\mathbb{R}^d) \times L_2(\mathbb{R}^d)) \longrightarrow \mathfrak{S}_2,$$
$$\Gamma: l_{\delta_0}(\mathbb{Z}^d, L_2(\Delta)) \times l_{\delta_0}(\mathbb{Z}^d, L_2(\Delta)) \longrightarrow \mathfrak{S}_{\delta_0}.$$

Now use Theorem 4.5 for $\mathscr{F} = \mathscr{G} = L_2(\Delta)$, $\theta = (\delta_0^{-1} - \delta^{-1})(\delta_0^{-1} - 1/2)^{-1}$. This proves all the assertions of Subsections 5.7, 5.8. Note that (5.6) was obtained in [S2, Theorem 4.6] but only for the case $1 < \delta < 2$.

5.10. Comments. 1°) There is an essential difference between the assertions of Subsections 5.2–5.5 and those of 5.7, 5.8. The former are true for arbitrary $T \in \mathscr{E}$ (see Theorems 4.8, 4.12). At the same time the proofs of the latter two are based on specific properties of the kernel $\exp(ix\xi)$; see the proof of Theorem 11.1 in [BS1].

2°) Compare the requirements we have put on a, b to estimate E_{ab} in $\mathfrak{S}_{\delta, s}$ for $\delta > 2$ on the one hand and for $\delta < 2$ on the other. The terms in which these requirements have been stated are quite different. The only result we have for the critical value $\delta = 2$ is the trivial equality (5.5). It would be interesting to find sharp conditions on a, b for the inclusion $E_{ab} \in \Sigma_2$.

3°) Let A be a selfadjoint operator bounded from below on a Hilbert space and let the part of its spectrum lying to the left of the point $\lambda \in \mathbb{R}$ be discrete. Let $N(\lambda; A)$ denote the number of eigenvalues of A in the interval $(-\infty, \lambda)$ counted according to their multiplicities. We shall be interested in this number for the Schrödinger-type operators on $L_2(\mathbb{R}^d)$:

$$A = A_l(\alpha V) = (-\Delta)^l - \alpha V, \tag{5.10}$$

with a nonnegative function V and "coupling constant" $\alpha > 0$. Further, let $\nu(\lambda; R)$ denote the quantity of s-numbers of an operator $R \in \mathfrak{S}_\infty$ situated to the right of a point $\lambda > 0$. The following identity is true (see [BS5, formulas (3.12), (3.13)]):

$$\begin{gathered} N(-\gamma; A_l(\alpha V)) = \nu(\alpha^{-1/2}; b(X)a_\gamma(D)), \\ \alpha, \gamma > 0, \quad b = \sqrt{V}, \quad a_\gamma(\xi) = (|\xi|^{2l} + \gamma)^{-1/2}. \end{gathered} \tag{5.11}$$

For $2l < d$, (5.11) remains true for $\gamma = 0$ as well.

The identity (5.11) enables one to use the results on the operators E_{ab} to estimate $N(-\gamma; A_l(\alpha V))$. In fact, Cwikel [C] obtained (5.2) precisely in connection with this problem. On the other hand, a lot of information on the operators given by (5.10) is known. By (5.11), one can use this information to check that the estimates stated in Subsections 5.2–5.8 are of sharp order. Therefore, the estimates from Subsections 5.2; 5.3, 1°, 3°; 5.7, 1° are sharp. See [BS5] for details.

4°) Estimate (5.9) for the case $\delta = 1$ is sharp as well. Indeed, if $0 \neq b(X)a(D) \in \mathfrak{S}_1$ then $v(a), v(b) \in l_1$. This was proved in [S2]. It is unknown whether or not (5.9) is sharp for $\delta \neq 1$.

REFERENCES

[A1] A. S. Andreev, *Hardy inequalities for anisotropic Sobolev classes*, Probl. Mat. Anal. **6** (1977), 3–10. (Russian)

[A2] ———, *Hardy inequalities for anisotropic Besov classes*, Probl. Mat. Anal. **7** (1979), 3–13; English transl. in Selecta Math. Sovietica **5** (1986), 257–266.

[BL] J. Berg and J. Löfström, *Interpolation spaces. An introduction*, Springer-Verlag, Berlin, Heidelberg, and New York, 1976.

[B1] M. S. Birman, *Scattering problems for differential operators with constant coefficients*, Funktsional. Anal. i Prilozhen. **3** (1969), no. 3, 1–16; English transl. in Functional Anal. Appl. **3** (1969), no. 3.

[B2] ———, *Discrete spectrum in a gap of the perturbed periodic operator at large coupling constants*, Proc. Conf. Exact Methods in Quantum Dynamics, Prague, 1990 (to appear).

[B3] ———, *On discrete spectrum in gaps of a perturbed periodic second order operator*, Funktsional. Anal. i Prilozhen. (to appear).

[BS1] M. S. Birman and M. Z. Solomyak, *Estimates for the singular numbers of integral operators*, Uspekhi Mat. Nauk **32** (1977), no. 1 (193), 17–84; English transl. in Russian Math. Surveys **32** (1977), no. 1, 15–89.

[BS2] ———, *Asymptotic behaviour of the spectrum of pseudodifferential operators with aniso tropically homogeneous symbols*. I, Vestnik Leningrad. Univ. Mat. Mekh. Astronom. **1977**, no. 13, 13–21; II, **1979**, no. 3, 5–10; English transl. in Vestnik Leningrad Univ. Math. **10** (1982); **12** (1980).

[BS3] ———, *Negative discrete spectrum of the Schrödinger operator with large coupling constant: a qualitative discussion*, Oper. Theory: Adv. Appl. **46** (1990), 3–16.

[BS4] ———, *Schrödinger operator. Estimates for the number of the bound states as a function-theoretic problem*, Lectures of the XIV School on Operator Theory and Function Spaces, I, Novgorod, 1990; English transl., Amer. Math. Soc., Providence, RI (to appear).

[BS5] ———, *Estimates for the number of negative eigenvalues of the Schrödinger operator and its generalizations*, this volume.

[C] M. Cwikel, *Weak type estimates for singular values and the number of bound states of Schrödinger operators*, Ann. Mat. **106** (1977), 93–100.

[KA] L. V. Kantorovich and G. P. Akilov, *Functional analysis*, 2nd ed., "Nauka", Moscow, 1977; English transl., Pergamon Press, Oxford, 1982.

[M] V. G. Maz′ya, *Sobolev spaces*, Izdat. Leningrad. Univ., Leningrad, 1985; English transl., Springer-Verlag, Berlin, 1985.

[MS] V. G. Maz′ya and T. O. Shaposhnikova, *Theory of multipliers in spaces of differentiable functions*, Izdat. Leningrad. Univ., Leningrad, 1986; English transl., Pitman (Advanced Publishing Program), Boston, MA, and London, 1985.

[N] S. M. Nikol′skiĭ, *Approximation of functions of several variables and embedding theorems*, "Nauka", Moscow, 1969; English transl., Springer-Verlag, Berlin and New York, 1975.

[O'N] R. O'Neyl, *Convolution operators and $L(p, q)$-spaces*, Duke Math. J. **30** (1963), 129–142.

[P] M. V. Paukshto, *Extension theorem for a certain class of function spaces*, Mat. Zametki **25** (1979), 825–834; English transl. in Math. Notes **25** (1979).

[S1] B. Simon, *Weak trace ideals and the number of bound states of Schrödinger operators*, Trans. Amer. Mat. Soc. **224** (1976), 367–380.

[S2] ———, *Trace ideals and their applications*, Cambridge Univ. Press, Cambridge, 1979.

[T] H. Triebel, *Interpolation theory. Function spaces. Differential operators*, VEB Deutscher Verlag, Berlin, 1978.

Translated by M. Z. SOLOMYAK

ADVANCES IN SOVIET MATHEMATICS
Volume 7, 1991

Reflection Operators and Their Applications to Asymptotic Investigations of Semiclassical Integral Equations

A. M. BUDYLIN AND V. S. BUSLAEV

ABSTRACT. We describe the asymptotic behavior of the solutions of the integral equations

$$\varepsilon^{-1}\int_{-1}^{1}\mathscr{A}(\varepsilon^{-1}(x-y))f(y)\,dy=g(x),$$

as $\varepsilon\to 0$. We include in our considerations the equations with symbols

$$a(\varepsilon\xi)=\int_{-\infty}^{+\infty}\exp(-ix\xi)\varepsilon^{-1}\mathscr{A}(\varepsilon^{-1}x)\,dx,$$

that can have jumps (or jumps of derivatives) and roots at some points. Our approach is based on general constructions which can be treated as constructions of the so-called method of projections. We combine four ideas: the classical Wiener-Hopf technique, the well-known Kozak formula, a certain new version of the alternating Schwarz method, and the notion of reflection operator (which may be a new notion in this field).

CONTENTS

1991 *Mathematics Subject Classification*. Primary 45A05, 47B99; Secondary 47A68.

§1. Introduction

1.1. Properties of symbols. There is an extensive literature devoted to integral equations of the form

$$\varepsilon^{-1}\int_{-1}^{1}\mathscr{A}(\varepsilon^{-1}(x-y))f(y)\,dy = g(x) \tag{1.1}$$

with small parameter ε. Frequently a different form of the same equations,

$$\int_{-\varepsilon^{-1}}^{\varepsilon^{-1}}\mathscr{A}(x-y)f(y)\,dy = g(x),$$

elementarily related to (1.1), is used. These equations have applications in quite different fields of mathematical physics; in particular they appear in the theory of elasticity, in hydrodynamics, electrostatics, diffraction theory, and quantum statistics. General indications about the literature and about the main applications can be obtained from [BaSh, Be, ChKo, Gau, Hu, LeLe, Nob, Nov, Sol, VoBa, VAB].

To describe the outline of this work, we must comment on the properties of the kernel $\mathscr{A}$. It is more convenient to discuss them using the corresponding symbol:

$$a(\varepsilon\xi) = \int_{-\infty}^{+\infty}\exp(-ix\xi)\varepsilon^{-1}\mathscr{A}(\varepsilon^{-1}x)\,dx. \tag{1.2}$$

Actually we shall consider more general symbols: $a(\xi, \varepsilon) = b(\xi) * v(\varepsilon\xi)$.

The most important are the three following properties of the symbol:

(1) its smoothness (which determines the velocity of decay of the kernel $\mathscr{A}$ at infinity);
(2) its asymptotic behavior at infinity (which can change the character of equation (1.1) from an almost differential equation to an integral equation of the first kind);
(3) the absence or presence of roots at some points (they determine the essential properties of the formally limiting equation).

In general, in theoretical work devoted to the investigation of equation (1.1) with small ε, it is usually assumed that the symbol is a smooth (frequenly even analytic) function, that it has no roots and has special power asymptotic behavior at infinity.

However, in the applications we meet symbols which have jumps (or jumps of derivatives) and/or roots at some points, the jump points and the roots sometimes coinciding.

Here our aim is to consider this more complicated situation. We study the case when the symbol has only one singular point, but it will be clear that our approach can be applied to more general cases. As to the influence of the asymptotic behavior of the symbol at infinity on the properties of equation (1.1), we view it as a separate question and limit our consideration to the case $a(\xi) \to a_{\pm}$ as $\xi \to \pm\infty$.

Of course our work is not an isolated paper in which the case of singular symbols is considered, see, for example, [Hu, Nov]. These articles were devoted to some special examples of equations with singular symbols (later we shall review these examples); they demonstrate that the problem is nontrivial. From our point of view, however, the structure of the asymptotic behavior of the solutions was not developed in a satisfactory form in these works.

We should note separately the series of papers by H. Widom [Wi1–Wi4] devoted to the principal terms of the asymptotic expansion of the trace $\operatorname{tr} f(A[\varepsilon])$. Widom considers certain smooth symbols and symbols with jumps; he uses a straightforward approach to the trace that cannot be applied to the investigation of the inverse operator $(A[\varepsilon])^{-1}$ directly.

The main terms of the asymptotic expansion of $\det A[\varepsilon]$ and some close functionals were considered in [IIKS, JMMS]. The authors of these papers use very special approaches, depending on the specific form of the symbol.

1.2. Reflection operators. We begin with some constructions which can be related to the general theory of linear operators. The initial object of study is a linear operator $\mathscr{A}$ in a linear space $\mathscr{X}$ and two linear projection operators P, P':

$$I = P + P', \tag{1.3}$$

determining the direct decomposition of the space $\mathscr{X}$ into two subspaces:

$$\mathscr{X} = X + X'. \tag{1.4}$$

We shall consider the operator:

$$A = P\mathscr{A}_X : X \longrightarrow X, \tag{1.5}$$

and represent the inverse operator A^{-1} in the form:

$$A^{-1} = (I - \Gamma)\mathscr{A}^{-1}|_X. \tag{1.6}$$

The first main object in our investigation is the operator Γ, which we call the reflection operator. The meaning of the operator A and of the subspaces X and X' is obvious in the case of equation (1.1). It turns out that the operator Γ is considerably more suitable for the purpose of our asymptotic investigation than the operator A^{-1} itself. Apparently the reason is that the operator Γ can be constructed as a projection operator. In fact it is the projection operator onto the subspace $\mathscr{A}^{-1}X'$ in the direct decomposition:

$$\mathscr{X} = X + \mathscr{A}^{-1}X'. \tag{1.7}$$

Further we shall suppose that, in its turn, the subspace X' can be represented as a direct sum of two subspaces X'_+ and X'_-:

$$X' = X'_+ + X'_-. \tag{1.8}$$

The meaning of the decomposition (1.8) is again clear in the case of equation (1.1).

The second idea is that the action of the operator $\mathscr{A}$ between X'_+ and X'_- can be modelled by means of a certain operator $\mathscr{B} : \mathscr{Y} \to \mathscr{Y}$ which can be effectively investigated; the model space $\mathscr{Y}$ can also be represented as a direct sum $\mathscr{Y} = Y + Y'$, for which we let Q and Q' denote the corresponding projection operators.

The connection between $\mathscr{A}$ and $\mathscr{B}$ is more complicated and cannot be described in terms of standard perturbation theory. Here a certain operator interpretation of the classical Wiener-Hopf technique must be introduced in our considerations to display this connection.

Let $\mathscr{A}$ and $\mathscr{B}$ be intertwined by the relation:

$$\mathscr{A} = L\mathscr{B}R, \tag{1.9}$$

where

$$PLQ' = 0, \qquad Q'RP = 0. \tag{1.10}$$

Basically, the formula for the Wiener-Hopf approach can be written as follows

$$A^{-1} = PR^{-1}B^{-1}L^{-1}|_X, \tag{1.11}$$

where B is the analog of A.

Using these ideas, it is possible to obtain an equation for the operator Γ:

$$CT = Q', \quad C = (Q'\mathscr{B}^{-1}Q') + D, \quad T = Q'TQ', \quad \Gamma = WTV, \tag{1.12}$$

where W and V are operators that can be constructed explicitly by means of the Wiener-Hopf approach and D is a small operator (in a certain appropriate sense) when ε is a small number.

Mainly, equation (1.12) is based on the well-known Kozak formula, connecting the restriction of the operator $\mathscr{A}$ to the subspace X and the restriction of the inverse operator $\mathscr{A}^{-1}$ to the complementary subspace X', see for example [BoSi, KoSi]. Here we can write this formula in the form:

$$\mathscr{A}\Gamma = (I - \Gamma')\mathscr{A}, \tag{1.13}$$

where Γ' is the Γ-type operator corresponding to the subspace X'.

In the case of integral equation (1.1) with smooth symbol (the kernel is generated by the quickly decreasing function $\mathscr{A}(x)$), we can choose the identity operator I as the modelling operator: $\mathscr{B} = I$, $\mathscr{Y} = \mathscr{X}$. In this special case the equation for the operator Γ can be rewritten in the more expressive form:

$$\begin{pmatrix} I & \Gamma_+ \\ \Gamma_- & I \end{pmatrix} \begin{pmatrix} \Gamma_{++} & \Gamma_{+-} \\ \Gamma_{-+} & \Gamma_{--} \end{pmatrix} = \begin{pmatrix} \Gamma_+ & 0 \\ 0 & \Gamma_- \end{pmatrix}, \qquad \Gamma = \sum_{\alpha,\beta} \Gamma_{\alpha\beta}. \tag{1.14}$$

Here Γ_+ and Γ_- correspond to the restrictions of the operator $\mathscr{A}$ to the subspaces:

$$X_+ = X + X'_-, \qquad X_- = X + X'_+, \tag{1.15}$$

respectively. It is clear that in our case the operators Γ_+ and Γ_- can be constructed explicitly. The products $\Gamma_+\Gamma_-$ and $\Gamma_-\Gamma_+$ are small operators when ε is a small number, so that equation (1.14) can be investigated effectively.

It is possible to relate the system (1.14) with the system of integral equations obtained earlier in [Nob]; see also [Sol, VoBa]. We can also view system (1.14) as a version of the classical alternating Schwarz method; see, for example [GlLu, Ma]. The operator approach to this system allows us to introduce the general modelling operator $\mathscr{B}$. It turns out that the case of the integral equation (1.1) with smooth symbol can be treated without passing to the equation for the operator Γ, see [BB1, BB2].

It is worth noting that the class of equations (1.1) considered in [Nob] is characteristic of applications to the diffraction problems [Be, Nob, Sol, VoBa, VAB]. After the normalization of the asymptotic behavior at infinity, the corresponding symbol acquires the form

$$a(\xi) = (1+\xi^2)^{1/2}(1-\xi^2)^{-1/2}. \tag{1.16}$$

The crucial velocity of the kernel decay (i.e., the function $\mathscr{A}$) at infinity, separating the simple case (the kernel decreases quickly) from the complicated case (the kernel decreases slowly), has the order

$$\mathscr{A}(x) = O(x^{-1}). \tag{1.17}$$

For the symbol (1.16), the velocity of decay is less crucial, but it is clear that the inverse symbol generates a quickly decreasing kernel. In fact, the formulas proposed in [Nob] reflect in undeveloped form the connection between the equation with the direct symbol on the interval $[-1, 1]$ and the equation with the inverse symbol outside the interval. The passage from the internal problem to the external one uses the operator (defined on the entire axis) generated by the symbol a^{-1}, therefore the roots of the symbol a play an essential role. The symbol (1.16) has no roots, and the passage to the external problem answers all the questions.

1.3. Applications. The general constructions described above form §§2 and 3 of this work. In §2 we mainly discuss the notion of reflection operator, in §3 we combine it with the ideas of the Wiener-Hopf factorization.

§4 is devoted to applications. Our approach leads to the following representation of the solutions of equation (1.1):

$$f(x, \varepsilon) = \int_{-1}^{1} F(x, y; \varepsilon) g(y)\, dy, \tag{1.18}$$

$$\begin{aligned} F(x, y; \varepsilon) = \iint_{|z|, |z'|\geqslant 1} & dz\, dz' \varepsilon^{-1} W(\varepsilon^{-1}(x-z))\varepsilon^{-1} V(\varepsilon^{-1}(z'-y)) \\ & \times \sum_{k=0,\ 0\leqslant l\leqslant k}^{K} T_{kl}(z, z')\varepsilon^{k}(\log \varepsilon^{-1})^{l} \\ & + O(\varepsilon^{K+1}(\log\varepsilon^{-1})^{K+1}). \end{aligned} \tag{1.19}$$

The norm implied in (1.19) depends on the class of the symbols. The functions W, V, and T_{kl} are described by certain explicit formulas; they generate the operators W and V introduced earlier, see (1.12).

The integrals in (1.18) admit considerable simplification as $\varepsilon \to 0$, but it turns out that their asymptotic behavior is quite different if the point x is close to the boundaries of the interval $[-1, 1]$ or is far from the boundaries. It is difficult to combine the derivation of the asymptotic formulas and their simplification near the boundaries and inside the interval, see [Hu, Nov]; this leads to very unwieldy constructions. We can derive the results of the work [Hu] from formula (1.18) by elementary computations; unfortunately, the lack of space does not permit us to reproduce these computations here.

We consider three classes of symbols. The first one consists of symbols generating quickly decreasing kernels. Earlier we described this case as simple and it is included in the work for completeness; our aim is also to demonstrate the technique used in a simple case.

We include in this class symbols of the form:

$$a(\xi, \varepsilon) = r(\varepsilon\xi), \tag{1.20}$$

where

$$r \in C^{\infty}(\mathbb{R}\setminus 0)\cap C(\overline{\mathbb{R}}), \quad r(\xi) = r(\infty) + O(\xi^{-\gamma}), \quad \gamma > 0, \quad r(\xi) \neq 0.$$[1]

Thus, the symbols of this class can have jumps of the derivatives at the point $\xi = 0$. The symbol

$$a(\varepsilon\xi) = 1 + \exp(-\varepsilon|\xi|) \tag{1.21}$$

belongs to the first class and generates the following integral equation:

$$f(x) + \frac{1}{\pi}\int_{-1}^{1} \frac{\varepsilon}{(x-y)^2 + \varepsilon^2} f(y)\, dy = g(x). \tag{1.22}$$

Equation (1.22) is known in electrostatics and in quantum statistics, see [Gau].

[1] Here we describe only the main properties of the symbols, additional details will be given later.

The second class consists of symbols of the form:

$$a(\xi, \varepsilon) = b(\xi) r(\varepsilon\xi), \tag{1.23}$$

where r has the above-mentioned properties and $b^{\pm 1}$ is a bounded function analytic in the angles: $|\arg\xi| < \varphi$, $|\arg\xi - \pi| < \varphi$, $\varphi > 0$. Note that the function b can have a jump at the point $\xi = 0$ and can change sign at this point. The symbol belonging to this class,

$$a(\xi, \varepsilon) = \operatorname{sgn}\xi[1 + \exp(-\varepsilon|\xi|)], \tag{1.24}$$

generates the equation

$$\frac{1}{\pi} \text{p.v.} \int_{-1}^{1} \left[\frac{1}{x-y} + \frac{x-y}{(x-y)^2 + \varepsilon^2}\right] f(y)\, dy = g(x), \tag{1.25}$$

which is known in hydrodynamics; see, for example, [BaSh, Nov].

The third class contains the symbols

$$a(\xi, \varepsilon) = b(\xi) \frac{\varepsilon\xi}{\sqrt{1 + (\varepsilon\xi)^2}} r(\varepsilon\xi), \tag{1.26}$$

where b and r have the previous properties. Note that $a(0, \varepsilon) = 0$; moreover, at the point $\xi = 0$ the symbol can lose its smoothness. Typical specimens of this class are the symbols

$$a(\xi, \varepsilon) = \operatorname{sgn}\xi[1 - \exp(-\varepsilon|\xi|)] \tag{1.27}$$

and

$$a(\xi, \varepsilon) = 1 - \exp(-\varepsilon|\xi|). \tag{1.28}$$

They generate the equations

$$\frac{i}{\pi} \text{p.v.} \int_{-1}^{1} \left[\frac{1}{x-y} - \frac{x-y}{(x-y)^2 + \varepsilon^2}\right] f(y)\, dy = g(x) \tag{1.29}$$

and

$$f(x) - \frac{1}{\pi} \int_{-1}^{1} \frac{\varepsilon}{(x-y)^2 + \varepsilon^2} f(y)\, dy = g(x) \tag{1.30}$$

respectively. The first of them is known in hydrodynamics [BaSh], the second one in electrostatics and in quantum statistics [Gau, Hu]. We should note that equation (1.30) with $g = 1$ can be treated as a concrete form of the representation of an electrostatic problem admitting direct characterization in terms of differential equations. As a problem of the theory of differential equations, it has been considered in [Sh, Soi].

The work also contains two appendices with the proofs of certain analytical facts.

We would like to express our gratitude to B. P. Belinskiĭ, M. Sh. Birman, N. J. Reshetikhin, E. K. Sklyanin, Ya. S. Soĭbel′man, M. Z. Solomyak, and H. Widom for helpful discussions.

§2. Reflection operators

2.1. General properties of reflection operators.

2.1.1. *Initial framework.* The general framework of our further considerations will be determined by the following objects.

Let $\mathscr{X}$ be a complex vector space. Let

$$\mathscr{X} = X + X'$$

be its decomposition in the direct sum of subspaces X and X'. Introduce the corresponding projection operators P and P',

$$I = P + P',$$

where I is the identity operator.

Let $\mathscr{A}$ be an operator in $\mathscr{X}$ (here and further *operator* always means *linear operator*). The decomposition of the space $\mathscr{X}$ described above allows us to consider the operator $\mathscr{A}$ as the following matrix,

$$\mathscr{A} = \begin{pmatrix} \mathscr{A}_{XX} & \mathscr{A}_{XX'} \\ \mathscr{A}_{X'X} & \mathscr{A}_{X'X'} \end{pmatrix},$$

with operator-valued entries:

$$\mathscr{A}_{XX}\colon X \longrightarrow X, \quad \mathscr{A}_{XX'}\colon X' \longrightarrow X, \quad \mathscr{A}_{X'X}\colon X \longrightarrow X', \quad \mathscr{A}_{X'X'}\colon X' \longrightarrow X'.$$

The block $\mathscr{A}_{XX}$ will be the main object of our further study; denote it by A, so that $A = \mathscr{A}_{XX}$. Also introduce the operator $\mathbf{A}\colon \mathscr{X} \to \mathscr{X}$:

$$\mathbf{A} = \begin{pmatrix} A & 0 \\ 0 & 0 \end{pmatrix}.$$

This operator has the following evident properties:

$$\mathbf{A} = P\mathbf{A}P = P\mathscr{A}P.$$

Consider an operator $G\colon \mathscr{X} \to \mathscr{X}$ satisfying the conditions:

1_G: $P'G = 0$,
2_G: $\mathbf{A}G = P$.

It is not difficult to understand that equations 1_G–2_G have a solution if and only if the operator A has a right inverse operator $A^{-1}\colon X \to X$ so that $AA^{-1} = I$. In this case the general solution of equations 1_G–2_G is given by the formula:

$$G = \begin{pmatrix} A^{-1} & n \\ 0 & 0 \end{pmatrix}, \tag{2.1}$$

where A^{-1} is an arbitrary right inverse operator and n is an arbitrary operator $X' \to X$ satisfying the condition $An = 0$.

We shall also consider the operator G characterized by the following conditions:

3_G: $GP' = 0$,
4_G: $G\mathbf{A} = P$.

These equations have a solution if and only if there exists a left inverse operator A^{-1} such that $A^{-1}A = I$. In this case

$$G = \begin{pmatrix} A^{-1} & 0 \\ m & 0 \end{pmatrix}, \tag{2.2}$$

where A^{-1} is an arbitrary left inverse operator and m is an arbitrary operator $X \to X'$ satisfying the condition $mA = 0$.

The system 1_G–4_G has a solution if and only if A is a bijection, i.e., A has two-sided inverse operator A^{-1}. In this case

$$G = \begin{pmatrix} A^{-1} & 0 \\ 0 & 0 \end{pmatrix}. \tag{2.3}$$

Note that systems 1_G–2_G and 3_G–4_G are equivalent if A is a bijection. If any of these systems has a unique solution, then A is a bijection (with all the consequences).

Both systems can be considered in a similar way; we prefer to deal with the right inverse operator.

In the case when the operator G obeys 1_G–2_G, let us define the operators:

$$N = GP', \qquad \Theta = P - G\mathbf{A}.$$

It is clear that

$$N = PNP', \qquad \Theta = P\Theta P,$$

and

$$N = \begin{pmatrix} 0 & n \\ 0 & 0 \end{pmatrix}, \qquad \Theta = \begin{pmatrix} \theta & 0 \\ 0 & 0 \end{pmatrix},$$

where

$$\theta = I - A^{-1}A.$$

Since the operator θ is a projection operator: $\theta^2 = \theta$, $\theta X = \operatorname{Ker} A$, the operator Θ is also a projection operator.

2.1.2. *Reflection operator.* Let G obey conditions 1_G–2_G. Introduce the *reflection operator*:

$$\Gamma = I - G\mathscr{A}. \tag{2.4}$$

The matrix of the operator Γ has the form:

$$\Gamma = \begin{pmatrix} \theta - n & \gamma - n \\ 0 & I \end{pmatrix}, \tag{2.5}$$

where

$$\gamma = -A^{-1}\mathscr{A}_{XX'}. \tag{2.6}$$

PROPOSITION 2.1.1. *The operator* Γ *satisfies the following relations*:

1_Γ: $P'(I - \Gamma) = 0$,

2_Γ: $P\mathscr{A}\Gamma = 0$.

PROOF. 1_Γ:

$$P'(I-\Gamma) = P'G\mathscr{A} = 0,$$

2_Γ:

$$\begin{aligned} P\mathscr{A}\Gamma &= P\mathscr{A}(I - G\mathscr{A}) \\ &= P\mathscr{A} - P\mathscr{A}G\mathscr{A} \\ &= P\mathscr{A} - P\mathscr{A}PG\mathscr{A} \\ &= P\mathscr{A} - \mathbf{A}G\mathscr{A} = 0. \quad \square \end{aligned}$$

We call the operator $\Gamma: \mathscr{X} \to \mathscr{X}$ obeying conditions 1_Γ–2_Γ *the generalized reflection operator.*

Further we shall often use the condition

(A): the operator $\mathscr{A}$ is a bijection.

At this point we suppose that condition (A) is fulfilled; only the last proposition (Proposition 2.1.5) will not depend on this assumption.

Under condition (A)

$$G = (I-\Gamma)\mathscr{A}^{-1}, \tag{2.7}$$

where Γ is a reflection operator. Hence

$$A^{-1}f = (I-\Gamma)\mathscr{A}^{-1}f, \qquad f \in X. \tag{2.8}$$

PROPOSITION 2.1.2. *Any generalized reflection operator Γ is a reflection operator.*

PROOF. It is sufficient to verify that the operator $G = (I-\Gamma)\mathscr{A}^{-1}$, where Γ is a generalized reflection operator, obeys relations 1_G–2_G:

1_G:

$$P'G = [P'(I-\Gamma)]\mathscr{A}^{-1} = 0,$$

2_G:

$$\begin{aligned} \mathbf{A}G &= P\mathscr{A}P(I-\Gamma)\mathscr{A}^{-1} \\ &= P\mathscr{A}(I-\Gamma)\mathscr{A}^{-1} \\ &= P - (P\mathscr{A}\Gamma)\mathscr{A}^{-1} = P. \quad \square \end{aligned}$$

PROPOSITION 2.1.3. *The reflection operator Γ satisfies the following relations:*

$\widetilde{3}_\Gamma$: $\Gamma P = \Theta - N\mathscr{A}P$,

$\widetilde{4}_\Gamma$: $(I-\Gamma)\mathscr{A}^{-1}P' = N$.

PROOF. Relation $\widetilde{4}_\Gamma$ follows from formula (2.7) and the definition of N. Relation $\widetilde{3}_\Gamma$ is an evident consequence of the matrix representation of the operator Γ. $\square$

PROPOSITION 2.1.4. *The operator A is a bijection if and only if the reflection operator Γ satisfies the relations:*

3_Γ: $\Gamma P = 0$,

4_Γ: $(I-\Gamma)\mathscr{A}^{-1}P' = 0$.

PROOF. The set of relations 1_Γ–4_Γ is equivalent to the set 1_G–4_G, which in turn implies that A is a bijection. □

Note that under conditions 1_Γ–4_Γ the matrix representation (2.5) of the operator Γ becomes simpler:

$$\Gamma = \begin{pmatrix} 0 & \gamma \\ 0 & I \end{pmatrix}. \tag{2.9}$$

In this case Γ is a projection.

The following more general statement is also true.

PROPOSITION 2.1.5. *Any operator Γ satisfying relations 1_Γ, 3_Γ, or 2_Γ, 4_Γ, is a projection operator.*

PROOF. Consider, for example, relations 1_Γ, 3_Γ:

$$\begin{aligned} \Gamma - \Gamma^2 &= \Gamma(I - \Gamma) \\ &= \Gamma(P + P')(I - \Gamma) \\ &= (\Gamma P)(I - \Gamma) + \Gamma[P'(I - \Gamma)] = 0. \quad \square \end{aligned}$$

Call the (generalized) reflection operator satisfying the additional condition 3_Γ a *projecting (generalized) reflection operator.*

Introducing the notion of projecting reflection operator, we prefer condition 3_Γ as the condition which leads to the projection property of Γ, since this property does not depend on condition (A).

Call the (generalized) reflection operator obeying conditions 1_Γ–4_Γ a *total reflection operator.*

It is evident that we can use this notion only if condition (A) is fulfilled.

Equations 1_Γ–4_Γ admit a simple geometric interpretation.

Conditions 1_Γ, 3_Γ imply that $I - \Gamma$ is a projection operator into the subspace X; 2_Γ, 4_Γ imply that Γ is a projection operator into the subspace $\mathscr{A}^{-1}X'$. Therefore the projection operator Γ can be characterized by the decomposition

$$\mathscr{X} = X + \mathscr{A}^{-1}X' \tag{2.10}$$

as the projection operator onto the second subspace $\mathscr{A}^{-1}X'$.

2.1.3. *Gauging.* Let G be a solution of system 1_G–2_G and Γ be the corresponding reflection operator. Consider the transformation $G \to G_1$:

$$G_1 = G(I - \Phi) + \Gamma\Psi, \tag{2.11}$$

where Φ and Ψ are operators in $\mathscr{X}$ and $P\Phi = 0$, $P'\Psi = 0$.

PROPOSITION 2.1.6. *The operator G_1 is a solution of system 1_G–2_G, and formula* (2.11) *gives the general solution of this system.*

PROOF. Let us verify relations 1_G–2_G.

1_G: $$P'G_1 = P'\Gamma\Psi = P'\Psi = 0,$$

$$2_G: \qquad \begin{aligned} \mathbf{A}G_1 &= \mathbf{A}[G(I-\Phi)+\Gamma\Psi] \\ &= P(I-\Phi)+P\mathscr{A}P\Gamma\Psi \\ &= P-P\mathscr{A}(I-P')\Gamma\Psi \\ &= P+P\mathscr{A}P'\Gamma\Psi \\ &= P+P\mathscr{A}(P'\Psi)=P. \end{aligned}$$

Now we discuss the second part of the proposition. Let

$$\Phi=-P'\mathscr{A}P\Psi, \quad \Psi=\begin{pmatrix}\psi & \psi' \\ 0 & 0\end{pmatrix},$$

then

$$\begin{aligned} G_1 &= G+GP'\mathscr{A}P\Psi+(I-G\mathscr{A})\Psi \\ &= G+GP'\mathscr{A}P\Psi+P\Psi-G\mathscr{A}P\Psi \\ &= G+(P-G\mathbf{A})\Psi=G+\Theta\Psi. \end{aligned}$$

This formula means that the nontrivial matrix elements of the operators G_1 and G are connected by the relations:

$$A_1^{-1}=A^{-1}+\theta\psi, \qquad n_1=n+\theta\psi'.$$

Since the operators ψ and ψ' are arbitrary, these formulas prove the second part of the proposition. □

Transformation (2.11) generates the corresponding transformation of the reflection operators:

$$\Gamma_1=\Gamma(I-\Psi_1)+G\Phi_1, \tag{2.12}$$

where $P\Phi_1=0$, $P'\Psi_1=0$ and $\Phi_1=\Phi\mathscr{A}$, $\Psi_1=\Psi\mathscr{A}$.

Under condition (A), the transformation (2.12) can be treated as the initial one.

Let us indicate two important special cases:

a) $\Psi=0 \implies G_1=G(I-\Phi)$; in this case $\Gamma_1=\Gamma+G\Phi_1$,
b) $\Phi=0 \implies \Gamma_1=\Gamma(I-\Psi_1)$; in this case $G_1=G+\Gamma\Psi$.

The transformation $\Gamma\to\Gamma_1=\Gamma(I-\Psi_1)$ can be considered as a transformation of the generalized reflection operators independently of the existence of the operator G.

Among transformations a) there is a transformation with $\Phi=P'$, so that $G_1=GP$ and therefore G_1 obeys condition 3_G: $G_1P'=0$. This means that the corresponding reflection operator satisfies condition 4_Γ.

Among transformations b) there is a transformation with $\Psi_1=P$, so that $\Gamma_1=\Gamma P'$ and therefore Γ_1 obeys condition 3_Γ.

In both cases the transformed operator Γ_1 is a projection operator.

2.1.4. *Additional subspace.* At this point we shall assume condition (A) to be fulfilled.

Similarly to the operator G, define the operator G' as the solution of the system:

$1'_G$: $PG' = 0$,
$2'_G$: $A'G' = P'$,

where

$$\mathbf{A}' = P'\mathscr{A}^{-1}P' = \begin{pmatrix} 0 & 0 \\ 0 & A' \end{pmatrix}, \qquad A' = (\mathscr{A}^{-1})_{X'X'}.$$

The operator G' can be considered as a G-type operator with the operator $\mathscr{A}$ replaced by the operator $\mathscr{A}^{-1}$ and the subspace X replaced by the subspace X'. G' generates the reflection operator $\Gamma' = I - G'\mathscr{A}^{-1}$, $G' = (I - \Gamma')\mathscr{A}$.

PROPOSITION 2.1.7. *The operators* G *and* G' *exist simultaneously and are related by the formulas*:

$$G' = \mathscr{A}\Gamma, \qquad \Gamma' = \mathscr{A}G. \tag{2.13}$$

Let us explain this proposition. Suppose the operator G exists; then the operator Γ is defined and the first formula (2.13) gives a solution of the system $1'_G$–$2'_G$, and the second formula describes the corresponding reflection operator. On the contrary, if the operator G' exists, formulas (2.13) allow us to recover the operators G and Γ.

PROOF. Let the operator G be defined; let us verify that the operator $G' = \mathscr{A}\Gamma$ is a solution of system $1'_G$–$2'_G$:

$1'_G$: $PG' = P\mathscr{A}\Gamma = 0$,
$2'_G$: $\mathbf{A}'G' = P'\mathscr{A}^{-1}P'\mathscr{A}\Gamma = P'\mathscr{A}^{-1}I\mathscr{A}\Gamma = P'\Gamma = P'$.

Now compute the operator Γ':

$$\Gamma' = I - G'\mathscr{A}^{-1} = I - \mathscr{A}\Gamma\mathscr{A}^{-1} = \mathscr{A}(I - \Gamma)\mathscr{A}^{-1} = \mathscr{A}G. \quad \square$$

Let us indicate some evident consequences of formulas (2.13):

$$\mathscr{A}\Gamma = (I - \Gamma')\mathscr{A}, \tag{2.14}$$

$$G = \mathscr{A}^{-1} - \mathscr{A}^{-1}G'\mathscr{A}^{-1}, \qquad G' = \mathscr{A} - \mathscr{A}G\mathscr{A}. \tag{2.15}$$

All formulas (2.13)–(2.15) can be viewed as different representations of the Kozak formula mentioned in the introduction.

Formula (2.14) implies that the set of relations 1_Γ–4_Γ can be written in the following symmetric form:

$$P'(I - \Gamma) = 0, \quad P(I - \Gamma') = 0, \quad \Gamma P = 0, \quad \Gamma'P' = 0. \tag{2.16}$$

2.2. Equations for the operator Γ.

2.2.1. *The equation for* Γ. Here we assume that condition (A) is fulfilled.

Using formula (2.13), we can consider the relations $1'_G$– $2'_G$ as the system of equations for the operator Γ:

E_1: $\mathbf{A}'\mathscr{A}\Gamma = P'$,
E_2: $P\mathscr{A}\Gamma = 0$.

Any operator Γ satisfying these equations is a reflection operator.

In fact all further constructions of this section and the next ones will be devoted to the description of different regularizations of equations E_1–E_2. The efficiency of all these regularizations will depend on certain concrete additional topological hypotheses; the significance of these conditions will be made clear during our further study of applications to integral equations.

The conditions mentioned above will be expressed in terms of certain additional algebraic structures in the space $\mathscr{X}$.

Let X' be a direct sum:

$$X' = X'_+ + X'_-,$$

of some subspaces X'_+ and X'_- and let P'_+, P'_- be the corresponding projection operators,

$$P' = P'_+ + P'_-.$$

Let us introduce the subspaces

$$X_\alpha = X + X'_{\alpha'}, \qquad \alpha \neq \alpha', \quad \alpha, \alpha' = +, -;$$

consider the decompositions

$$\mathscr{X} = X_\alpha + X'_\alpha,$$

and the corresponding projection operators

$$I = P_\alpha + P'_\alpha.$$

In accordance to the decompositions $\mathscr{X} = X_\alpha + X'_\alpha$, the operator $\mathscr{A}$ generates A-type operators $A_\alpha = P_\alpha \mathscr{A}|_{X_\alpha} : X_\alpha \to X_\alpha$. At first we assume the corresponding operators Γ_α to be total reflection operators.

Let G and Γ be defined and consider equations E_1–E_2. Now we shall describe a regularization of these equations, assuming that the operator $\mathscr{A}$ connects the subspaces X'_+ and X'_- weakly.

In accordance with subsection 2.1.3, it is possible to reinforce condition E_2 and assume that the condition

(C): $\mathscr{A}\Gamma = P'(\mathscr{A}\Gamma)P'$

is true. Consider equation E_1 jointly with condition (C).

Introduce the operators $\Gamma_{\alpha\beta}: \mathscr{X} \to \mathscr{X}$, $\alpha, \beta = +, -$, by the formulas

$$\mathscr{A}\Gamma_{\alpha\beta} = P'_\alpha(\mathscr{A}\Gamma)P'_\beta. \tag{2.17}$$

Owing to (C), it is possible to recover Γ if the $\Gamma_{\alpha\beta}$ are known:

$$\Gamma = \sum_{\alpha, \beta} \Gamma_{\alpha\beta}. \tag{2.18}$$

Rewrite equation E_1 as a matrix relation:

$$\begin{pmatrix} P'_+\mathscr{A}^{-1}P'_+ & P'_+\mathscr{A}^{-1}P'_- \\ P'_-\mathscr{A}^{-1}P'_+ & P'_-\mathscr{A}^{-1}P'_- \end{pmatrix} \begin{pmatrix} \mathscr{A}\Gamma_{++} & \mathscr{A}\Gamma_{+-} \\ \mathscr{A}\Gamma_{-+} & \mathscr{A}\Gamma_{--} \end{pmatrix} = \begin{pmatrix} P'_+ & 0 \\ 0 & P'_- \end{pmatrix}.$$

Multiply both sides of this relation by the matrix $\operatorname{diag}(\Gamma_+, \Gamma_-)$ and take into account the formulas $\mathscr{A}\Gamma_\alpha(P'_\alpha\mathscr{A}^{-1}P'_\alpha) = P'_\alpha$:

$$\begin{pmatrix} \mathscr{A}^{-1}P'_+ & \Gamma_+\mathscr{A}^{-1}P'_- \\ \Gamma_-\mathscr{A}^{-1}P'_+ & \mathscr{A}^{-1}P'_- \end{pmatrix} \begin{pmatrix} \mathscr{A}\Gamma_{++} & \mathscr{A}\Gamma_{+-} \\ \mathscr{A}\Gamma_{-+} & \mathscr{A}\Gamma_{--} \end{pmatrix} = \begin{pmatrix} \Gamma_+ & 0 \\ 0 & \Gamma_- \end{pmatrix}.$$

Using formulas (2.17), one can omit the projection operators P'_α, $\alpha = +, -$, and cancel the factors $\mathscr{A}$, $\mathscr{A}^{-1}$:

$$E_3: \qquad \begin{pmatrix} I & \Gamma_+ \\ \Gamma_- & I \end{pmatrix} \begin{pmatrix} \Gamma_{++} & \Gamma_{+-} \\ \Gamma_{-+} & \Gamma_{--} \end{pmatrix} = \begin{pmatrix} \Gamma_+ & 0 \\ 0 & \Gamma_- \end{pmatrix}.$$

This is the desired equation.

Sometimes it will be useful to write this equation in the following form:

$$\Gamma_{\alpha\beta} = \Gamma_\alpha(\delta_{\alpha\beta}I - \Gamma_{\alpha'\beta}). \tag{2.19}$$

2.2.2. *Properties of equation* E_3. Equation E_3 does not lose its significance if condition (A) is violated and the operators Γ_α are only generalized reflection operators.

PROPOSITION 2.2.1. *Let* Γ_α *be generalized reflection operators and* $(\Gamma_{\alpha,\beta})$ *be a solution of* E_3; *then*

$$P_\alpha\mathscr{A}\Gamma_{\alpha\beta} = 0, \tag{2.20}$$

and $\Gamma = \sum_{\alpha,\beta}\Gamma_{\alpha\beta}$ *is a generalized reflection operator.*

PROOF. Use equation E_3 in the form (2.19):

$$P'_\alpha\Gamma_{\alpha\beta} = P'_\alpha\Gamma_\alpha(\delta_{\alpha\beta}I - \Gamma_{\alpha'\beta}) = P'_\alpha(\delta_{\alpha\beta}I - \Gamma_{\alpha'\beta}).$$

Hence $P'_\alpha(\Gamma_{\alpha\beta} + \Gamma_{\alpha'\beta}) = P'_\alpha\delta_{\alpha\beta}$. Summing with respect to β gives $P'_\alpha\Gamma = P'_\alpha$ and, as a result, 1_Γ: $P'(I - \Gamma) = 0$.

Let us verify 2_Γ:

$$P_\alpha\mathscr{A}\Gamma_{\alpha\beta} = (P_\alpha\mathscr{A}\Gamma_\alpha)(\delta_{\alpha\beta}I - \Gamma_{\alpha'\beta}) = 0.$$

Owing to $P = P_{\alpha'}P_\alpha$:

$$P\mathscr{A}\Gamma_{\alpha\beta} = 0 \quad \text{and} \quad P\mathscr{A}\Gamma = 0. \quad \square$$

PROPOSITION 2.2.2. *Let* Γ_α *be a projecting reflection operator and let* $(\Gamma_{\alpha\beta})$ *obey equation* E_3; *then*

$$\Gamma_{\alpha\beta} = \Gamma_\alpha\Gamma_{\alpha\beta}, \qquad \Gamma_\alpha(I - \Gamma) = 0, \tag{2.21}$$

the matrix $(\Gamma_1)_{\alpha\beta} = \Gamma_{\alpha\beta}\Gamma_\beta$ *is also a solution of equation* E_3, $(\Gamma_1)_{\alpha\beta}P_\beta = 0$, *and* $\Gamma_1 = \sum_{\alpha,\beta}(\Gamma_1)_{\alpha\beta}$ *is a projecting generalized reflection operator.*

This proposition can be derived from equation E_3 in the form (2.19).

As a consequence of Proposition 2.2.2, we can conclude that equation E_3 is equivalent to the equation

$$E_4: \qquad \begin{pmatrix} I & \Gamma_+\Gamma_- \\ \Gamma_-\Gamma_+ & I \end{pmatrix}\begin{pmatrix} \Gamma_{++} & \Gamma_{+-} \\ \Gamma_{-+} & \Gamma_{--} \end{pmatrix} = \begin{pmatrix} \Gamma_+ & 0 \\ 0 & \Gamma_- \end{pmatrix}$$

if Γ_α are projecting generalized reflection operators.

Now let us return to the case when condition (A) is fulfilled and Γ_α are total reflection operators.

PROPOSITION 2.2.3. *If equation* E_3 *(or* E_4*) has a unique solution, then the corresponding operator* $\Gamma = \sum_{\alpha,\beta}\Gamma_{\alpha\beta}$ *is a total reflection operator and* $(I-\Gamma)\Gamma_\alpha = 0$.

Under the conditions of the proposition, equations $1'_G$–$2'_G$ have a unique solution; this means that equations 1_G–2_G also have a unique solution, therefore the operator Γ is a total reflection operator. The formula $(I-\Gamma)\Gamma_\alpha = 0$ can be obtained from the relation

$$\Gamma_{\alpha\beta} = (\delta_{\alpha\beta}I - \Gamma_{\alpha\beta'})\Gamma_\beta. \tag{2.22}$$

This relation is an analog of (2.19) and, in its turn, can be obtained from E_3 by uniqueness of the solution.

§3. Factorization

3.1. The modelling operator.

3.1.1. *Comparison of two operators.* In §2 we assumed that the operator $\mathscr{A}$ connected the subspaces X'_+ and X'_- weakly. Further we shall suppose that there exists an operator $\mathscr{B}: \mathscr{Y} \to \mathscr{Y}$, admitting an effective investigation, which models the action of the operator between X'_+ and X'_- in an appropriate sense.

We shall suppose that the space $\mathscr{Y}$ possesses all the structures which exist in the space $\mathscr{X}$, i.e., we have the decompositions

$$\mathscr{Y} = Y + Y', \quad Y' = Y'_+ + Y'_-, \quad Y_\alpha = Y + Y'_{\alpha'}, \quad \mathscr{Y} = Y_\alpha + Y'_\alpha, \quad \alpha = +, -,$$

and the corresponding projection operators

$$I = Q + Q', \quad Q' = Q'_+ + Q'_-, \quad Q_\alpha = Q + Q'_{\alpha'}, \quad I = Q_\alpha + Q'_\alpha.$$

We shall use the natural analogs of the operators A, $\mathbf{A}$ and the corresponding notation for them; the analogs of the operators G and Γ will be denoted by H and Δ respectively. If necessary, these symbols will be supplied with the natural indices.

We shall always assume *the operator* $\mathscr{B}$ *to be a bijection.*

Suppose that the operators $\mathscr{A}$ and $\mathscr{B}$ are related by the formula

$$\mathscr{A} = L\mathscr{B}R, \tag{3.1}$$

where $L: \mathscr{Y} \to \mathscr{X}$ and $R: \mathscr{X} \to \mathscr{Y}$ are bijections satisfying the conditions

$$PLQ' = 0, \tag{3.2}$$

$$QL^{-1}P' = 0, \tag{3.3}$$

$$Q'RP = 0, \tag{3.4}$$

$$P'R^{-1}Q = 0. \tag{3.5}$$

Note the following evident consequences of these conditions:

$$PL = PLQ, \quad LQ' = P'LQ', \quad QL^{-1} = QL^{-1}P, \quad L^{-1}P' = Q'L^{-1}P',$$
$$Q'R = Q'RP', \quad RP = QRP, \quad P'R^{-1} = P'R^{-1}Q', \quad R^{-1}Q = PR^{-1}Q.$$

Further we shall use these last formulas without special comments.

Let the operator $B: Y \to Y$ be a bijection; then all the operators H, H', Δ, Δ' are defined. Under all these conditions we have the following set of relations:

$$G = R^{-1}HL^{-1}, \tag{3.6}$$

$$G' = LH'R, \tag{3.7}$$

$$\Gamma = R^{-1}\Delta R, \tag{3.8}$$

$$\Gamma' = L\Delta' L^{-1}, \tag{3.9}$$

which also show that G, G', Γ, and Γ' are defined and that the operator A is a bijection.

All these relations can be viewed as representations of the classical Wiener-Hopf approach.

The simplest way to prove the given relations is by direct computation. For example, let us verify the formula for the operator G:

1_G: $P'G = P'R^{-1}HL^{-1} = P'R^{-1}Q'HL^{-1} = 0$,

2_G: $\mathbf{A}G = P\mathscr{A}PR^{-1}HL^{-1} = P\mathscr{A}R^{-1}QHL^{-1} = PLQ\mathscr{B}QHL^{-1} = PLQL^{-1} = P$,

3_G: $GP' = R^{-1}HL^{-1}P' = R^{-1}HQ'L^{-1}P' = 0$,

4_G: $G\mathbf{A} = R^{-1}HL^{-1}P\mathscr{A}P = R^{-1}HQL^{-1}\mathscr{A}P = R^{-1}HQ\mathscr{B}QRP = R^{-1}QRP = P$.

Note the following special case. If $\mathscr{B} = I$ and $\mathscr{X} = \mathscr{Y}$, we have

$$\mathscr{A} = LR, \tag{3.10}$$

$$H = P, \quad H' = P', \quad \Delta = P', \quad \Delta' = P. \tag{3.11}$$

In this case

$$G = R^{-1}PL^{-1}, \quad G' = LP'R, \quad \Gamma - R^{-1}P'R, \quad \Gamma' = LPL^{-1}. \tag{3.12}$$

3.1.2. *Generalizations.* In fact we shall not be able to use the formulas of the previous subsection for applications to integral equations precisely because in all interesting cases the operators $\mathscr{B}$, L, and R do not possess all the properties (3.2)–(3.9).

PROPOSITION 3.1.1. *Let the operators $\mathscr{A}$ and $\mathscr{B}$ be connected by relation* (3.1), *where L, R satisfy conditions* (3.2), (3.5) *and R is a bijection. Let Γ be given by formula* (3.8).

(1) *If Δ is a generalized reflection operator, then Γ is also a generalized reflection operator.*
(2) *If R satisfies condition* (3.4) *additionally and Δ is a projecting generalized reflection operator, then Γ is also a projecting generalized reflection operator.*
(3) *If L, R satisfy conditions* (3.3), (3.4) *and Δ is a total reflection operator, then Γ is also a total reflection operator.*

PROOF. In case (1) the operator Γ satisfies equations 1_Γ, 2_Γ. In case (2) the operator Γ additionally satisfies equation 3_Γ. In case (3) this operator also satisfies equation 4_Γ. We verify these statements in order, using the corresponding assumptions (1), (2), and (3):

$$1_\Gamma: \qquad \begin{aligned} P'(I-\Gamma) &= P'R^{-1}(I-\Delta)R \\ &= P'R^{-1}(Q+Q')(I-\Delta)R \\ &= (P'R^{-1}Q)(I-\Delta)R + P'R^{-1}[Q'(I-\Delta)]R = 0, \end{aligned}$$

$$2_\Gamma: \qquad \begin{aligned} P\mathscr{A}\Gamma &= PL\mathscr{B}RR^{-1}\Delta R \\ &= PL(Q+Q')\mathscr{B}\Delta R \\ &= (PLQ')\mathscr{B}\Delta R + PL(Q\mathscr{B}\Delta)R = 0, \end{aligned}$$

$$3_\Gamma: \qquad \begin{aligned} \Gamma P &= R^{-1}\Delta RP \\ &= R^{-1}\Delta(Q+Q')RP \\ &= R^{-1}(\Delta Q)RP + R^{-1}\Delta(Q'RP) = 0, \end{aligned}$$

4_Γ:

$$\begin{aligned} (I-\Gamma)\mathscr{A}^{-1}P' &= R^{-1}(I-\Delta)RR^{-1}\mathscr{B}^{-1}L^{-1}P' \\ &= R^{-1}(I-\Delta)\mathscr{B}^{-1}(Q+Q')L^{-1}P' \\ &= R^{-1}[(I-\Delta)\mathscr{B}^{-1}Q']L^{-1}P' + R^{-1}(I-\Delta)\mathscr{B}^{-1}(QL^{-1}P') \\ &= 0. \quad \square \end{aligned}$$

The following proposition can be considered as the converse of the previous one.

PROPOSITION 3.1.2. *Let* R *be a bijection and suppose the projecting generalized reflection operators* Δ *and* Γ *are connected by relation* (3.8); *then* R *satisfies conditions* (3.4) *and* (3.5). *If additionally* Δ *is a total reflection operator, then the operators* $\mathscr{A}$ *and* $\mathscr{B}$ *are connected by formula* (3.1), *the operator* L *satisfying condition* (3.2). *If, besides,* Γ *is a total reflection operator, then* L *is a bijection and* L^{-1} *satisfies condition* (3.3).

PROOF. Verify the first statement:

$$Q'RP = (Q'\Delta)RP = Q'R(\Gamma P) = 0,$$
$$P'R^{-1}Q = (P'\Gamma)R^{-1}Q = P'R^{-1}(\Delta Q) = 0.$$

The second statement:

$$PLQ' = P(\mathscr{A}R^{-1}\mathscr{B}^{-1})Q' = P\mathscr{A}R^{-1}(\Delta\mathscr{B}^{-1}Q') = (P\mathscr{A}\Gamma)R^{-1}\mathscr{B}^{-1}Q' = 0.$$

The third statement, i.e., condition (3.3), can be treated as formula (3.2) with respect to the relation $\mathscr{B} = L^{-1}\mathscr{A}R^{-1}$. □

3.1.3. *New approach to regularization.* The initial point of the new approach is the pair of the relations

$$\mathscr{A} = L_\alpha \mathscr{B} R_\alpha, \tag{3.13}$$

where R_α are bijections and R_α, L_α satisfy conditions (3.2), (3.4), (3.5).

Introduce the operators

$$V = \sum_\alpha Q'_\alpha R_\alpha P'_\alpha, \quad V_1 = \sum_\alpha P'_\alpha R_\alpha^{-1} Q'_\alpha, \quad W = \sum_\alpha R_\alpha^{-1}\mathscr{B}^{-1}Q'_\alpha. \tag{3.14}$$

PROPOSITION 3.1.3. *The operators* $V : \mathscr{X} \to \mathscr{Y}$, $V_1 : \mathscr{Y} \to \mathscr{X}$, *and* W: $\mathscr{Y} \to \mathscr{X}$ *satisfy the following relations*:

a) $V = Q'VP'$,
b) $\mathscr{A}W = P'\mathscr{A}WQ'$,
c) $VV_1 = Q'$, $V_1V = P'$,
d) *if additionally condition* (A) *is satisfied and the operators* L_α *possess properties* (3.3), *then*

$$\mathscr{A}WW_1 = P', \qquad W_1\mathscr{A}W = Q',$$

where

$$W_1 = \sum_\alpha Q'_\alpha L_\alpha^{-1} P'_\alpha. \tag{3.15}$$

PROOF. Formulas a) and c) are evident directly and formulas b) and d) are evident consequences of the relation

$$\mathscr{A}W = \sum_\alpha P'_\alpha L_\alpha Q'_\alpha. \quad \square$$

The main idea of this work is the use of the operators V *and* $\mathscr{A}W$ *as good approximations of the operators* $Q'RP'$ *and* $P'LQ'$.

At first note that the formula $G' = LH'R$ can be rewritten in the form $G' = P'LQ'H'Q'RP'$. Our approach is to try to study the operator G' using the representation $G' = \mathscr{A}WTV$, or

$$\Gamma = WTV. \tag{3.16}$$

Return to the main equations E_1–E_2. Now we have:

E_1: $\mathbf{A}'\mathscr{A}WTV = P'$ or $P'\mathscr{A}^{-1}P'\mathscr{A}WTV = P'$,
E_2: $P\mathscr{A}WTV = 0$.

The second equation is satisfied due to the general properties of the operator W. Transform the first one to more explicit form. Having multiplied it by V from the left and by V_1 from the right, we get:

E_5: $CT = Q'$,

where

$$C = VP'\mathscr{A}^{-1}P'\mathscr{A}W$$

or

$$C = VW. \tag{3.17}$$

Equation E_5 together with representation (3.17) will be the main tool in the investigation of the integral equations.

Compute the operator C explicitly:

$$C = \sum_{\alpha,\beta} Q'_\alpha R_\alpha R_\beta^{-1}\mathscr{B}^{-1}Q'_\beta = \sum_{\alpha,\beta} Q'_\alpha \mathscr{B}^{-1}L_\alpha^{-1}L_\beta Q'_\beta.$$

Therefore

$$C = B' + D, \tag{3.18}$$

where

$$D = \sum_\alpha D_\alpha, \tag{3.19}$$

$$D_\alpha = Q'_\alpha(R_\alpha R_{\alpha'}^{-1} - I)\mathscr{B}^{-1}Q'_{\alpha'} = Q'_\alpha\mathscr{B}^{-1}(L_\alpha^{-1}L_{\alpha'} - I)Q'_{\alpha'}. \tag{3.20}$$

Without any losses we can consider this equation jointly with the condition

E_6: $T = Q'TQ'$.

We can now forget about the process of deriving E_5 and consider it as our initial object.

PROPOSITION 3.1.4. *Let an operator* $T: \mathscr{Y} \to \mathscr{Y}$ *satisfy equation* E_5 *with* V *and* W *possessing properties* a), b), *and* c) *from Proposition* 3.1.3. *Then the operator* $\Gamma = WTV$ *is a projecting generalized reflection operator.*

PROOF. Let us verify the properties of Γ. We have:

$$VP'(I - \Gamma) = VP' - (VWT)V = VP' - Q'V = 0,$$

hence c) implies 1_Γ.

2_Γ: $P\mathscr{A}\Gamma = (P\mathscr{A}W)TV = 0$.
3_Γ: $\Gamma P = WT(VP) = 0$. □

PROPOSITION 3.1.5. *Let* $\mathscr{A}$ *satisfy condition* (A) *and* V, W *possess all properties* a), b), c), *and* d) *from Proposition* 3.1.3. *If the operator* T *is the unique solution of equations* E_5–E_6, *then the corresponding operator* Γ *is a total reflection operator.*

PROOF. Taking into account Proposition 3.1.4, we need only prove property 4_Γ of the operator Γ. The assumptions of the proposition mean that besides E_5 we also have the relation $TC = Q'$; hence

$$\begin{aligned}(I-\Gamma)\mathscr{A}^{-1}P'\mathscr{A}W &= (I-\Gamma)\mathscr{A}^{-1}I\mathscr{A}WQ' \\ &= (I-\Gamma)WQ' \\ &= WQ' - W(TVW)Q' \\ &= WQ' - WQ' = 0.\end{aligned}$$

This implies 4_Γ. □

3.2. Degenerate cases.

3.2.1. *Simple degenerate case.* We know that under condition (A) any generalized reflection operator turns out to be a reflection operator, i.e., formula (2.7) generates the operator G satisfying equations 1_G–2_G. However, if the operator Γ is a generalized reflection operator, it can generate a solution of equations 1_G–2_G even in some cases when condition (A) is not satisfied.

In the general case we shall admit that $\operatorname{Ker}\mathscr{A} \neq 0$ and $\operatorname{Im}\mathscr{A} \neq \mathscr{X}$, $\operatorname{Im}\mathscr{A} = \mathscr{A}\mathscr{X}$. Introduce the *linear* operator $\mathscr{A}^{-1}$ which satisfies the relation $\mathscr{A}\mathscr{A}^{-1}f = f$, $f \in \operatorname{Im}\mathscr{A}$.

PROPOSITION 3.2.1. *Let there exist an operator* $K : \mathscr{X} \to \mathscr{X}$ *which satisfies conditions*:

(1) $\operatorname{Im}K \subset \operatorname{Im}\mathscr{A}$ (*in this case it is possible to consider the product* $\mathscr{A}^{-1}K$) *and*
(2) $PK = P$.

Let some operator Γ *be a generalized reflection operator; then the operator*

$$G = (I-\Gamma)\mathscr{A}^{-1}K \tag{3.21}$$

satisfies system 1_G–2_G.

PROOF. Equation 1_G is a direct consequence of 1_Γ; to get 2_G, we do a simple computation:

$$\begin{aligned}\mathbf{A}G &= (P\mathscr{A}P)(I-\Gamma)\mathscr{A}^{-1}K \\ &= P\mathscr{A}(I-P')(I-\Gamma)\mathscr{A}^{-1}K \\ &= P\mathscr{A}I\mathscr{A}^{-1}K - (P\mathscr{A}\Gamma)\mathscr{A}^{-1}K - P\mathscr{A}[P'(I-\Gamma)]\mathscr{A}^{-1}K \\ &= PK - P.\end{aligned}$$ □

3.2.2. *Degenerate equation.* Generally speaking, if the operator $C = VW$, considered as an operator in Y', is not a bijection, the equations E_5–E_6 have no solutions or may have many solutions. Nevertheless this does not mean that equations 1_G–2_G also have no solutions. In the applications, we shall meet a case when there exists a (unique) operator T satisfying the relations

$$TC = Q', \qquad CT = Q'', \tag{3.22}$$

where Q'' is the projection operator subjected to the projection operator Q':

$$Q'Q'' = Q''Q' = Q''. \tag{3.23}$$

Thus, the operator T satisfies equation E_5 with a defect that can be characterized by the projection operator

$$\triangle Q = Q' - Q''. \tag{3.24}$$

It turns out that in this case we can also study the operator G in the form (3.21):

$$G = (I - \Gamma)(\mathscr{A}^{-1}K), \qquad \Gamma = WTV. \tag{3.25}$$

As for the operator K, we shall assume that the conditions of the previous subsection are fulfilled:

(1) the operator $\mathscr{A}^{-1}K$ is defined,
(2) $PK = P$.

Substitute representation (3.25) in equations 1_G–2_G. Instead of equation 1_G: $P'G = 0$, it is sufficient to verify the relation $VP'G = 0$:

$$VP'G = VP'(I - WTV)(\mathscr{A}^{-1}K) = (VP' - Q''V)(\mathscr{A}^{-1}K) = \triangle QV(\mathscr{A}^{-1}K).$$

Equation 1_G will be satisfied if

$(\triangle)$: $\triangle QV(\mathscr{A}^{-1}K) = 0$.

The last relation $(\triangle)$ can be treated as an equation for K. Let us show that if this relation is true, the operator G also obeys equation 2_G:

$$\begin{aligned}\mathbf{A}G &= P\mathscr{A}PG = P\mathscr{A}IG = P\mathscr{A}(I - WTV)(\mathscr{A}^{-1}K)\\ &= P\mathscr{A}(\mathscr{A}^{-1}K) = PK = P.\end{aligned}$$

For a given K equation $(\triangle)$ can be invalid, so let us change the operator K to the operator K_1:

$$K \longrightarrow K_1, \qquad K_1 = K + K', \tag{3.26}$$

such that the operator K_1 has properties (1), (2) and satisfies condition $(\triangle)$. In other words, the operator K' must obey the conditions: $\operatorname{Im} K' \subset \operatorname{Im} \mathscr{A} \cap X'$ and

$$\triangle QV\mathscr{A}^{-1}K' = -\triangle QV(\mathscr{A}^{-1}K),$$

or

$$\triangle QV\mathbf{A}'K' = -\triangle QV(\mathscr{A}^{-1}K). \tag{3.27}$$

Now consider the special case when $\triangle Q$ is a one-dimensional projection operator:

$$\triangle Q f = l(f)\psi. \tag{3.28}$$

Here ψ is a fixed element of $\operatorname{im} \triangle Q$ and l is some functional, $l(\psi) = 1$.

The operator K' can be found as a one-dimensional operator:

$$K' f = \mu(f)\psi', \tag{3.29}$$

where ψ' is a fixed vector from the subspace $\operatorname{im} \mathscr{A} \cap X'$. The operator K_1 possesses properties (1), (2), and condition ($\triangle$) leads to the equation:

$$l(V\mathscr{A}^{-1}\psi')\mu(f) = -l(V\mathscr{A}^{-1}Kf), \tag{3.30}$$

where $f \in \mathscr{X}$ is arbitrary. This equation can be solved if the element ψ' can be chosen such that the condition

$$l(V\mathscr{A}^{-1}\psi') \neq 0 \tag{3.31}$$

is fulfilled.

3.3. Solvability of the equations.

3.3.1. *About the solvability of equation E_4.* We have stressed earlier that in the case of integral equations (1.1) with symbols listed in §1, the main equations E_4 and E_5 are generated by operators which are close to the identity operator or to invertible operators. To formalize this idea, let us limit the framework of the consideration.

At first we assume that the space $\mathscr{X}$ is a Banach space and the projection operators P, P'_+, P'_- onto the corresponding subspaces $\mathscr{X} = X + X'_+ + X'_-$ are bounded operators. Then let $\mathscr{A}$ and the corresponding projecting generalized reflection operators Γ_α be defined and be bounded operators. Finally suppose the operator K from Proposition 3.2.1 exists and the operator $\mathscr{A}^{-1}K$ is also bounded.

PROPOSITION 3.3.1. *Let*

$$\max\{\|\Gamma_+\Gamma_-\|, \|\Gamma_-\Gamma_+\|\} \leqslant h < 1; \tag{3.32}$$

then equation E_3 has a (unique) solution. This solution satisfies the estimates

$$\left\|\Gamma_{\alpha\alpha} - \sum_{l=0}^{k-1}(\Gamma_\alpha\Gamma_{\alpha'})^l\Gamma_\alpha\right\| \leqslant h^k(1-h)^{-1}\|\Gamma_\alpha\|, \tag{3.33}$$

$$\left\|\Gamma_{\alpha\alpha'} - \sum_{l=1}^{k-1}(\Gamma_\alpha\Gamma_{\alpha'})^l\right\| \leqslant h^k(1-h)^{-1}, \tag{3.34}$$

$$\left\|\Gamma - \Gamma_+ - \Gamma_- + \sum_\alpha \sum_{l=1}^{k-1}(\Gamma_\alpha\Gamma_{\alpha'})^l(I - \Gamma_\alpha)\right\| \leqslant h^k(1-h)^{-1}(2 + \|\Gamma_+\| + \|\Gamma_-\|); \tag{3.35}$$

in particular,

$$\|\Gamma\| \leqslant (1-h)^{-1}(2h + \|\Gamma_+\| + \|\Gamma_-\|). \tag{3.36}$$

Note that under the conditions of Proposition 3.3.1 the projection operator Γ is the sum of the convergent series:

$$\Gamma = \Gamma_+ + \Gamma_- - \Gamma_+\Gamma_- - \Gamma_-\Gamma_+ + \Gamma_+\Gamma_-\Gamma_+ + \Gamma_-\Gamma_+\Gamma_- - \cdots. \tag{3.37}$$

Using this representation it is not difficult to show that the following limit relation between Γ and Γ_α is also true:

$$I - \Gamma = \lim_{n\to\infty}[(I-\Gamma_+)(I-\Gamma_-)]^n = \lim_{n\to\infty}[(I-\Gamma_-)(I-\Gamma_+)]^n. \tag{3.38}$$

It is natural to treat this last formula as an analog of the classical alternating Schwarz method; its abstract version can be found, for example, in [GlLu, Ma]. In short, formula (3.38) is a slightly more general result, since all projection operators within the framework of the standard Schwarz method are orthogonal, but in our case the three operators Γ, Γ_+, and Γ_- are not orthogonal. We are not going to discuss this generalization in more detail in the present work.

The main object of our consideration is the operator A^{-1}:

$$A^{-1}f = Gf = (I-\Gamma)\mathscr{A}^{-1}Kf, \qquad f \in X. \tag{3.39}$$

Under the conditions of Proposition 3.3.1, the operator A has a right inverse operator A^{-1} which admits the effective estimate:

$$\|A^{-1}\| \leqslant \{1 + (1-h)^{-1}(2h + \|\Gamma_+\| + \|\Gamma_-\|)\}\|\mathscr{A}^{-1}K\|. \tag{3.40}$$

Taking into account estimate (3.35), we can say that the complete asymptotic expansion of A^{-1} as $h \to 0$ has been obtained.

If additional condition (A) is satisfied and the operators Γ_α are total reflection operators, then A is a bijection and it is possible to set $K = I$.

3.3.2. *About the solvability of equation* E_5. Let $\mathscr{A}$ and $\mathscr{B}$ be operators in the spaces $\mathscr{X} = X + X'_+ + X'_-$ and $\mathscr{Y} = Y + Y'_+ + Y'_-$ connected by relations (3.13). Formulas (3.14) allow us to construct the operators V and W, thus we can introduce equation E_5. We shall consider equation E_5, provided that the operators $I - R_\alpha R_{\alpha'}^{-1}$ connect the subspaces $Y'_{\alpha'}$ and Y'_α weakly, in other words the operators D_α, see (3.20), are small in an appropriate sense.

Assume that the spaces $\mathscr{X}$, $\mathscr{Y}$ are Banach spaces and the projection operators $P'_{\alpha'}$, Q'_α are bounded. Then suppose that the operators $\mathscr{A}$, $\mathscr{A}^{-1}K$ (the operator K is described in Proposition 3.2.1), $\mathscr{B}^{\pm 1}$, and $R_\alpha^{\pm 1}$ are bounded operators. In this case the operators B', $\mathbf{B}'$, and D are also bounded. Finally, suppose that the operator B' has a bounded right inverse operator $B'^{-1}: Y' \to Y'$ such that $B'B'^{-1} = I$. Select the operator H' in the following special form:

$$H' = \begin{pmatrix} 0 & 0 \\ 0 & B'^{-1} \end{pmatrix}. \tag{3.41}$$

PROPOSITION 3.3.2. *If*

$$\|H'D\| \leqslant h < 1, \tag{3.42}$$

then equation E_5 *has a solution* T *satisfying the condition* E_6, *moreover*

$$\left\| T - \sum_{l=0}^{k-1} (H'D)^l H' \right\| \leqslant (1-h)^{-1} h^k \|H'\|. \tag{3.43}$$

If the operator B *(simultaneously* B'*) is a bijection, then the solution* T *(obeying condition* E_6*) is unique.*

The last proposition means that under its assumptions the operator A has a right inverse operator A^{-1}:

$$A^{-1}f = (I - WTV)\mathscr{A}^{-1}Kf, \qquad f \in X, \tag{3.44}$$

which admits the effective estimate:

$$\|A^{-1}\| \leqslant \{1 + 2(1-h)^{-1}\|W\|\,\|V\|\,\|H'\|\}\|\mathscr{A}^{-1}K\|. \tag{3.45}$$

We should also note that the following estimates are true:

$$\begin{aligned} &\|V\| \leqslant \sum_\alpha \|Q'_\alpha\|\,\|R_\alpha\|, \\ &\|W\| \leqslant \left(\sum_\alpha \|R_\alpha^{-1}\|\,\|Q'_\alpha\|\right)\|\mathscr{B}^{-1}\|, \\ &\|H'\| \leqslant \|B'^{-1}\|\,\|Q'\|. \end{aligned} \tag{3.46}$$

Formula (3.43) leads to the complete asymptotic expansion of the operator A^{-1} as $h \to 0$.

If condition (A) and relations (3.3) for L_α are satisfied and the operator B is a bijection, then the operator A is also a bijection and it is possible to set $K = I$ in the above formulas.

In subsection 3.2.2 we considered the case when the operator C has no right inverse operator T in the space Y', but has a left inverse. Suppose that the operator B' has a left inverse operator B'^{-1}: $B'^{-1}B' = I$, which is bounded. It can be used instead of the right inverse operator to construct the operator H', see formula (3.41). Let us suppose that the conditions of Proposition 3.3.2 are satisfied with this operator H' and introduce the left inverse operator T: $TC = Q'$, which exists in this case. Provided that the operator K obeys the additional conditions indicated in subsection 3.2.2, the operator A^{-1} has all properties described in Proposition 3.3.2.

§4. Integral equations

4.1. Quickly decreasing kernels.

4.1.1. *Preliminary notions.* We are now going to consider integral equations of the form

$$\varepsilon^{-1}\int_{\Omega}\mathscr{A}(\varepsilon^{-1}(x-y))f(y)\,dy=g(x),\qquad x\in\Omega=(-1\,,\,1)\,,\tag{4.1}$$

where $\mathscr{A}$ is a given smooth complex-valued function on the axis and ε is a small positive parameter. We also consider the more general equations:

$$\int_{\Omega}\mathscr{A}(x-y\,,\,\varepsilon)f(y)\,dy=g(x)\,,\qquad x\in\Omega\,;\tag{4.2}$$

the character of dependence of the function $\mathscr{A}(*\,,\,\varepsilon)\colon\mathbb{R}\to\mathbb{C}$ on the parameter ε has been described in the Introduction in terms of the corresponding symbol. According to the general outline, let us introduce the integral operators $\mathscr{A}[\varepsilon]$ and $A[\varepsilon]$:

$$(\mathscr{A}[\varepsilon]f)(x)=\int_{\mathbb{R}}\mathscr{A}(x-y\,,\,\varepsilon)f(y)\,dy\,,\qquad x\in\mathbb{R}\,,\quad f\colon\mathbb{R}\to\mathbb{C}\,,$$

$$(A[\varepsilon])(x)=\int_{\Omega}\mathscr{A}(x-y\,,\,\varepsilon)f(y)\,dy\,,\qquad x\in\Omega\,,\quad f\colon\Omega\to\mathbb{C}.$$

The operators $\mathscr{A}[\varepsilon]$ and $A[\varepsilon]$ are connected by the general construction of §2:

$$A[\varepsilon]f=P\mathscr{A}[\varepsilon]f\,,\qquad f\colon\Omega\to\mathbb{C}.$$

Equations (4.1) and (4.2) can be written in the shorter form

$$A[\varepsilon]f=g.$$

The exact description of the space $\mathscr{X}$ of functions $\mathbb{R}\to\mathbb{C}$ will be given later; it will depend on the classes of symbols indicated in the Introduction. But in all cases the projection operators P and P'_α are given by the formulas:

$$(Pf)(x)=\omega(x)f(x)\,,\qquad (P'_\alpha f)(x)=\omega'_\alpha(x)f(x)\,,\tag{4.3}$$

where ω is the characteristic function of the interval Ω and ω'_+ (ω'_-) is the characteristic function of the half-axis $[1\,,\,\infty)$ ($(-\infty\,,\,-1]$). It is clear that $\omega+\omega'_++\omega'_-=1$.

Passing to Fourier transforms, we now limit ourselves to a list of formal relations without specifying the classes of functions under consideration.

We shall use the following notation:

$$Ff=\hat f\,,\qquad f=F^{-1}\hat f\,,$$

$$\hat f(\xi)=\int_{\mathbb{R}}f(x)\exp(-ix\xi)\,dx\,,\qquad f(x)=(2\pi)^{-1}\int_{\mathbb{R}}\hat f(\xi)\exp(ix\xi)\,d\xi.$$

The standard definition of the *symbol* is $a=F\mathscr{A}$. The relation

$$\mathscr{A}[\varepsilon]=F^{-1}a[\varepsilon]F\tag{4.4}$$

holds, $a[\varepsilon]$ being the multiplication operator: $(a[\varepsilon]f)(\xi) = a(\xi, \varepsilon)f(\xi)$ ($= a(\varepsilon\xi)f(\xi)$ in case (4.1)).

The projection operators $\widehat{P}'_\alpha = FP'_\alpha F^{-1}$, $\widehat{P}_\alpha = FP_\alpha F^{-1}$ are given by the formal expressions:

$$\widehat{P}'_\alpha = E_\alpha p_{\alpha'} E^*_\alpha, \qquad \widehat{P}_\alpha = E_\alpha p_\alpha E^*_\alpha. \tag{4.5}$$

Here E_α are the multiplication operators: $(E_\alpha f)(\xi) = E_\alpha(\xi)f(\xi)$, $E_\alpha(\xi) = \exp(-i\alpha\xi)$, $\alpha = +, -$, and the operator p_α can be considered as the integral operator with kernel:

$$p_\alpha \sim \frac{-\alpha}{2\pi i}\frac{1}{\xi - \eta + i\alpha 0}. \tag{4.6}$$

In the case of appropriate classes of functions, the operators p_+ and p_- are complementary projection operators onto the sets of functions that are the limit values of holomorphic functions in the upper and lower half-planes, respectively.

The kernel of the operator $\widehat{P} = FPF^{-1}$ is given by the formula:

$$\widehat{P} \sim \frac{1}{\pi}\frac{\sin(\xi - \eta)}{\xi - \eta}. \tag{4.7}$$

The operator $A[\varepsilon]$ can now be represented in the form:

$$(A[\varepsilon]f)(x) = (2\pi)^{-1}\int_{\mathbb{R}} a(\xi, \varepsilon)\exp(i\xi x)\hat{f}(\xi)\,d\xi.$$

Introduce the class S of functions r satisfying the following conditions:

s_1) $r \in C(\overline{\mathbb{R}}) \cap C^\infty(\mathbb{R})$,

s_2) $|r(\xi) - r(\infty)| \leqslant c(1 + |\xi|)^{-\gamma}$,
$|r^{(n)}(\xi)| \leqslant c_n(1 + |\xi|)^{-\gamma}$, $\gamma > 1/2$, $n \in \mathbb{N}$,

s_3) $r(\xi) \neq 0$, $\xi \in \mathbb{R}$, $\operatorname{ind}(r, \mathbb{R}) = 0$, $r(0) = 1$.

It is known that all these functions admit the representation:

$$r = r_+ r_-. \tag{4.8}$$

In this representation the $r_\pm$ are holomorphic functions in the half-planes $\mathbb{C}_\pm = \{\xi \mid \operatorname{im}\xi \gtrless 0\}$. The functions r_α are continuous and have no roots in the closed half-planes, moreover $r_\alpha \in C^\infty(\mathbb{R})$ and both functions admit the asymptotic estimates

$$r_\alpha(\xi) - r_\alpha(\infty) = O(|\xi|^{-\gamma'}), \qquad r_\alpha^{(n)}(\xi) = O(|\xi|^{-\gamma'}), \tag{4.9}$$

as $\xi \to \infty$, $n \in \mathbb{N}$, $\gamma' < \gamma$. Let us recall the well-known explicit formulas for r_α:

$$r_\alpha(\xi) = c_\alpha \exp\left\{\frac{-\alpha}{2\pi i}\int \log\left[\frac{r(\eta)}{r(\infty)}\right]\frac{d\eta}{\xi - \eta + i\alpha 0}\right\}. \tag{4.10}$$

All the above assertions concerning r_α can be verified simply by using formulas (4.10). Since the constants c_α are arbitrary, it is possible to assume that

$$r_\alpha(0) = 1. \tag{4.11}$$

We also introduce the more general class S_0 of functions r satisfying condition $s_1^0)$ instead of condition $s_1)$:

$s_1^0)$ $r \in C(\overline{\mathbb{R}}) \cap C^\infty(\mathbb{R} \setminus 0)$.

All formulas (4.8)–(4.11) remain valid in this general case, but now the relation $r_\alpha \in C^\infty(\mathbb{R})$ must be replaced by the relation $r_\alpha \in C^\infty(\mathbb{R}\setminus 0)$ and the functions r_α do not belong to $C^\infty(\mathbb{R})$, they obey only the Hölder condition at the point $\xi = 0$ (of arbitrary order less than 1). As $\xi \to 0$, $\xi \in \overline{\mathbb{C}}_\alpha$, the functions r_α admit the following asymptotic representation

$$r_\alpha(\xi) \sim 1 + \sum_{1 \leqslant k,\ 0 \leqslant l \leqslant k} r_{kl}^\alpha \xi^k \log^l \xi^{-1}. \tag{4.12}$$

It is possible to differentiate this representation any number of times.

4.1.2. *Inclusion into the framework of the general outline.* Now consider the operator $\mathscr{A}[\varepsilon]$ with symbol of class S_0. The most convenient space is $\mathscr{X} = L_2(\mathbb{R})$. The projection operators P and P'_α have been described earlier; of course, they are bounded operators. The operator $\mathscr{A}$ is a bijection, it is bounded together with the inverse operator $\mathscr{A}^{-1}$. In this case we need not introduce any special modelling operator $\mathscr{B}$; in fact this means that $\mathscr{Y} = \mathscr{X}$, $\mathscr{B} = I$, $Q = P$, and $Q'_\alpha = P'_\alpha$. As a result, $\Delta_\alpha = P'_\alpha$.

The analytic factorization (4.8) of the symbol a,

$$a = a_+ a_-,$$

generates operator factorization:

$$\mathscr{A} = L_\alpha R_\alpha, \qquad R_\alpha = F^{-1} a_\alpha F, \qquad L_\alpha = F^{-1} a_{\alpha'} F.$$

All the operators R_α and L_α are bijections, they all possess the characteristic properties (3.2)–(3.5). As an example, consider the operator $P'_+ R_+ P_+$:

$$P'_+ R_+ P_+ = F^{-1} E_+ p_- r_+ p_+ E_+^* F = 0.$$

Here we used the relation $p_- r_+ p_+ = 0$ implied by the fact that the function r_+ is a multiplier in the space $p_+ L_2(\mathbb{R})$, in other words $r_+ p_+ L_2(\mathbb{R}) \subset p_+ L_2(\mathbb{R})$. Recalling Proposition 3.1.3, we can assert that the operators

$$\Gamma_\alpha = R_\alpha^{-1} P'_\alpha R_\alpha$$

are total reflection operators.

Actually, the last assertion can be treated as a new interpretation of the classical Wiener-Hopf approach to integral equations.

We can now consider equation E_4 for the reflection operator Γ.

We shall systematically use the following proposition from calculus.

LEMMA 4.1.1. *Let two functions f, $g\colon (0, \infty) \to \mathbb{C}$ have the properties*:

f) $f \in C^2([0, +\infty))$ *and admits the analytic continuation into the angle* $0 < \arg\xi < \varphi$, $\varphi > 0$, *where*

$$|f(\xi)| \leqslant c_1 \exp(c_2 |\operatorname{im} \xi|), \qquad c_2 < 2;$$

g) $g \in C^{\infty}(0, \infty)$ *and* g *admits the asymptotic representations*:

$$1_g: \qquad g(\xi) \sim \sum_{0 \leqslant k, 0 \leqslant l \leqslant k} g_{kl} \xi^k \log^l \xi^{-1}, \qquad \xi \to 0, \tag{4.13}$$

where the series is differentiable (arbitrary number of times), and

$$2_g: \qquad g(\xi) = g(\infty) + O(\xi^{-\gamma}), \qquad g^{(n)}(\xi) = O(\xi^{\gamma}),$$
$$\xi \to \infty, \quad \gamma > 2^{-1}, \quad n \in \mathbb{N}.$$

Under these conditions, the integral

$$I(\xi, \eta; \varepsilon) = \int_0^{\infty} \frac{f(\zeta) g(\varepsilon\zeta) \exp(2i\zeta)}{(\xi - \zeta - i0)(\zeta - \eta + i0)} \, d\zeta, \tag{4.14}$$

where $\xi, \eta \in \mathbb{R}$, $\varepsilon > 0$, *admits the asymptotic representation*:

$$I(\xi, \eta; \varepsilon) = \sum_{\substack{0 \leqslant k \leqslant K \\ 0 \leqslant l \leqslant k}} I_{kl}(\xi, \eta) \varepsilon^k \log^l \varepsilon^{-1} + I_k(\xi, \eta; \varepsilon), \qquad K \geqslant 2, \tag{4.15}$$

as $\varepsilon \to 0$. *Here*

$$I_{kl}(\xi, \eta) = \sum_{j=l}^{k} g_{kl} \binom{j}{l} J_{k, j-l}(\xi, \eta), \tag{4.16}$$

$$J_{kl}(\xi, \eta) = \int_L \frac{f(\zeta) \zeta^k \log^l \zeta^{-1} \exp[2i\zeta]}{(\xi - \zeta)(\zeta - \eta)} \, d\zeta, \tag{4.17}$$

the contour L *is the ray* $\arg \zeta = \varphi'$, $0 < \varphi' < \varphi$. *The remainder* $I_K(\xi, \eta; \varepsilon)$ *possesses the estimate*

$$|I_K(\xi, \eta; \varepsilon)| \leqslant C_K \frac{\varepsilon^K}{1 + |\xi - \eta|} \left(\frac{1}{1 + |\varepsilon\xi|^{\gamma'}} + \frac{1}{1 + |\varepsilon\eta|^{\gamma'}} \right), \qquad \gamma' < \gamma. \tag{4.18}$$

If $k \geqslant 2$, *the integrals* J_{kl} *possess the estimates*

$$|J_{kl}(\xi, \eta)| \leqslant c_{kl} (1 + |\xi|)^{-1} (1 + |\eta|)^{-1}. \tag{4.19}$$

The proof of the lemma will be given in Appendix 5.1. Note that the meaning of the lemma is rather simple. In order to get the series (4.15), one must replace the function $g(\varepsilon\xi)$ in the integral $I(\xi, \eta; \varepsilon)$ by its asymptotic expansion at the point $\varepsilon\xi = 0$ and interchange the order of integration and summation. The integrals obtained are divergent and have to be regularized by replacing the integration path $\mathbb{R}_+$ by the ray L.

4.1.3. *Investigation of equation* E_4. Let us verify the conditions of solvability of the equation E_3 given by Proposition 3.3.1. Suppose that the parameter ε is sufficiently small.

In accordance with Proposition 3.3.1, we must estimate the norms of the operators $\Gamma_+\Gamma_-$ and $\Gamma_-\Gamma_+$, treated as operators in $L_2(\mathbb{R})$. The procedures are the same in both cases, therefore we only consider the operator $\Gamma_+\Gamma_-$:

$$\begin{aligned}\Gamma_+\Gamma_- &= R_+^{-1}P_+'R_+R_-^{-1}P_-'R_- \\ &= R_+^{-1}F^{-1}E_+p_-E_+^*r_+r_-^{-1}E_-p_+E_-^*FR_- \\ &= -R_+^{-1}F^{-1}E_+m_+E_-^*FR_-,\end{aligned} \tag{4.20}$$

where

$$m_+ = -p_-r_+r_-^{-1}E_-^2p_+ = p_-(1-r_+r_-^{-1})E_-^2p_+. \tag{4.21}$$

It is evident that $R_\alpha^{\pm 1}$, $F^{\pm 1}$, and $E_\alpha^{\pm 1}$ are uniformly bounded operators, so that it is sufficient to estimate the norm of the operator m_+.

Its kernel has the form:

$$m_+ \sim \frac{1}{4\pi^2}\int_{\mathbb{R}} \frac{[1-r_+(\varepsilon\zeta)r_-^{-1}(\varepsilon\zeta)]\exp(2i\zeta)}{(\xi-\zeta-i0)(\zeta-\eta+i0)}\,d\zeta. \tag{4.22}$$

Let us use Lemma 4.1.1. It is clear that the last integral is the sum of the integrals $I(\xi, \eta; \varepsilon)$ and $\overline{I(-\eta, -\xi; \varepsilon)}$, the function f being equal to 1.

Let us discuss all the analytic objects characterized by Lemma 4.1.1 from the operator point of view. All functions of the variables $\xi, \eta \in \mathbb{R}$ can be considered as kernels of the integral operators. From estimates (4.18) and (4.19) it is evident that the kernels $I_{kl}(\xi, \eta)$, $k \geqslant 2$, and $I_K(\xi, \eta; \varepsilon)$ generate operators I_{kl} and $I_K[\varepsilon]$ of Hilbert-Schmidt class. Specifically, the Hilbert-Schmidt norm $|*|_2$ of the operator $I_K[\varepsilon]$ admits the estimate:

$$|I_K[\varepsilon]|_2 \leqslant c_K\varepsilon^{K-1/2}. \tag{4.23}$$

The kernels $I_{10}(\xi, \eta)$ and $I_{11}(\xi, \eta)$ also determine operators of Hilbert-Schmidt class. Let us show this for the kernel $J_{11}(\xi, \eta)$:

$$\begin{aligned}|J_{11}|_2^2 &= \left(\iint d\xi\,d\eta\,|J_{11}(\xi, \eta)|^2\right)^{1/2} \\ &= \left(\iint d\xi\,d\eta\left|\int_L d\zeta\frac{\zeta\log\zeta\exp(2i\zeta)}{(\xi-\zeta)(\zeta-\eta)}\right|^2\right)^{1/2} \\ &\leqslant \int_L |d\zeta||\zeta\log\zeta|\exp(-2\,\mathrm{im}\,\zeta)\left(\iint\frac{d\xi d\eta}{|\xi-\eta|^2|\zeta-\eta|^2}\right)^{1/2} \\ &\leqslant c\int_L |d\zeta|\frac{|\zeta\log\zeta|}{\mathrm{im}\,\zeta}\exp(-2\,\mathrm{im}\,\zeta).\end{aligned}$$

In this computation we have used the Minkowski integral inequality.

It is now possible to assert that the operator $m_+[\varepsilon]$ is an operator of Hilbert-Schmidt class and admits the following asymptotic representation:

$$m_+[\varepsilon] = \sum_{1\leqslant k,\ 0\leqslant l\leqslant k}^{K} m_{kl}^+\varepsilon^k\log^l\varepsilon^{-1} + m_K^+[\varepsilon], \tag{4.24}$$

where

$$|m_K^+[\varepsilon]|_2 \leqslant c_K \varepsilon^{K+1} \log^{K+1} \varepsilon^{-1}. \tag{4.25}$$

Thus if the parameter ε is sufficiently small, the assumptions of Proposition 3.3.1 are fulfilled with

$$h = O(\varepsilon \log \varepsilon^{-1}). \tag{4.26}$$

Therefore the operator Γ and its components $\Gamma_{\alpha\beta}$ satisfy all the conclusions of Proposition 3.3.1; and formula (3.37) gives, in fact, the asymptotic description of the operator Γ.

Since the operator $\mathscr{A}$ is a bijection, the operator Γ is a total reflection operator. Thus we can conclude that the complete asymptotic description of the operator A^{-1} has been obtained.

Let us make a short remark about the case $a \in S$. In this case

$$|m_\alpha|_2 = O(\varepsilon^\infty). \tag{4.27}$$

This can be understood by considering formulas (4.22) for the kernels of the operators m_α. The remark after Lemma 4.1.1 asserts that the asymptotic expansions of these kernels can be obtained by substituting the Taylor series at the point $\varepsilon\xi = 0$ for the expressions in brackets of the integrands. If $a \in S$, these series do not contain any logarithmic terms, therefore all the regularized integrals which arise after interchanging the order of integration and summation vanish.

As a result, if $a \in S$, we have the representation:

$$\Gamma = \Gamma_+ + \Gamma_- + O(\varepsilon^\infty) \tag{4.28}$$

(in the Hilbert-Schmidt norm).

4.2. Slowly decreasing kernels.

4.2.1. *Operators with discontinuous symbols.* In subsection 4.2 we shall consider the equations generated by discontinuous symbols. From the very beginning it is worth noting that generally the discontinuity of the symbol is not compatible with the choice of $L_2(\mathbb{R})$-topology as the topology in the space $\mathscr{X}$ (or $\mathscr{Y}$). To clarify this circumstance and to prepare the basis for the following considerations, it is useful to recall certain basic properties of the singular Hilbert operator on a bounded interval:

$$(Sf)(x) = \frac{i}{\pi} \int_\Omega \frac{f(y)}{x-y}\, dx, \qquad x \in \Omega, \quad f\colon \Omega \to \mathbb{C};$$

here the integral is treated as a Cauchy singular integral. The operator S arises as the block $P\mathscr{S}P$ of the classical singular Hilbert operator $\mathscr{S}$ on the entire axis:

$$(\mathscr{S}f)(x) = \frac{i}{\pi} \int_\mathbb{R} \frac{f(y)}{x-y}\, dy, \qquad x \in \mathbb{R}, \quad f\colon \mathbb{R} \to \mathbb{C}.$$

Its symbol is

$$s(\xi) = \operatorname{sgn} \xi.$$

Let us indicate some well-known properties of the operator S, mainly following [GoKr, MiPr, Tr]. The space $L_2(\Omega)$ is not suitable for the operator S: in this space the operator S has no inverse operator with satisfactory properties. It is more convenient to consider the operator S in the Hilbert space $L_2(\rho, \Omega)$ with an appropriate singular weight ρ. Depending on the weight ρ, S can have a two-sided inverse operator or only the single-sided (left or right) inverse operator, or may have no inverse operator at all.

Let us give an example of weight ρ for which the operator S has right inverse operator S^{-1} in the space $L_2(\rho, \Omega)$. This weight can be chosen in the form:

$$\rho(x) = \left|\frac{x^2-1}{x^2+1}\right|^{\nu}, \tag{4.29}$$

$0 < \nu < 1$. In this case S and S^{-1} are bounded operators. The right inverse operator is given by the expression:

$$(S^{-1}f)(x) = \frac{1}{\sqrt{1-x^2}}\frac{i}{\pi}\int_{\Omega}\sqrt{1-y^2}\frac{f(y)}{x-y}\,dy. \tag{4.30}$$

The kernel subspace of S, $\operatorname{Ker} S = \{f : Sf = 0\}$, is generated by the function $(1-x^2)^{-1/2}$.

If the weight ρ is

$$\rho(x) = \left|\frac{x+1}{x-1}\right|^{\nu}, \qquad 0 < |\nu| < 1, \tag{4.31}$$

then the operator S has a two-sided inverse operator. Both S and S^{-1} are bounded operators in this case. If $\nu > 0$, the operator S^{-1} is given by the formula:

$$(S^{-1}f)(x) = \sqrt{\frac{1-x}{1+x}}\frac{i}{\pi}\int_{\Omega}\sqrt{\frac{1+y}{1-y}}\frac{f(y)}{x-y}\,dy. \tag{4.32}$$

This operator transforms a smooth function f into the function $S^{-1}f$ which is bounded at the point $+1$ and, generally, is unbounded at the point -1. If $\nu < 0$, the points $+1$ and -1 interchange their roles and formula (4.32) must be replaced by its natural analog.

We shall treat the space $L_2(\rho, \Omega)$ as a subspace of $L_2(\rho) = L_2(\rho, \mathbb{R})$. It is worth noting that the weights ρ in both cases (4.29) and (4.31) tend to 1 as $x \to \infty$. This is a useful property of the weights, because we shall use Fourier transforms in the space $L_2(\rho)$.

We also need some information about the operator $\mathscr{S}$ in the space $L_2(\rho)$, where ρ is given by (4.29) or (4.31). It can be shown that the operator $\mathscr{S}$ has a two-sided inverse operator $\mathscr{S}^{-1}$ which coincides with $\mathscr{S} : \mathscr{S}^{-1} = \mathscr{S}$, and the operator $\mathscr{S}$ is bounded, see [GoKr, MiPr].

4.2.2. *Fourier multipliers.* The Fourier transformation F takes the operator $\mathscr{S}$ to the multiplication operator: $f(\xi) \to s(\xi)f(\xi)$. The last remark of the previous subsection means that this is a bounded operator in the space $FL_2(\rho)$, in other words, the function s is a Fourier multiplier in the space $L_2(\rho)$.

Further we shall give other examples of Fourier multipliers. We need some preliminary remarks. We assume that the weights ρ are given by formulas (4.29) or (4.31).

We have the following system of continuous embeddings:

$$L_2(\mathbb{R}) \cap L_{r'}(\mathbb{R}) \subset L_2(\rho) \subset L_2(\Omega_2') \oplus L_r(\Omega_2), \tag{4.33}$$

where $\Omega_2 = (-2, 2)$, $\Omega_2' = \mathbb{R}\backslash\Omega_2$. In these formulas $1 < r < 2(1+\nu)^{-1}$, and r' is the conjugate index, i.e., $r^{-1} + r'^{-1} = 1$. For the norm of the function f in the space $L_2(\mathbb{R}) \cap L_{r'}(\mathbb{R})$ we can take the expression $\|f\|_2 + \|f\|_{r'}$, and in the space $L_2(\Omega_2') \oplus L_r(\Omega_2)$, the sum $\|f\|_{2,\Omega_2'} + \|f\|_{r,\Omega_2}$.

Let us verify relations (4.33). Let χ be the indicator of the interval Ω_2 and $\chi' = 1 - \chi$, then

$$\begin{aligned} \|f\| &\leqslant \|\chi f\| + \|\chi' f\| = \|\chi\rho|f|^2\|_1^{1/2} + \|\chi' f\| \\ &\leqslant c(\|f^2\|_s^{1/2}\|\chi\rho\|_{s'}^{1/2} + \|f\|_2) \leqslant c(\|f\|_{r'} + \|f\|_2), \end{aligned}$$

where $2s = r'$, $\nu s' = \nu r'(r'-2)^{-1} = \nu r(2-r)^{-1} < 1$, s and s' are conjugate indices. Moreover

$$\begin{aligned} \|\chi' f\|_2 + \|\chi f\|_r &\leqslant c(\|f\| + \|\chi|f|^r\|_1^{1/r}) \\ &\leqslant c(\|f\| + \|\,|f|^r\rho^{1/t}\|_t^{1/r}\|\chi\rho^{-1/t}\|_{t'}^{1/r}) \\ &\leqslant c(\|f\| + \|\,|f|^{rt}\rho\|_1^{1/rt}) = 2c\|f\|, \end{aligned}$$

where $rt = 2$, $\nu t' t^{-1} = \nu r(2-r)^{-1} < 1$, t and t' are conjugate indices. Thus the relations (4.33) are proved.

Since $FL_r(\mathbb{R}) \subset L_{r'}(\mathbb{R})$, $r, r' > 1$, $r < 2$, we have the following system of embeddings:

$$\begin{array}{ccccc} L_2(\mathbb{R}) \cap L_{r'}(\mathbb{R}) & \subset & L_2(\rho) & \subset & L_2(\Omega_2') \oplus L_r(\Omega_2) \\ F\big\downarrow\big\uparrow F^{-1} & \cup\big\uparrow F^{-1} & F\big\downarrow\big\uparrow F^{-1} & & \cap\big\downarrow F \qquad \cap\big\downarrow F \\ L_2(\mathbb{R}) \cap L_r(\mathbb{R}) & \subset & FL_2(\rho) & \subset & L_2(\mathbb{R}) + L_{r'}(\mathbb{R}) \end{array} \tag{4.34}$$

It is almost evident that a similar system also exists for the space $L_2(\rho^{-1})$ with weight ρ^{-1} instead of ρ. We shall use this remark below.

Let us introduce the notation

$$\check{\mu} f = F^{-1}\mu\hat{f}, \qquad (\mu\hat{f})(\xi) = \mu(\xi)\hat{f}(\xi), \qquad \mu\colon \mathbb{R} \to \mathbb{C}. \tag{4.35}$$

PROPOSITION 4.2.1. *The operator $\check{\mu}$ is bounded in the space $L_2(\rho)$ if $\mu \in L_\infty(\mathbb{R}) \cap L_p(\mathbb{R})$, where $1 < p < \nu^{-1}$, and $\|\check{\mu}\| \leqslant c(\|\mu\|_\infty + \|\mu\|_p)$.*

PROOF. Let $p = rr'(r'-r)^{-1}$, $1 < r < 2$. The condition $p < \nu^{-1}$ implies $1 < r < 2(1+\nu)^{-1}$. Now let us show that the operator $\check{\mu}$ is bounded as an operator from $L_2(\Omega_2') \oplus L_r(\Omega_2)$ to $L_2(\mathbb{R}) \cap L_{r'}(\mathbb{R})$. Using the diagram (4.34), we see that it is sufficient to prove the inequality

$$\|\mu\hat{f}\|_2 + \|\mu\hat{f}\|_r \leqslant c_\mu(\|\widehat{\chi' f}\|_2 + \|\widehat{\chi f}\|_{r'}).$$

This inequality can be obtained as the result of the following chain of computations:

$$\begin{aligned}\|\mu\hat{f}\|_2 + \|\mu\hat{f}\|_r &\leqslant c(\|\mu(\widehat{\chi' f})\|_2 + \|\mu(\widehat{\chi f})\|_2 + \|\mu(\widehat{\chi' f})\|_r + \|\mu(\widehat{\chi f})\|_r) \\ &\leqslant c(\|\mu\|_\infty \|\widehat{\chi' f}\|_2 + \|\mu\|_q \|\widehat{\chi f}\|_{r'} + \|\mu\|_q \|\widehat{\chi' f}\|_2 + \|\mu\|_p \|\widehat{\chi f}\|_{r'}) \\ &\leqslant c(\|\mu\|_\infty + \|\mu\|_p)(\|\widehat{\chi' f}\|_2 + \|\widehat{\chi f}\|_{r'}),\end{aligned}$$

$q = 2r'(r'-2)^{-1} = 2r(2-r)^{-1} > p$. Here we used the well-known interpolation inequality:

$$\|\mu\|_q \leqslant \|\mu\|_p^{p/q} \|\mu\|_\infty^{1-p/q} \leqslant \|\mu\|_p + \|\mu\|_\infty.$$

Now let us give a simple sufficient condition for the boundedness of an integral operator in the space $FL_2(\rho)$.

PROPOSITION 4.2.2. *Let $\mathscr{M}(\xi, \eta)$ be the kernel of an integral operator $\mathscr{M}$ in the space $FL_2(\rho)$. If*

$$\begin{aligned}\|\!|\mathscr{M}|\!\| \equiv &\left(\int d\xi \int d\eta\, |\mathscr{M}(\xi, \eta)|^2\right)^{1/2} \\ &+ \left(\int d\xi \int d\eta\, |\mathscr{M}(\xi, \eta)|^r\right)^{1/r} \\ &+ \left(\int d\eta \left(\int d\xi\, |\mathscr{M}(\xi, \eta)|^r\right)^{2/r}\right)^{1/2} \\ &+ \left(\int d\eta \left(\int d\xi\, |\mathscr{M}(\xi, \eta)|^2\right)^{r/2}\right)^{1/r} < \infty,\end{aligned}$$

where $1 < r < 2(1+\nu)^{-1}$, then the operator $\mathscr{M}$ is an operator of Hilbert-Schmidt class and $|\mathscr{M}|_2 \leqslant \|\!|\mathscr{M}|\!\|$.

PROOF. Consider the norm of the operator $\mathscr{M}$:

$$\begin{aligned}|\mathscr{M}|_2 = &\left(\iint \rho(x)\rho^{-1}(y)|(F^{-1}\mathscr{M}F)(x, y)|^2\, dx\, dy\right)^{1/2} \\ \leqslant c &\left\{\left(\int dx\, \rho(x) \int dy\, |(F^{-1}\mathscr{M}F)(x, y)|^2\right)^{1/2}\right. \\ &\left. + \left(\int dx\, \rho(x) \left(\int dy\, |(F^{-1}\mathscr{M}F)(x, y)|^{r'}\right)^{2/r'}\right)^{1/2}\right\}\end{aligned}$$

$$
\begin{aligned}
&\leqslant c\left\{\left(\int dx\,\rho(x)\int d\eta\|(F^{-1}\mathscr{M})(x,\eta)|^2\right)^{1/2}\right.\\
&\qquad\left.+\left(\int dx\,\rho(x)\left(\int d\eta\,|(F^{-1}\mathscr{M})(x,\eta)|^r\right)^{2/r}\right)^{1/2}\right\}\\
&\leqslant c\left\{\left(\int dx\int d\eta|(F^{-1}\mathscr{M})(x,\eta)|^2\right)^{1/2}\right.\\
&\qquad+\left(\int dx\left(\int d\eta\,|(F^{-1}\mathscr{M})(x,\eta)|^2\right)^{r'/2}\right)^{1/r'}\\
&\qquad+\left(\int dx\left(\int d\eta\,|(F^{-1}\mathscr{M})(x,\eta)|^r\right)^{2/r}\right)^{1/2}\\
&\qquad\left.+\left(\int dx\left(\int d\eta\,|(F^{-1}\mathscr{M})(x,\eta)|^r\right)^{r'/r}\right)^{1/r'}\right\}\\
&\leqslant c\left\{\left(\iint|\mathscr{M}(\xi,\eta)|^2\,d\xi\,d\eta\right)^{1/2}\right.\\
&\qquad+\left(\int d\eta\left(\int dx\,|(F^{-1}\mathscr{M})(x,\eta)|^{r'}\right)^{2/r'}\right)^{1/2}\\
&\qquad+\left(\int d\eta\left(\int dx\,|(F^{-1}\mathscr{M})(x,\eta)|^2\right)^{r/2}\right)^{1/r}\\
&\qquad\left.+\left(\int d\eta\left(\int dx\,|(F^{-1}\mathscr{M})(x,\eta)|^{r'}\right)^{r/r'}\right)^{1/r}\right\}\\
&\leqslant c\left\{\left(\iint|\mathscr{M}(\xi,\eta)|^2\,d\xi\,d\eta\right)^{1/2}\right.\\
&\qquad+\left(\int d\eta\left(\int d\xi\,|\mathscr{M}(\xi,\eta)|^r\right)^{2/r}\right)^{1/2}\\
&\qquad+\left(\int d\eta\left(\int d\xi\,|\mathscr{M}(\xi,\eta)|^2\right)^{r/2}\right)^{1/r}\\
&\qquad\left.+\left(\int d\eta\int d\xi\,|\mathscr{M}(\xi,\eta)|^r\right)^{1/r}\right\}\\
&=c\|\mathscr{M}\|.
\end{aligned}
$$

Here we used diagram (4.34) and the Minkowski integral inequality.

4.2.3. *Description of a class of kernels.* We can now describe the class of integral operators that will be the object of investigation in subsection 4.2. We write $a \in J_0$ if

$$a(\xi, \varepsilon) = b(\xi) r(\varepsilon\xi),$$

where $r \in S_0$ and b satisfies the following conditions:

j_1) $b^{\pm 1}(\xi)$ are bounded functions on $\mathbb{R} \setminus 0$;

j_2) $b^{-1}(\xi)$ is an analytic function in the angles

$$|\arg \xi| < \varphi, \qquad |\arg \xi - \pi| < \varphi,$$

φ is a positive number, and

$$|b^{-1}(\xi)| \leqslant c_1 \exp(c_2 |\operatorname{im} \xi|), \qquad c_2 < 2;$$

j_3) the functions $b^{\pm 1}(\xi)$ are Fourier multipliers in the space $L_2(\rho, \mathbb{R})$.

As an example of such functions, we can propose

$$b(\xi) = a + b \operatorname{sgn} \xi + c(\xi), \tag{4.36}$$

where a and b are some constants, $a^2 \neq b^2$, c admits an analytical continuation into the angles described in j_2), and (in these angles) we have

$$c(\xi) \neq -(a \pm b), \qquad |c(\xi)| \leqslant c(1 + |\xi|)^{-\sigma}, \qquad \nu < \sigma. \tag{4.37}$$

We also need some information concerning the operators $\check{r}[\varepsilon]$ and $\check{r}_{\pm}[\varepsilon]$ in the space $L_2(\rho)$.

Let γ be the parameter from the condition s_2) and ν be the parameter characterizing the weight function ρ.

PROPOSITION 4.2.3. *If $\nu < \gamma$, then the operators $\check{r}^{\pm 1}[\varepsilon]$, $\check{r}_{\pm}[\varepsilon]$ and $\check{r}_{\pm}^{-1}[\varepsilon]$ are bounded and their norms are $O(\varepsilon^{-\nu'})$, where $\nu' > \nu$.*

The proof can be obtained as an immediate consequence of Proposition 4.2.1.

The operator $\mathscr{A}[\varepsilon]$ will be considered as an operator in the space $\mathscr{X} = L_2(\rho)$, its symbol $a(\xi, \varepsilon)$ belonging to the class J_0, and $\nu < \gamma$. The operator $\mathscr{B}$ (it does not depend on ε) with symbol $b(\xi)$ will act in the same space, $\mathscr{Y} = \mathscr{X}$. The operator $\mathscr{B}$ is a bijection and both $\mathscr{B}$ and $\mathscr{B}^{-1}$ are bounded. Proposition 4.2.3 asserts that $\mathscr{A}[\varepsilon]$ is also bijective and the norms of $\mathscr{A}[\varepsilon]$ and $\mathscr{A}^{-1}[\varepsilon]$ are $O(\varepsilon^{-\nu'})$, $\nu' > \nu$. The operators $P'_\alpha = Q'_\alpha$ have the same meaning as in subsection 4.1.

The operator B is bounded but is not necessarily invertible. We assume that B has a bounded right or two-sided inverse.

In the case of symbols

$$b(\xi) = a + b \operatorname{sgn} \xi, \tag{4.38}$$

it is possible to give effective criteria for the bounded right- or two-sided invertability of the operator B. These conditions can be written as explicit relations between the parameters a, b, and ν, see [GoKr, MiPr]. We are not going to reproduce these relations here. Under the above-mentioned conditions, the operator B with a more general symbol (4.36) also admits a bounded right- or two-sided inverse if the function c, more exactly the constant c in estimate (4.37), is sufficiently small.

In the conclusion of this subsection let us make a remark about a special case. Suppose the line segment connecting the points $a+b$ and $a-b$ in the complex plane does not contain the point 0. In this case the operator B with symbol (4.38) is two-sided invertible in the standard space $L_2(\Omega)$ and it is possible to use $L_2(\mathbb{R})$ as the space $\mathscr{X}=\mathscr{Y}$, formally $\nu=0$. This assertion is also true for the operator with symbol (4.36) if the constant c in estimate (4.37) is sufficiently small. All further results of this subsection admit the same simplification for symbols of this type; practically we can set ν, $\nu'=0$ in all further expressions and formulas.

4.2.4. *Investigation of equation* E_4. Let us introduce the factorization $r=r_+r_-$ and consider the operators $R_\alpha[\varepsilon]=\check{r}_\alpha[\varepsilon]$. As a result, we get

$$\mathscr{A}[\varepsilon]=L_\alpha[\varepsilon]\mathscr{B}R_\alpha[\varepsilon], \tag{4.39}$$

where $L_\alpha[\varepsilon]=\check{r}_{\alpha'}[\varepsilon]$. Taking into account Proposition 4.2.3 and repeating the computations of subsection 4.1, we can verify relations (3.2)–(3.5).

In accordance with Proposition 3.1.3, the operators V and W (introduced by (3.14)) possess properties a), b), and c). Let us also note the estimates of their norms $\|V\|$, $\|W\|=O(\varepsilon^{-\nu'})$, that follow from (3.14), and estimates from the previous subsection.

Thus to construct the reflection operator Γ we can use equation E_4. Let us recall Proposition 3.3.2. Since the operator B does not depend on ε and has a bounded inverse operator (right- or two-sided) the assumptions of Proposition 3.3.2 will be satisfied for sufficiently small ε if $\|D[\varepsilon]\|=o(1)$ as $\varepsilon\to 0$. In fact we can give an estimate of the Hilbert-Schmidt norm of D:

$$|D[\varepsilon]\|_2\leqslant c\varepsilon\log\varepsilon^{-1}. \tag{4.40}$$

Let us prove this estimate. We can get such estimates for operators $D_\alpha=-F^{-1}E_\alpha m_\alpha E_\alpha^* F$, where now

$$m_\alpha=p_{\alpha'}(1-r_\alpha r_{\alpha'}^{-1})E_{\alpha'}^2 b^{-1}p_\alpha.$$

In particular the kernel of the operator m_+ has the form

$$m_+\sim\frac{1}{4\pi^2}\int_{\mathbb{R}}\frac{[1-r_+(\varepsilon\zeta)r_-^{-1}(\varepsilon\zeta)]b^{-1}(\zeta)\exp(2i\zeta)}{(\xi-\zeta\quad i0)(\zeta-\eta+i0)}\,d\zeta, \tag{4.41}$$

compare (4.22). As in subsection 4.1, the last integral can be represented as the sum of the integrals $I(\xi, \eta; \varepsilon)$ and $\overline{I(-\eta, -\xi; \varepsilon)}$, see Lemma 4.1.1, the function f being equal to b^{-1}.

Let us estimate the Hilbert-Schmidt norm of the operator m_+ using Lemma 4.1.1 and Proposition 4.2.2. Let us discuss the objects introduced in Lemma 4.1.1 from the operator point of view.

Consider the operators I_{kl} and $I_K[\varepsilon]$ described by the corresponding kernels. The norms $\|\!| * |\!\|$ of I_{kl}, $k \geqslant 2$, are finite, therefore I_{kl} are operators of Hilbert-Schmidt class with respect to the norm of $FL_2(\rho)$. We discuss the estimates of the norms of the operators I_{1l} and $I_K[\varepsilon]$ in more detail.

To demonstrate the procedure, consider the operator J_{11}. First note that the norm $\|\!| * |\!\|$ is the sum of norms of the form

$$\|\mathscr{M}\|_{s,t} = \left(\int d\eta \left(\int d\xi \, |\mathscr{M}(\xi, \eta)|^s\right)^t\right)^{1/st},$$

where s, $st = 2, r$. It is clear that

$$\begin{aligned}
\|J_{11}\|_{s,t} &= \left(\int d\eta \left(\int d\xi \left|\int_L d\zeta \frac{f(\zeta)\zeta \log \zeta^{-1} \exp(2i\zeta)}{(\xi - \zeta)(\zeta - \eta)}\right|^s\right)^t\right)^{1/st} \\
&\leqslant \left(\int d\eta \left(\int_L |d\zeta| \left|\frac{f(\zeta)\zeta \log \zeta^{-1} \exp(2i\zeta)}{\zeta - \eta}\right| \left(\int \frac{d\xi}{|\xi - \zeta|^s}\right)^{1/s}\right)^{st}\right)^{1/st} \\
&\leqslant \int_L |d\zeta| |f(\zeta)\zeta \log \zeta^{-1} \exp(2i\zeta)| \left(\int \frac{d\eta}{|\zeta - \eta|^{st}}\right)^{1/st} \left(\int \frac{d\xi}{|\xi - \zeta|^s}\right)^{1/s} \\
&\leqslant c \int_L |d\zeta| e^{-(2-c_2)\operatorname{im}\zeta} |\zeta \log \zeta| \cdot |\operatorname{im}\zeta|^{-1/(st)' - 1/s'} < \infty.
\end{aligned}$$

Now we estimate $I_K[\varepsilon]$:

$$\begin{aligned}
\|I_K[\varepsilon]\|_{s,t} &\leqslant C_K \varepsilon^K \left(\int d\eta \left(\int d\xi \left\{\frac{1}{1 + |\xi - \eta|} \right.\right.\right. \\
&\qquad \left.\left.\left. \times \left[\frac{1}{1 + |\varepsilon\xi|^{\gamma'}} + \frac{1}{1 + |\varepsilon\eta|^{\gamma'}}\right]\right\}^s\right)^t\right)^{1/st} \\
&\leqslant C_K \varepsilon^K \left\{\int d\eta \left(\int d\xi \, (1 + |\xi - \eta|)^{-s} (1 + |\varepsilon\xi|)^{-\gamma' s}\right)^t \right. \\
&\qquad \left. + \int d\eta \, (1 + |\varepsilon\eta|)^{-\gamma' st} \left(\int d\xi \, (1 + |\xi - \eta|)^{-s}\right)^t\right\}^{1/st} \\
&\leqslant C_K \varepsilon^K \left(\int d\eta \, (1 + |\varepsilon\eta|))^{-\gamma' r}\right)^{1/r} \leqslant C_K \varepsilon^{K - 1/r}.
\end{aligned}$$

Here we used the assumption $\gamma r > 1$, hence $\gamma' r > 1$.

We can now repeat the argument of subsection 4.1 and assert that under the condition $\nu < 2\gamma + 1$ (a consequence of the condition $\gamma r > 1$) the operator $D[\varepsilon]$ is an operator of Hilbert-Schmidt class which admits the asymptotic representation:

$$D[\varepsilon] = \sum_{k=1,\ 0\leqslant l\leqslant k}^{K} D_{kl}\varepsilon^k \log^l \varepsilon^{-1} + D_K[\varepsilon], \tag{4.42}$$

where

$$|D_K[\varepsilon]|_2 \leqslant c_K \varepsilon^{K+1} \log^{K+1} \varepsilon^{-1}. \tag{4.43}$$

Thus, all the assumptions of Proposition 3.3.2 are fulfilled with

$$h = O(\varepsilon \log \varepsilon^{-1}). \tag{4.44}$$

Therefore, in accordance to Propositions 3.1.4 and 3.1.5,

1) the operator $\Gamma = WTV$ is a projection reflection operator if B is a right invertible operator, and
2) the operator Γ is a total reflection operator if B is a two-sided invertible operator.

In both cases we have obtained a total asymptotic expansion for the operator Γ and have proved that it satisfies the estimates:

$$\begin{gathered} \|\Gamma\| = O(\varepsilon^{-2\nu'}), \\ |\Gamma - WH'V|_2 = O(\varepsilon^{1-2\nu'} \log \varepsilon^{-1}). \end{gathered} \tag{4.45}$$

We can now conclude that an exhaustive asymptotic description of the operator A^{-1} has been obtained.

4.3. Degenerate symbols.

4.3.1. *Factorization of degenerate symbols.* In subsection 4.3 we shall investigate the asymptotic properties of operators generated by smooth and nonsmooth symbols with simple root at the point $\xi = 0$. A typical symbol of this class has the form

$$a(\xi, \varepsilon) = b(\xi) z(\varepsilon\xi) r(\varepsilon\xi), \qquad z(\xi) = \frac{\xi}{\sqrt{1+\xi^2}}, \tag{4.46}$$

where $r \in S_0$ and b satisfies conditions j_1) and j_2). This time the corresponding operator $\mathscr{A}[\varepsilon]$ is not a bijection in the standard spaces.

The factor $z(\varepsilon\xi)$ in the representation (4.46), considered as a symbol, generates an operator whose restriction to the interval Ω cannot be investigated directly. We shall include this factor in the factorization procedure, but this time the analytic factorization problem has no solutions with the properties used earlier.

The standard factorization of the function r: $r = r_+ r_-$, generates the following factorization of the symbol zr:

$$zr = l_\alpha v_\alpha, \tag{4.47}$$

where

$$v_\alpha(\xi) = r_\alpha(\xi)(1 - i\alpha\xi)^{-1/2}. \tag{4.48}$$

These functions are analytic in the corresponding half-planes $\mathbb{C}_\alpha$. It is clear that

$$l_\alpha(\xi) = \xi r_{\alpha'}(\xi)(1 + i\alpha\xi)^{-1/2}. \tag{4.49}$$

The operators v_α are not bijections in the space $L_2(\mathbb{R})$. This forces us to distinguish the spaces $\mathscr{X}$ and $\mathscr{Y}$. We choose the Sobolev space H_σ as the space $\mathscr{X}$. Let us recall that this space is characterized by the norm

$$\|f\|_\sigma^2 = \int_{\mathbb{R}} (1 + \xi^2)^\sigma |\hat{f}(\xi)|^2 \, d\xi.$$

If we set $\mathscr{Y} = H_\tau$, $\tau = \sigma + 1/2$, the operator $R_\alpha[\varepsilon]\colon \mathscr{X} \to \mathscr{Y}$ generated by the symbol $v_\alpha(\varepsilon\xi)$ turns out to be a bijection. Moreover

$$\|R_\alpha[\varepsilon]\| = O(\varepsilon^{-1/2}), \qquad \|R_\alpha^{-1}[\varepsilon]\| = O(1), \qquad \varepsilon \to 0. \tag{4.50}$$

The operator $L_\alpha[\varepsilon]\colon \mathscr{Y} \to \mathscr{X}$ generated by the symbol $l_\alpha(\varepsilon\xi)$ is bounded.

The symbol $b(\xi)$ generates the operator $\mathscr{B}$ (in $\mathscr{Y}$) which is a bijection; $\mathscr{B}$ is bounded together with the inverse operator $\mathscr{B}^{-1}$.

Relation (4.39) is again true. To derive the standard properties of the operators R_α and L_α, we must discuss the properties of the projection operators P'_α, Q'_α in the spaces $\mathscr{X}$, $\mathscr{Y}$ in more detail. In the framework of the previous considerations, their boundedness was evident, but now we have to prove it. The problem is reduced to proving the boundedness of the operator $\mathscr{S}$ in the spaces $L_2(\rho_\sigma)$, $L_2(\rho_\tau)$, $\tau = \sigma + 1/2$,

$$\rho_\sigma(\xi) = (1 + \xi^2)^\sigma.$$

This operator has been investigated in [GoKr, MiPr]. It was found that $\mathscr{S}$ is a bounded operator in $L_2(\rho_\sigma)$ if $|\sigma| < 1/2$. This means that we must set $-1/2 < \sigma < 0$.

Properties (3.2), (3.4), and (3.5) of the operators $R_\alpha[\varepsilon]$ and $L_\alpha[\varepsilon]$ are now evident. It is worth noting that property (3.3) has no straightforward meaning in the present case.

4.3.2. *Investigation of equation* E_4. We need some additional information about the operator $B\colon Y \to Y$. Further we assume that B has a bounded right inverse B^{-1}. An elementary example is given by the expression:

$$b(\xi) = 1 + c(\xi),$$

if $\sup|c|$ is sufficiently small.

The symbol $b(\xi) = s(\xi)$ $(= \operatorname{sgn}\xi)$ gives a more interesting example. We shall consider this example at the end of the section separately because it requires modifications of the general construction.

To prove the solvability of equation E_4, we shall again derive the estimate

$$|D[\varepsilon]|_2 \leqslant c\varepsilon \log \varepsilon^{-1}. \tag{4.51}$$

In fact we can repeat the general line of reasoning from subsection 4.2.4, but some details of the forthcoming formulas will be new. In the present case

$$D_\alpha[\varepsilon] = -F^{-1}E_\alpha m_\alpha[\varepsilon]E^*_{\alpha'}F, \qquad m_\alpha = p_{\alpha'}(1 - v_\alpha v_{\alpha'}^{-1})E^2_{\alpha'}b^{-1}p_\alpha.$$

The main tool is again Lemma 4.1.1, and the problem is to treat all the analytical objects of the lemma from the operator point of view. The Hilbert-Schmidt norms of the operators I_{kl}, $k \geqslant 2$, are evidently finite and we only consider I_{1l}, $I_K[\varepsilon]$. Setting $0 < \tau < 1/2$, we get:

$$\begin{aligned}
|J_{11}|_2 &= \left(\iint d\xi\, d\eta \left(\frac{1+\xi^2}{1+\eta^2}\right)^\tau \left|\int_L d\zeta \frac{f(\zeta)\zeta \log \zeta^{-1} \exp(2i\zeta)}{(\xi-\zeta)(\zeta-\eta)}\right|^2\right)^{1/2} \\
&\leqslant \int_L |d\zeta|\,|f(\zeta)\zeta\log\zeta^{-1}\exp(2i\zeta)| \left(\iint \left(\frac{1+\xi^2}{1+\eta^2}\right)^\tau \frac{d\xi\, d\eta}{|\xi-\zeta|^2|\zeta-\eta|^2}\right)^{1/2} \\
&\leqslant c\int_L |d\zeta|\, e^{-(2-c_2)\operatorname{im}\zeta}|\zeta\log\zeta^{-1}|\frac{1}{\operatorname{im}\zeta} < \infty
\end{aligned}$$

and

$$\begin{aligned}
|I_K(\varepsilon)|_2^2 &\leqslant C_K\varepsilon^{2K}\iint d\xi\, d\eta \left(\frac{1+\xi^2}{1+\eta^2}\right)^\tau \frac{1}{1+|\xi-\eta|^2}\left(\frac{1}{1+|\varepsilon\xi|^{\gamma'}} + \frac{1}{1+|\varepsilon\eta|^{\gamma'}}\right)^2 \\
&\leqslant C_K\varepsilon^{2K}\left\{\iint \left(\frac{1+\xi^2}{1+\eta^2}\right)^\tau \frac{d\xi\, d\eta}{(1+|\xi-\eta|^2)(1+|\varepsilon\xi|^{2\gamma'})}\right. \\
&\qquad \left. + \iint \left(\frac{1+\xi^2}{1+\eta^2}\right)^\tau \frac{d\xi\, d\eta}{(1+|\xi-\eta|^2)(1+|\varepsilon\eta|^{2\gamma'})}\right\} \\
&\leqslant C_K\varepsilon^{2K}\int \frac{d\xi}{(1+|\varepsilon\xi|)^{2\gamma'}} \leqslant C_K\varepsilon^{2K-1}.
\end{aligned}$$

As consequences we have formulas (4.42)–(4.44). Of course the meaning of these formulas is now somewhat different.

Thus we can assert that the operator $\Gamma = WTV$ is a projecting generalized reflection operator; it admits a total asymptotic expansion and the estimates

$$\begin{aligned}
\|\Gamma\| &= O(\varepsilon^{-1/2}), \\
|\Gamma - WH'V|_2 &= O(\varepsilon^{1/2}\log\varepsilon^{-1}).
\end{aligned} \tag{4.52}$$

4.3.3. *The operator* A^{-1}. Since the operator $\mathscr{A}[\varepsilon]$ is not a bijection, we cannot use the standard construction to get the operator A^{-1} and must turn our attention to the results of subsection 3.2.1. In accordance to Proposition 3.2.1, we try to construct a bounded operator $K: \mathscr{X} \to \mathscr{X}$ possessing the properties

(1) $\operatorname{Im} K \subset \operatorname{Im} \mathscr{A}$,
(2) $PK = P$,
(3) the operator $\mathscr{A}^{-1}K$ is bounded.

In the last condition, the operator $\mathscr{A}^{-1}$ is the left inverse operator; in the case considered it exists, but of course is not a bounded operator.

Let $\varphi \in X'$ be a rapidly decreasing function and $\hat{\varphi}(0) = 1$. Let us set

$$Kf = Pf - \widehat{Pf}(0) \cdot \varphi, \qquad f \in \mathscr{X}. \tag{4.53}$$

Since Pf is a finite function, the functional $\widehat{Pf}(0)$ is well defined. Properties (1) and (2) are evidently true so that we must only verify (3).

It is clear that if the functional $l: X \to \mathbb{C}$, $l(f) = \hat{f}(0)$, is bounded, then the functional $f \to \widehat{Pf}(0)$ is also bounded, so we now consider the functional l.

The function $f \in X$ can be characterized by the relations:

$$\int_{\mathbb{R}} (1+\xi^2)^\sigma |\hat{f}(\xi)|^2\, d\xi < \infty, \qquad \hat{f}(\xi) = \frac{1}{\pi}\int_{\mathbb{R}} \frac{\sin(\xi-\eta)}{\xi-\eta}\hat{f}(\eta)\, d\eta.$$

Then

$$\begin{aligned} |\hat{f}(0)|^2 &= \frac{1}{\pi^2}\left|\int \frac{\sin\xi}{\xi}\hat{f}(\xi)\, d\xi\right|^2 \\ &\leqslant \int (1+\xi^2)^\sigma |\hat{f}(\xi)|^2\, d\xi\, \frac{1}{\pi^2}\int (1+\xi^2)^{-\sigma}\left(\frac{\sin\xi}{\xi}\right)^2 d\xi \\ &= c\|f\|_\sigma^2. \end{aligned}$$

Let us consider the operator $\mathscr{A}^{-1}K$. Since $KP = K$, it is sufficient to take the norm $\|\mathscr{A}^{-1}Kf\|_\sigma$ for $f \in X$:

$$\begin{aligned} \|\mathscr{A}^{-1}Kf\|_\sigma^2 &= \int (1+\xi^2)^\sigma \left|\frac{\hat{f}(\xi) - \hat{f}(0)\hat{\varphi}(\xi)}{a(\xi,\varepsilon)}\right|^2 d\xi \\ &\leqslant c\int (1+\xi^2)^\sigma \left|\frac{\sqrt{1+(\varepsilon\xi)^2}}{\varepsilon\xi}[\hat{f}(\xi) - \hat{f}(0)\hat{\varphi}(\xi)]\right|^2 d\xi \\ &\leqslant c\left[\int_{|\varepsilon\xi|>1} \cdots + \int_{|\varepsilon\xi|<1} \cdots\right] \\ &\leqslant c\|f\|_\sigma^2 + \frac{c}{\varepsilon^2}\int (1+\xi^2)^\sigma \left|\frac{\hat{f}(\xi) - \hat{f}(0)\hat{\varphi}(\xi)}{\xi}\right|^2 d\xi. \end{aligned}$$

Let us estimate the integral:

$$\left(\int (1+\xi^2)^\sigma \left|\frac{\hat{f}(\xi)-\hat{f}(0)\hat{\varphi}(\xi)}{\xi}\right|^2 d\xi\right)^{1/2}$$

$$= \left(\int (1+\xi^2)^\sigma \left|\frac{1}{\xi}\int d\eta \left[\frac{\sin(\xi-\eta)}{\xi-\eta} - \frac{\sin\eta}{\eta}\hat{\varphi}(\xi)\right]\hat{f}(\eta)\right|^2\right)^{1/2}$$

$$\leqslant \int d\eta\, |\hat{f}(\eta)| \left(\int d\xi\,(1+\xi^2)^\sigma \left|\frac{\frac{\sin(\eta-\xi)}{\eta-\xi} - \frac{\sin\eta}{\eta}\hat{\varphi}(\xi)}{\xi}\right|^2\right)^{1/2}$$

$$\leqslant |f|_\sigma \left[\int (1+\eta^2)^{-\sigma} d\eta \int d\xi\,(1+\xi^2)^\sigma \left|\frac{\frac{\sin(\eta-\xi)}{\eta-\xi} - \frac{\sin\eta}{\eta}\hat{\varphi}(\xi)}{\xi}\right|^2\right]^{1/2}.$$

Since the last integral converges, we can assert that

$$\|\mathscr{A}^{-1}K\| \leqslant c\varepsilon^{-1}. \tag{4.54}$$

We can now present the final formula for the right inverse operator

$$A^{-1}f = (I-\Gamma)(\mathscr{A}^{-1}K)f, \qquad f \in X. \tag{4.55}$$

Taking into account the asymtotic description of the operator Γ, we can say that we have obtained an effective asymptotic description of the operator A^{-1}.

Let us make some short remarks concerning the special case $\mathscr{B} = I$. In this case it is possible to use equation E_4 written in terms of the reflection operators Γ_α,

$$\Gamma_\alpha = R_\alpha^{-1} Q'_\alpha R_\alpha.$$

They are projecting generalized reflection operators. To investigate equation E_4, we can use Proposition 3.3.1 and estimate (4.51). If the factor r in (4.46) belongs to the class S, the estimate (4.51) can be improved:

$$m_\alpha = O(\varepsilon^\infty), \tag{4.56}$$

see (4.27). As a result we get

$$\Gamma = \Gamma_+ + \Gamma_- + O(\varepsilon^\infty). \tag{4.57}$$

4.3.4. *About the symbols with a root of $|\varepsilon\xi|$-type.* As we have already mentioned, the case $b(\xi) = \operatorname{sgn}\xi$ requires some modifications. The reason is that the operators $\mathbf{B}$ and $\mathbf{B}'$ now have only left inverse operators. More exactly, the operator $\mathbf{B}'$ satisfies the following relations

$$H'\mathbf{B}' = Q', \qquad \mathbf{B}'H' = Q' - q, \tag{4.58}$$

where q is a one-dimensional projection operator in $\mathscr{Y}$. In explicit form

$$(\mathbf{B}'f)(x) = \frac{i}{\pi}\int_{\Omega'} \frac{\omega'(x)}{x-y} f(y)\,dy\,, \qquad f\colon \mathbb{R}\to\mathbb{C}\,,$$

$$(H'f)(x) = \frac{\sqrt{x^2-1}}{|x|}\frac{i}{\pi}\int_{\Omega'} \frac{\omega'(x)}{x-y}\frac{|y|}{\sqrt{y^2-1}} f(y)\,dy\,, \tag{4.59}$$

$$qf = l_0(f)\psi_0\,, \quad \psi_0(x) = \frac{\omega'(x)}{x}\,, \quad l_0(f) = \frac{1}{\pi}\int_{\Omega'} \frac{\operatorname{sgn} y}{\sqrt{y^2-1}} f(y)\,dy. \tag{4.60}$$

The analytic approach to these formulas can be found in [Tr, Ga, Mu]. To give operator treatment of these formulas we must prove the following

LEMMA 4.3.1. *Formula* (4.59) *defines a bounded operator in* H_τ.

The proof of the lemma can be found in Appendix 5.2.

The results of subsection 4.3.2 show that the operator $C = \mathbf{B}' + D$ has the same qualities as the operator $\mathbf{B}'$. This leads us to turn our attention to the results of subsection 3.2.2. In accordance to subsection 3.2.2, the operator A^{-1} can be described by the formula

$$A^{-1}f = (I-\Gamma)(\mathscr{A}^{-1}K)f\,, \qquad \Gamma = WTV\,, \quad f\in X\,, \tag{4.61}$$

for an appropriate operator K. As compared with the previous subsection, we must prove additionally that there exists an element $\psi_1 \in \operatorname{Im}\mathscr{A}\cap X'$ such that

$$l(V\mathscr{A}^{-1}\psi_1) \neq 0.$$

Under this condition, the functional l is defined by the defect of the equation E_5:

$$CT = Q' - \Delta Q\,, \qquad \Delta Qf = l(f)\psi. \tag{4.62}$$

The results of subsection 4.3.2 say that

$$C = \mathbf{B}' + O(\varepsilon\log\varepsilon^{-1})\,, \qquad T = H' + O(\varepsilon\log\varepsilon^{-1})\,;$$

therefore

$$\Delta Q = q + O(\varepsilon\log\varepsilon^{-1}). \tag{4.63}$$

Now we can assume that $\psi = \psi_0 + O(\varepsilon\log\varepsilon^{-1})$, so that

$$l = l_0 + O(\varepsilon\log\varepsilon^{-1}). \tag{4.64}$$

Let us construct an element ψ_1 with the properties

$$|l_0(V\mathscr{A}^{-1}\psi_1)| \geqslant c_0\varepsilon^{-1}\,, \qquad c_0 \neq 0\,, \tag{4.65}$$

$$|V\mathscr{A}^{-1}\psi_1|_\tau \leqslant c_1\varepsilon^{-1}\,; \tag{4.66}$$

then

$$|l(V\mathscr{A}^{-1}\psi_1)| \geqslant c\varepsilon^{-1}\,, \qquad c\neq 0. \tag{4.67}$$

The element ψ_1 does not depend on ε.

Let ψ_1 be a smooth odd rapidly decreading function equal to 0 when $|x| \leqslant 1$ and positive when $x > 1$. It is clear that $\psi_1 \in \operatorname{Im}\mathscr{A} \cap X'$. Let us verify (4.65) and (4.66).

The estimate $|R_\alpha \mathscr{A}^{-1}\psi_1|_\tau \leqslant c\varepsilon^{-1}$ is evident, it immediately leads to (4.66). Moreover

$$\varepsilon R_\alpha \mathscr{A}^{-1}\psi_1 - \varkappa\psi_1 \longrightarrow 0, \qquad \varepsilon \to 0,$$

where $\varkappa$ is the operator with the symbol $|\xi|^{-1}$. The element is well defined because $\hat{\psi}_1(0) = 0$. The last formula implies

$$l_0(V\mathscr{A}^{-1}\psi_1) \sim \varepsilon^{-1} l_0(Q'\varkappa\psi_1) = \varepsilon^{-1} l_0(\varkappa\psi_1).$$

The element $\varkappa\psi_1$ is described by the formula

$$\varkappa = \mathscr{S}J, \qquad (Jf)(x) = \int_{-\infty}^{x} f(y)\,dy,$$

hence

$$\begin{aligned}
l_0(\varkappa\psi_1) &= \frac{1}{\pi}\int \frac{\operatorname{sgn} x \cdot \omega'(x)}{\sqrt{x^2-1}}\,dx \left(\frac{1}{\pi}\int \frac{dy}{x-y}\int_{-\infty}^{y} \psi_1(z)\,dz\right) \\
&= -\frac{1}{\pi}\int dx \left(\frac{1}{\pi}\int \frac{1}{x-y}\,\frac{\operatorname{sgn} y \cdot \omega'(y)}{\sqrt{y^2-1}}\,dy\right)\int_{-\infty}^{x} \psi_1(z)\,dz \\
&= -\frac{1}{\pi}\int dx\,\frac{\omega(x)}{\sqrt{1-x^2}}\int_{-\infty}^{x} \psi_1(z)\,dz \\
&= -\frac{1}{\pi}\int_\Omega \frac{dx}{\sqrt{1-x^2}}\int_{-\infty}^{-1} \psi_1(z)\,dz \\
&\neq 0.
\end{aligned}$$

Thus we have proved formula (4.65) and can now claim that we have obtained a justification of the asymptotic constructions of the present subsection in the particular case $b(\xi) = \operatorname{sgn}\xi$.

§5. Appendices

5.1. Proof of Lemma 4.1.1. Let g be the function from the statement of the Lemma, possessing the asymptotic expansion (4.13). Let N be a sufficiently large positive integer. Put

$$g_0(\zeta) = \left[\sum_{\substack{1\leqslant k\leqslant N \\ 0\leqslant l\leqslant k}} g_{kl}\zeta^k \log^l \zeta^{-1}\right](1+\zeta^{N+1})^{-1}.$$

The function g_0 is an analytic function in the angle $|\arg\zeta| < \pi/(N+1)$; it admits the asymptotic representation

$$g_0(\zeta) = \sum_{\substack{1\leqslant k\leqslant N \\ 0\leqslant l\leqslant k}} g_{kl}\zeta^k \log^l \zeta^{-1} + O(\zeta^{N+2}\log\zeta^{-1}), \qquad \zeta \to 0,$$

and the estimates $|g_0^{(r)}| \leqslant c_r(1+|\zeta|)^{-\gamma_0}$, where $\gamma_0 < 1$. The function

$$g_\infty(\zeta) = g(\infty)\frac{\zeta^{N+1}}{1+\zeta^{N+1}}$$

is also analytic in the same angle and $g_\infty(\zeta) = O(\zeta^{N+1})$, $\zeta \to 0$.

Now let us set

$$g = g_0 + g_\infty + h.$$

The function h belongs to $C^\infty(0, \infty)$, vanishes (together with its derivatives) as $O(|\zeta|^{-\gamma})$, $\gamma > 1/2$, if $\zeta \to \infty$, and as $O(\zeta^{N+1})$ if $\zeta \to 0$.

Let I_0 and I_∞ be the integrals given by (4.14) but with g_0 and g_∞ instead of g. Take into account the fact that f, g_0, and g are analytic and replace the integration path by the ray $\arg\zeta = \delta < \pi/(N+1)$. Using this new representation, it is easy to show that the integrals I_0 and I_∞ admit the asymptotic description

$$I_0 = \sum_{\substack{1\leqslant k\leqslant K \\ 0\leqslant l\leqslant k}} I_{kl}(\xi, \eta)\varepsilon^k \log^l \varepsilon^{-1} + O\left(\frac{\varepsilon^{K+1}\log^{K+1}\varepsilon^{-1}}{(1+|\xi|)(1+|\eta|)}\right),$$

$$I_\infty = o\left(\frac{\varepsilon^{K+1}\log^{K+1}\varepsilon^{-1}}{(1+|\xi|)(1+|\eta|)}\right),$$

where K is an integer depending on N and tending to $+\infty$ as $N \to \infty$.

Introduce the function k

$$h(\zeta) = \frac{\zeta^{K+1}}{1+\zeta^{K+1}}k(\zeta),$$

consider its Fourier transform:

$$K(x) = \frac{1}{2\pi}\int_0^\infty k(\zeta)e^{i\zeta x}\,d\zeta$$

and decomposition:

$$k(\varepsilon\zeta) = k_1(\zeta, \varepsilon) + k_2(\zeta, \varepsilon),$$

$$k_1 = \int_{\varepsilon^{-\nu}}^\infty K(x)e^{-i\varepsilon\zeta x}\,dx, \qquad k_2 = \int_\infty^{\varepsilon^{-\nu}} K(x)e^{-i\varepsilon\zeta x}\,dx, \qquad 0 < \nu < 1.$$

The function k_2 is an analytic function in $\mathbb{C}_+$ possessing the estimate

$$|k_2(\zeta, \varepsilon)| \leqslant c\exp[\varepsilon^{1-\nu}\operatorname{im}\zeta].$$

Consider the integral I_2, which is the integral in (4.14) but with the function

$$\frac{(\varepsilon\zeta)^{K+1}}{1+(\varepsilon\zeta)^{K+1}}k_2(\zeta, \varepsilon)$$

replacing g, and change the integration path to the ray L. Then obviously

$$I_2 = O\left(\frac{\varepsilon^{K+1}}{(1+|\xi|)(1+|\eta|)}\right).$$

The definition of k_1 and the representation

$$k_1 = \frac{1}{2\pi i}\int \frac{k(\tau)\exp[-\varepsilon^{-\nu}(\varepsilon\zeta-\tau)]}{\varepsilon\zeta-\tau-i0}\,d\tau$$

lead to the estimate

$$|k_1^{(l)}(\zeta, \varepsilon)| \leqslant c\frac{\varepsilon^{K+1}}{(1+|\varepsilon\zeta|)^{\gamma_1}},$$

where $l \leqslant 2$ and $\gamma_1 < \gamma$. Consider the integral I_1 generated by the function k_1:

$$\begin{aligned} I_1 &= \int_0^\infty \frac{f(\zeta)(\varepsilon\zeta)^{K+1}}{1+(\varepsilon\zeta)^{K+1}}\frac{k_1(\zeta, \varepsilon)\exp(2i\zeta)}{(\xi-\zeta-i0)(\zeta-\eta+i0)}\,d\zeta \\ &= \frac{1}{\xi-\eta}\left\{\int_0^\infty \frac{f(\zeta)(\varepsilon\zeta)^{K+1}}{1+(\varepsilon\zeta)^{K+1}}\frac{k_1(\zeta, \varepsilon)e^{2i\zeta}}{\xi-\zeta-i0}\,d\zeta\right. \\ &\qquad \left. -\int_0^\infty \frac{f(\zeta)(\varepsilon\zeta)^{K+1}}{1+(\varepsilon\zeta)^{K+1}}\frac{k_1(\zeta, \varepsilon)e^{2i\zeta}}{\eta-\zeta-i0}\,d\zeta\right\}. \end{aligned}$$

This representation shows that

$$|I_1| \leqslant c\frac{\varepsilon^{K+1}}{1+|\xi-\eta|}\left(\frac{1}{1+|\varepsilon\xi|^{\gamma'}}+\frac{1}{1+|\varepsilon\eta|^{\gamma'}}\right), \qquad \gamma' < \gamma.$$

Hence the lemma is proved.

5.2. Proof of Lemma 4.3.1. Consider the Fourier transform of the operator H'. Its kernel is given by

$$(FH'F^{-1})(\xi, \eta) = \frac{i}{2\pi^2}\iint_{\Omega'\Omega'} e^{i\xi x}\frac{\sqrt{x^2-1}}{|x|}\frac{1}{x-y}\frac{|y|}{\sqrt{y^2-1}}e^{-i\eta y}\,dx\,dy.$$

It is clear that we can restrict our consideration to the term

$$N(\xi, \eta) = \int_1^\infty\int_1^\infty e^{i\xi x}\frac{\sqrt{x^2-1}}{x}\frac{1}{x-y}\frac{y}{\sqrt{y^2-1}}e^{-i\eta y}\,dx\,dy.$$

Substitute the limit values for the main value, using the formulas

$$\text{p.v. } (x-y)^{-1} - \sigma i\pi\delta(x-y) + (x-y+\sigma i0)^{-1},$$

where $\sigma = +1, -1$. We choose $\sigma = \operatorname{sgn}\xi = \operatorname{sgn}\eta$ if $\operatorname{sgn}\xi * \operatorname{sgn}\eta = 1$. What σ is does not matter if $\operatorname{sgn}\xi * \operatorname{sgn}\eta = -1$. The delta-function term yields the kernel N_0:

$$N_0(\xi, \eta) = c\exp[i(\xi - \eta)](\xi - \eta + i0)^{-1},$$

where c depends on $\operatorname{sgn}\xi * \operatorname{sgn}\eta$. This kernel generates a bounded operator in $L_2(\rho_\tau)$.

In the integral N_1 with the limit values, we can deform the integration paths in the complex plane to separate the variables x and y. Using this new representation of N_1, we can conclude that

$$|N_1(\xi, \eta)| \leqslant cJ(\xi, \eta),$$

where

$$J(\xi, \eta) = \int_0^\infty \int_0^\infty \sqrt{\frac{x(1+y)}{y(1+x)}} \frac{\exp(-|\xi|x - |\eta|y)}{x+y}\, dx\, dy.$$

Now it is sufficient to prove that the operator J with kernel $J(\xi, \eta)$ is bounded in the space $L_2(\rho_\tau)$, $0 < \tau < 2^{-1}$. Let us give appropriate estimates of the kernel J, assuming $\xi, \eta > 0$. Set

$$J = \sum_{k=1}^4 J_k = \int_0^1 dx \int_0^1 dy \cdots + \int_1^\infty dx \int_0^1 dy \cdots + \int_0^1 dx \int_1^\infty dy \cdots + \int_1^\infty dx \int_1^\infty dy \cdots.$$

Then we have:

$$J_1 \leqslant c\int_0^1\int_0^1 \sqrt{\frac{x}{y}} e^{-\xi x - \eta y} \frac{dx\, dy}{x+y} = c\int_0^1 dx \int_0^{1/x} e^{-(\xi+\eta u)x} \frac{du}{\sqrt{u}(1+u)}$$
$$\leqslant c\int_0^\infty \frac{du}{\sqrt{u}(1+u)} \int_0^1 e^{-(\xi+\eta u)x}\, dx \leqslant \frac{c}{\sqrt{1+\xi}(\sqrt{1+\xi} + \sqrt{1+\eta})},$$

$$J_2 \leqslant c\int_1^\infty dx \int_0^1 \frac{1}{\sqrt{y}} e^{-\xi x - \eta y} \frac{dy}{x+y} \leqslant c\int_1^\infty e^{-\xi x} \frac{dx}{x} \int_0^1 e^{-\eta y} \frac{dy}{\sqrt{y}}$$
$$= c\int_\xi^\infty e^{-u} \frac{du}{u} \frac{1}{\sqrt{\eta}} \int_0^\eta e^{-v} \frac{dv}{v} \leqslant \frac{c|\log \xi^{-1}|}{1+\exp\xi} \frac{1}{\sqrt{1+\eta}},$$

$$J_3 \leqslant c\int_0^1 dx \sqrt{x} \int_1^\infty e^{-\xi x - \eta y} \frac{dy}{x+y} \leqslant c\int_0^1 \sqrt{x} e^{-\xi x}\, dx \int_1^\infty e^{-\eta y} \frac{dy}{y}$$
$$= c\xi^{-3/2} \int_0^\xi \sqrt{u} e^{-u}\, du \int_\eta^\infty e^{-v} \frac{dv}{v} \leqslant \frac{c}{(1+\xi)^{3/2}} \frac{|\log \eta|}{1+\exp \eta},$$

$$J_4 \leqslant c\int_1^\infty \int_1^\infty e^{-\xi x - \eta y} \frac{dx\, dy}{x+y} = ce^{-\xi-\eta} \int_0^\infty \int_0^\infty e^{-\xi u - \eta v} \frac{du\, dv}{2+u+v}$$
$$\leqslant ce^{-\xi-\eta} \int_0^\infty \frac{dw}{1+w} \int_0^\infty e^{-(\xi+\eta w)v}\, dv \leqslant ce^{-\xi-\eta} \frac{\log\xi - \log\eta}{\xi - \eta}.$$

It is clear that J_2 and J_3 are bounded operators of Hilbert-Schmidt class. To estimate J_1 and J_4, we shall use the well-known Schur test: the operator K with kernel $K(\xi, \eta)$ is bounded in $L_2(\mathbb{R})$ if

$$\sup_\xi \int |K(\xi, \eta)| \left|\frac{\xi}{\eta}\right|^\beta d\eta < \infty, \qquad \sup_\eta \int |K(\xi, \eta)| \left|\frac{\eta}{\xi}\right|^\beta d\xi < \infty,$$

where β is some positive number.

Let $0 < \beta < 1$; then

$$\begin{aligned}\sup_{\xi>0} \int_0^\infty \left|\frac{\xi+i}{\eta+i}\right|^\tau J_4(\xi, \eta) \left(\frac{\xi}{\eta}\right)^\beta d\eta &\leqslant c \sup_{\xi>0} \int_0^\infty \left(\frac{\xi}{\eta}\right)^\beta \log\frac{\xi}{\eta} \cdot \frac{d\eta}{\xi-\eta} \\ &= c \int_0^\infty \frac{\log\zeta\, d\zeta}{(\zeta-1)\zeta^\beta} < \infty,\end{aligned}$$

$$\begin{aligned}\sup_{\eta} \int_0^\infty \left|\frac{\xi+i}{\eta+i}\right|^\tau J_4(\xi, \eta) \left(\frac{\eta}{\xi}\right)^\beta d\xi &\leqslant c \sup_{\eta} \int_0^\infty \left(\frac{\eta}{\xi}\right)^\beta \log\frac{\xi}{\eta} \cdot \frac{d\xi}{\xi-\eta} \\ &= c \int_0^\infty \frac{\log\zeta\, d\zeta}{(\zeta-1)\zeta^\beta} < \infty.\end{aligned}$$

Finally,

$$\begin{aligned}\sup_{\xi>0} \int_0^\infty \left|\frac{\xi+i}{\eta+i}\right|^\tau J_1(\xi, \eta) \left(\frac{\xi}{\eta}\right)^\beta d\eta &\leqslant c \sup_{\xi\geqslant 1} \int_1^\infty \left(\frac{\xi}{\eta}\right)^{\tau+\beta} \frac{d\eta}{\sqrt{\xi}(\sqrt{\xi}+\sqrt{\eta})} \\ &\leqslant c \sup_{\xi>0} \int_0^\infty \left(\frac{\xi}{\eta}\right)^{\tau+\beta} \frac{d\eta}{\sqrt{\xi}(\sqrt{\xi}+\sqrt{\eta})} \\ &= c \int_0^\infty \frac{du}{u^{\tau+\beta}(1+\sqrt{u})},\end{aligned}$$

$$\begin{aligned}\sup_{\eta>0} \int_0^\infty \left|\frac{\xi+i}{\eta+i}\right|^\tau J_1(\xi, \eta) \left(\frac{\eta}{\xi}\right)^\beta d\xi &\leqslant c \sup_{\eta\geqslant 1} \int_1^\infty \left(\frac{\xi}{\eta}\right)^{\tau-\beta} \frac{d\xi}{\sqrt{\xi}(\sqrt{\xi}+\sqrt{\eta})} \\ &\leqslant c \sup_{\eta>0} \int_0^\infty \left(\frac{\xi}{\eta}\right)^{\tau-\beta} \frac{d\xi}{\sqrt{\xi}(\sqrt{\xi}+\sqrt{\eta})} \\ &= c \int_0^\infty \frac{du}{u^{1/2-\tau+\beta}(1+\sqrt{u})}.\end{aligned}$$

These integrals converge if $2^{-1} < \beta + \tau < 1$, $0 < \beta - \tau < 2^{-1}$. The set of solutions of this system of inequalities is not empty. Hence the lemma is proved.

References

[BaSh] M. A. Basin and V. P. Shadrin, *Hydroaerodynamics of wind near the boundary of the medium*, "Sudostroenie", Leningrad, 1980. (Russian)

[Be] B. P. Belinskiĭ, *Integral equations of stationary problems of short wave diffraction by barriers of segment type*, Zh. Vychisl. Mat. i Mat. Fiz. **13** (1973), 373–384; English transl. in U.S.S.R. Comput. Math. and Math. Phys. **13** (1973).

[BoSi] A. Böttcher and B. Silbermann, *Analysis of Toeplitz operators*, Akademie-Verlag, Berlin, 1989.

[BB1] A. M. Budylin and V. S. Buslaev, *On the asymptotic behavior of spectral characteristics of an integral operator with difference kernel on expanding regions*, Dokl. Akad. Nauk SSSR **287** (1986), 1529–531; English transl. in Soviet Math. Dokl. **33** (1986), 400–403.

[BB2] ______, *On the asymptotic behavior of spectral characteristics of integral operators with difference kernels on expanding regions*, Probl. Mat. Fiz. **14** (1991). (Russian)

[ChKo] W. Chew and J. Kong, *Asymptotic formula for the capacitance of two oppositely charged discs*, Math. Proc. Cambridge Philos. Soc. **89** (1981), no. 2, 373–384.

[Ga] F. D. Gakhov, *Boundary value problems*, "Nauka", Moscow, 1977; English transl. of 1st ed., Pergamon Press, Oxford; Addison-Wesley, Reading, MA, 1966.

[Gau] M. Gaudin, *La fonction d'onde de Bethe*, Masson, Paris, 1983.

[GlLu] I. Glazman and U. Lyubich, *Finite-dimensional linear analysis: A systematic presentation in problem form*, "Nauka", Moscow, 1969; English transl., M.I.T. Press, Cambridge, MA, 1974.

[GoKr] I. Gokhberg and N. Ya. Krupnik, *Introduction to the theory of one-dimensional singular integral operators*, "Shtiintsa", Kishinev, 1973; German transl., Birkhäuser, Basel, 1978.

[Hu] V. Hutson, *The circular plate condenser at small separations*, Math. Proc. Cambridge Philos. Soc. **59** (1963), no. 1, 211–224.

[IIKS] A. R. It·s, A. G. Izergin, V. E. Korepin, and N. A. Slavnov, *Differential equations for quantum correlation functions*, Internat. J. Modern Phys. B **4** (1990), 1003–1037.

[JMMS] M. Jimbo, T. Miwa, Y. Mori, and M. Sato, *Density matrix of an impenetrable Bose gas and the fifth Painlevé transcendent*, Physica **10** (1980), 80–158.

[KoSi] A. Kozak and I. Simonenko, *Projection methods for the solutions of multi-dimensional discrete convolution equations*, Sibirsk. Mat. Zh. **21** (1980), no. 2, 119–127; English transl. in Siberian Math. J. **21** (1980).

[LeLe] F. Leppington and H. Levin, *On the capacity of the circular disk condenser at small separation*, Math. Proc. Cambridge Philos. Soc. **68** (1970), no. 4, 236–254.

[Ma] K. Maurin, *Methods of Hilbert spaces*, PWN, Warsaw, 1959; English transl., 1967.

[MiPr] S. G. Mikhlin and S. Prössdorf, *Singular integral operators*, Akademie-Verlag, Berlin, 1986.

[Mu] N. I. Muskhelishvili, *Singular integral equations*, OGIZ, Moscow, 1946; 2nd ed., Fizmatiz, Moscow, 1962; 3rd ed., "Nauka", Moscow, 1968; English transl., Noordhoff, Groningen, 1953; reprinted, 1972.

[Nob] B. Noble, *Methods based on the Wiener-Hopf technique*, Pergamon Press, London, 1958.

[Nov] V. Yu. Novokshenov, *A singular integral equation with small parameter on a finite interval*, Mat. Sb. **105** (1978), 543–573; English transl. in Math. USSR Sb. **34** (1978), 475–502.

[Sh] S. Shaw, *Circular-disk viscometer and related electrostatic problems*, Phys. Fluids **13** (1970), 1935–1947.

[Soi] Ya. S. Soĭbel′man, *The asymptotic behavior of the capacity of a condenser with disks of arbitrary forms*, Sibirsk. Mat. Zh. **25** (1984), no. 6, 167–181; English transl. in Siberian Math. J. **25** (1984).

[Sol] V. G. Sologub, *On the solution of a convolution-type integral equation on a finite segment*, Zh. Vychisl. Mat. i Mat. Fiz. **11** (1971), no. 4, 837–854; English transl. in U.S.S.R. Comput. Math. and Math. Phys. **11** (1971).

[Tr] F. G. Tricomi, *Integral equations*, Interscience, New York, 1957.

[VoBa] I. I. Vorovich and V. A. Babeshko, *Dynamical mixed problem of the theory of elastisity for nonclassical domains*, "Nauka", Moscow, 1979. (Russian)

[VAB] I. I. Vorovich, V. M. Aleksandrov, and V. A. Babeshko, *Nonclassical mixed problems of the theory of elasticity*, "Nauka", Moscow, 1974. (Russian)

[Wi1] H. Widom, *A trace formula for Wiener-Hopf operators*, J. Operator Theory **8** (1982), 279–298.

[Wi2] ______, *Asymptotic expansions for pseudodifferential operators on bounded domains*, Lecture Notes in Math., vol. 1152, Springer-Verlag, Berlin and New York, 1986.

[Wi3] ______, *On a class of integral operators with discontinuous symbol*, Oper. Theory: Adv. Appl. **4** (1982), 477–500.

[Wi4] ______, *On a class of integral operators on a half-space with discontinuous symbol*, J. Funct. Anal. **88** (1990), no. 1, 166–193.

DEPARTMENT OF MATHEMATICAL PHYSICS
INSTITUTE OF PHYSICS
Current address: Leningrad University, Ulyanov str. 1,
Leningrad, Petrodvorets 198904 USSR

Translated by V. S. BUSLAEV

ADVANCES IN SOVIET MATHEMATICS
Volume 7, 1991

Weyl Asymptotics for the Discrete Spectrum of the Perturbed Hill Operator

A. V. SOBOLEV

ABSTRACT. We study discrete eigenvalues of a one-dimensional periodic operator (namely, the Hill operator) perturbed by a decreasing potential $gq(x)$, in a fixed gap of its essential spectrum. Here $g > 0$ is the coupling constant. Let λ, λ_1, λ_2 ($\lambda_1 < \lambda_2$) be certain points in the gap. Let $N(\lambda_1, \lambda_2, gq)$ be the number of eigenvalues lying in the interval (λ_1, λ_2). For a function q that does not change its sign, let $M(\lambda, gq)$ be the number of eigenvalues crossing the point λ while the coupling constant varies from 0 to g. For $N(\lambda_1, \lambda_2, gq)$ and $M(\lambda, gq)$ as $g \to \infty$ we establish Weyl type asymptotic formulas provided that $q(x)$ has a power-like behavior as $|x| \to \infty$. In particular, for $N(\lambda_1, \lambda_2, gq)$ we prove that

$$N(\lambda_1, \lambda_2, gq) \approx \pi^{-1} \iint dk(\lambda)\, dx, \qquad g \to \infty,$$

where the nondecreasing function $k(\lambda)$ denotes the so-called global quasi-momentum for the unperturbed Hill operator and the double integral is taken over the classically allowed domain

$$\{\lambda, x \mid \lambda_1 \leqslant \lambda + gq(x) \leqslant \lambda_2\}.$$

A similar result holds for $M(\lambda, gq)$.

§1. Introduction

Let us consider the Hill operator

$$h(gq) = -\frac{d^2}{dx^2} + p(x) + gq(x),$$

$x \in \mathbb{R}$, where p and q are real-valued functions, $p(x+1) = p(x)$ and $q(x) \to 0$ as $x \to \infty$, $g > 0$ is the coupling constant. It is well known that the continuous spectrum of $h(gq)$ (which coincides with the spectrum of $h_0 = h(0)$) consists of bands separated by gaps. We study properties of the discrete spectrum of $h(gq)$ in a fixed gap $(\lambda_1^0, \lambda_2^0)$ as $g \to \infty$. Discrete eigenvalues are continuous functions of g and if $q \geqslant 0$ (resp. $q \leqslant 0$), they increase (resp. decrease) monotonously as g grows. When q does not change

1991 *Mathematics Subject Classification*. Primary 34L20.

sign, we introduce the so-called counting function $M(\lambda, gq)$, defined as the number of eigenvalues of $h(g'q)$ crossing a point $\lambda \in (\lambda_1^0, \lambda_2^0)$ while g' varies from 0 to g. The values of $M(\lambda, gq)$ at the points λ_i^0, $i = 1, 2$, are defined as limits as $\lambda \to \lambda_i^0$ (see §2 for details). In the power scale a sufficient condition for both limits to be finite (which is equivalent to the finiteness of the discrete spectrum in the gap) is

$$q(x) = O(|x|^{-\alpha}), \tag{1.1}$$

where $\alpha > 2$.

Our purpose is to study the behavior of $M(\lambda, gq)$ as $g \to \infty$. By the splitting principle, we may restrict ourselves to the case when the operator $h(gq)$ is defined on $\mathbb{R}_+$ with a Dirichlet condition at the point $x = 0$. As mentioned in [DH], the asymptotics of $M(\lambda, gq)$ depends essentially on the sign of q. For nonpositive potentials q obeying (1.1) with $\alpha > 2$ and for λ lying strictly inside the gap, it was proved in [DH] that $M(\lambda, gq)$ is governed by the usual semiclassical formula:

$$\lim g^{-\frac{1}{2}} M(\lambda, gq) = \pi^{-1} \int |q(x)|^{\frac{1}{2}}\, dx, \qquad g \to \infty. \tag{1.2}$$

(Here and below integrals with no indication of limits mean integration over $\mathbb{R}_+$.) Moreover, if $q \geqslant 0$, then for such λ and α the bound $M(\lambda, gq) = O(g^{\frac{1}{\alpha}})$ was established in [DH].

The present paper generalizes the results of [DH] in several respects. First of all, if q decreases sufficiently quickly (in particular, as in (1.1) with $\alpha > 2$), our results are valid for *all* λ, including the endpoints λ_1^0, λ_2^0. For instance, the relation (1.2) for $q \leqslant 0$ holds uniformly in $\lambda \in [\lambda_1^0, \lambda_2^0]$ (see Theorem 3.3), the assumptions on q being a little bit broader than in [DH].

Then we study the function $M(\lambda, gq)$ as $q \leqslant 0$ and $\alpha < 2$ in (1.1) or $q \geqslant 0$ and α is an arbitrary positive number. In these cases the behavior of $M(\lambda, gq)$ depends explicitly on the asymptotics of $q(x)$ as $x \to \infty$ (for $q \geqslant 0$, $\alpha > 2$ this is seen from [DH]) and contains the global quasi-momentum $k(\lambda)$ for the free operator h_0 (see §2). Suppose that

$$q(x) = q_0 x^{-\alpha} + o(x^{-\alpha}), \qquad x \to \infty, \tag{1.3}$$

with some $q_0 \in \mathbb{R}$, $\alpha > 0$ and denote

$$A(\lambda; q_0, \alpha) = \begin{cases} \pi^{-1} \int [k(\lambda) - k(\lambda - q_0 t^{-\alpha})]\, dt, & q_0 \geqslant 0; \\ \pi^{-1} \int [k(\lambda - q_0 t^{-\alpha}) - k(\lambda)]\, dt, & q_0 \leqslant 0. \end{cases} \tag{1.4}$$

The following relation (see Theorems 3.5, 3.6)

$$\lim g^{-\frac{1}{\alpha}} M(\lambda, gq) = A(\lambda; q_0, \alpha), \qquad g \to \infty, \tag{1.5}$$

holds, provided that $q \geqslant 0$, $\alpha > 0$ or $q \leqslant 0$, $\alpha < 2$.

One should mention that the case of strictly inner λ is in a sense easier to study than the general one. In particular, as can be seen from [B1], the proof of (1.2) for such λ reduces to the case $p = 0$ treated in [BB] in which only the semi-infinite gap $(-\infty, 0)$ exists. One can also infer from [B1] that the asymptotics (1.2) with inner λ still holds true for an arbitrary bounded (not necessarily periodic) function p. We emphasize again that we prove (1.5) when $\alpha > 2$ and (1.2) for all $\lambda \in [\lambda_1^0, \lambda_2^0]$, including the endpoints λ_1^0 and λ_2^0. It is exactly here that the periodicity of p is used. In the multi-dimensional case for $q \leqslant 0$ the semiclassical asymptotics of $M(\lambda, gq)$ was obtained in [H] for inner λ without assuming the periodicity of p (see also the paper [B1], which allows us to weaken assumptions on q as compared to [H]). An extension of this result to all $\lambda \in [\lambda_1^0, \lambda_2^0]$ for periodic p under a certain additional assumption on the unperturbed operator was announced in [B2].

A multi-dimensional analog of (1.5) for an arbitrary bounded p and strictly inner λ was proved in [ADH] under the condition that the unperturbed operator has a so-called integrated density of states. Let us recall that for periodic p the density of states automatically exists and in the one dimensional case it coincides with $k(\lambda)$ up to a constant factor (see (2.15)).

The next result (see Theorems 3.7, 3.8) is related to the asymptotics of the total multiplicity of eigenvalues $N(\lambda_1, \lambda_2, gq)$ lying in the interval $(\lambda_1, \lambda_2) \subseteq [\lambda_1^0, \lambda_2^0]$. Now the function q is not assumed to have a fixed sign. Under condition (1.3), we obtain the asymptotics

$$\lim g^{-\frac{1}{\alpha}} N(\lambda_1, \lambda_2, gq) = A(\lambda_1, \lambda_2; q_0, \alpha), \qquad g \to \infty, \tag{1.6}$$

where

$$A(\lambda_1, \lambda_2; q_0, \alpha) = \pi^{-1} \int [k(\lambda_2 - q_0 t^{-\alpha}) - k(\lambda_1 - q_0 t^{-\alpha})]\, dt. \tag{1.7}$$

Similarly to (1.2) or (1.5), one may take $\lambda_1 = \lambda_1^0$, $\lambda_2 = \lambda_2^0$ if $\alpha > 2$.

The asymptotic relations (1.2), (1.5), (1.6) can be rewritten in a "semi-classical" form. For instance, by virtue of (1.7), the asymptotics (1.6) admits the representation

$$N(\lambda_1, \lambda_2, gq) \approx \pi^{-1} \iint dk(\lambda)\, dx, \qquad g \to \infty, \tag{1.8}$$

where the double integral is taken over the domain

$$\{\lambda \in \mathbb{R}, x \geqslant 0 \mid \lambda_1 \leqslant \lambda + gq(x) \leqslant \lambda_2\}.$$

Equality (1.8) is similar to the semiclassical Weyl asymptotics describing the behavior of discrete spectra in terms of the phase space volume. But the momentum is replaced by the quasimomentum in (1.8). So it is natural to think of (1.8) as of an extension of Weyl asymptotics to the Hill operator. Further discussion is given in §3.

Let us give a sketch of the main ideas. Assume, as above, that $h(gq)$ is defined on $\mathbb{R}_+$ with a Dirichlet condition at $x = 0$. The proofs of (1.2), (1.5), (1.6) are based on a simple argument, specific for the one-dimensional case, known as the splitting principle. Namely, we use the fact that the numbers $M(\lambda, gq)$ or $N(\lambda_1, \lambda_2, gq)$ will change at most by one if we put an additional Dirichlet condition on the operator $h(gq)$ at a suitable point (which may depend on g in general). This reduces the problem to two problems. The first is to study the spectral asymptotics of the Hill operator on a compact interval. Here we basically apply results due to [BS1] (proof of (1.2)) and [ADH] (proofs of (1.5), (1.6)). The most difficult part is to verify that the infinite "tail" does not contribute to the asymptotics of $M(\lambda, gq)$ (or $N(\lambda_1, \lambda_2, gq)$). We rely on the upper bound for $M(\lambda, gq)$ established in Theorem 3.2. As in [H] and [ADH], to prove it, we apply the Birman-Schwinger principle, which allows us to study the spectral properties of the compact operator $T(\lambda) = |q|^{\frac{1}{2}}(h_0 - \lambda)^{-1}|q|^{\frac{1}{2}}$ instead of those of $h(gq)$. To treat $T(\lambda)$, we construct an explicit representation for the integral kernel of $(h_0 - \lambda)^{-1}$ via the eigenfunctions of h_0. This allows us to use a result from [BS2], where eigenvalue estimates for a compact operator were given in terms of the smoothness of its kernel. This yields the desired estimate for $M(\lambda, gq)$. Strictly speaking, the eigenfunction expansion of h_0 is essential for the endpoints λ_1^0 or λ_2^0 only. For the other λ, the boundedness of $(h_0 - \lambda)^{-1}$ suffices (see §6).

The paper is organized as follows. Section 2 contains necessary information on the theory of compact operators and spectral analysis for the Hill operator. The main results are collected in §3. The crucial upper bound for $M(\lambda, gq)$ (Theorem 3.2) and the asymptotics (1.2) are proved in §4. In §5 we derive an estimate of the number $N(\lambda_1, \lambda_2, gq)$ for the Hill operator defined on a compact interval, needed for proving (1.6). Finally, asymptotics (1.5) and (1.6) are proved in §6. By C and c (with or without indices) we denote various positive constants, whose exact values are of no importance.

§2. Preliminaries

1. We need some elementary facts from the theory of compact operators (see [BS3]). Let $\mathfrak{S}_\infty$ be the class of compact operators, λ_n^+ $(-\lambda_n^-)$, $n \geqslant 1$, be positive (resp. negative) eigenvalues of a selfadjoint operator $A \in \mathfrak{S}_\infty$, arranged in descending order, counting multiplicity, $n_\pm(\lambda, A) = \#\{\lambda_n^\pm \mid \lambda_n^\pm > \lambda\}$ be their distribution function. For any $A \in \mathfrak{S}_\infty$ define the singular values (s-values) by $s_n(A) = [\lambda_n^+(A^*A)]^{\frac{1}{2}} = [\lambda_n^+(AA^*)]^{\frac{1}{2}}$, $n \geqslant 1$, and their distribution function by $n(s, \lambda) = \#\{s_n \mid s_n > s\}$. It is clear that

$$n(s, A) = n_+(s^2, A^*A) = n_+(s^2, AA^*). \tag{2.1}$$

Denote by $\nu(s, A)$ any of the functions $n_+(s, A)$, $n_-(s, A)$, $n(s, A)$. The inequality

$$\nu(s_1+s_2, A_1+A_2) \leqslant \nu(s_1, A_1) + \nu(s_2, A_2),$$
$$s_1>0, \quad s_2>0; \quad A_1, A_2 \in \mathfrak{S}_\infty, \tag{2.2}$$

is valid. By $\mathfrak{S}_1 \subset \mathfrak{S}_\infty$ we denote the trace class with norm $\|A\|_1 = \sum s_n(A)$. One has the bound

$$n(s, A) \leqslant s^{-1}\|A\|_1. \tag{2.3}$$

2. Let $p=p(x)$, $x\in\mathbb{R}$, be a bounded periodic function, $p(x)=p(x+1)$, $\Delta\subseteq\mathbb{R}$ be an interval (compact or not), $h_0 = -\frac{d^2}{dx^2}+p(x)$ be a selfadjoint operator in $L^2(\Delta)$ with Dirichlet boundary conditions at the endpoints of Δ, $\sigma(h_0)$ be the spectrum of h, $r_0(\lambda)=(h_0-\lambda)^{-1}$, $\lambda\notin\sigma(h_0)$ be its resolvent; for $p=0$ set $h_{00}=h_0$, $r_{00}(\lambda)=r_0(\lambda)$. Let $q\in L^1_{\mathrm{loc}}(\Delta)$ be a real-valued function decreasing as $|x|\to\infty$. The perturbed operator $h(q)=h_0+q$ may have at most a discrete spectrum on any interval (λ_1, λ_2), disjoint with the essential spectrum of h_0. Denote by $N(\lambda_1, \lambda_2, q)$ the total multiplicity of the eigenvalues lying on (λ_1, λ_2). If $\lambda_1=-\infty$, set $N(\lambda, q)=N(\lambda_1, \lambda, q)$. We introduce the coupling constant g into the notations and, for potentials of fixed sign, we define the function $M(\lambda, gq)$ as in §1. If $p=0$, we write $\mathring{N}(\lambda_1, \lambda_2, q)$ and $\mathring{M}(\lambda, q)$ in place of $N(\lambda_1, \lambda_2, q)$ and $M(\lambda, q)$ respectively.

When studying the function $M(\lambda, q)$, we use its relationship with the spectrum of the compact operator $T(\lambda)=T(\lambda, q)=|q|^{\frac12} r_0(\lambda)|q|^{\frac12}$; $\mathring{T}(\lambda)=T(\lambda)$ if $p=0$. If $\lambda\in(\lambda_1, \lambda_2)$ and λ is not an eigenvalue of h_0, then

$$M(\lambda, q) = n_+(1, T(\lambda, q)), \qquad q\leqslant 0; \tag{2.4$_+$}$$
$$M(\lambda, q) = n_-(1, T(\lambda, q)), \qquad q\geqslant 0. \tag{2.4$_-$}$$

By means of (2.4) one can show that for $q\leqslant 0$ ($q\geqslant 0$) the function $M(\lambda, q)$ is left- (right-) semicontinuous. For bounded Δ the following simple equalities hold

$$\begin{aligned} M(\lambda, q) &= N(\lambda, q) - N(\lambda, 0), && q\leqslant 0;\\ M(\lambda, q) &= \lim_{\lambda'\to\lambda+0}[N(\lambda', 0) - N(\lambda', q)], && q\geqslant 0. \end{aligned} \tag{2.5}$$

Note also the following relation between $M(\lambda, q)$ and $N(\lambda_1, \lambda_2, q)$:

$$\Big|N(\lambda_1, \lambda_2, q) - |M(\lambda_2, q) - M(\lambda_1, q)|\Big| \leqslant N(\lambda_1, \lambda_2, 0) + 1. \tag{2.6}$$

Sometimes we shall specify the dependence of various quantities on Δ in the notation: for example, $h(q;\Delta)$, $M(\lambda, q;\Delta)$, etc.

Set $\Delta_1=(a, b)$, $\Delta_2=[b, c)$, $-\infty\leqslant a<b<c\leqslant\infty$, and $\Delta_0=\Delta_1\cup\Delta_2$. The *splitting principle* holds:

$$\Big|N(\lambda_1, \lambda_2, q;\Delta_0) - [N(\lambda_1, \lambda_2, q;\Delta_1) + N(\lambda_1, \lambda_2, q;\Delta_2)]\Big| \leqslant 1, \tag{2.7}$$

$$\Big|M(\lambda, q;\Delta_0) - [M(\lambda, q;\Delta_1) + M(\lambda, q;\Delta_2)]\Big| \leqslant 1. \tag{2.8}$$

The first inequality is well known. The second one follows directly from (2.4) if λ is not an eigenvalue of $h_0(\Delta_i)$, $i = 0, 1, 2$. For the other λ it is proved by using left- or right-semicontinuity of $M(\lambda, q)$ in λ.

3. Here we recall some well-known results of spectral analysis for the the Hill operator $h_0 = h_0(\mathbb{R})$ (see [RS]). Let us consider the following set of regular boundary problems on the interval $[0, 1)$:

$$-u'' + pu = \lambda u, \qquad \frac{u(1)}{u(0)} = \frac{u'(1)}{u'(0)} = e^{ik}, \tag{2.9}$$

where $k \in \mathbb{R}$ is the quasimomentum. Let $\psi_s = \psi_s(x, k)$, $\lambda_s = \lambda_s(k)$, $\lambda_1 \leqslant \lambda_2 \leqslant \cdots$, $s = 1, 2, \dots$, be normalized eigenfunctions and eigenvalues of the problem (2.9). Obviously $\psi_s(x, k) = \overline{\psi_s(x, -k)}$, $\lambda_s(k) = \lambda_s(-k)$. The functions $\psi_s(x, k)$, $\lambda_s(k)$ are periodic in k with period 2π, continuous on $\mathbb{R}$ and real-analytic on the open intervals $(-\pi, 0)$, $(0, \pi)$. Note that

$$(-1)^{s-1}\frac{d\lambda_s}{dk} > 0, \qquad k \in (0, \pi), \quad s = 1, 2, \dots. \tag{2.10}$$

The union $\sigma = \bigcup_s \sigma_s$ of intervals

$$\sigma_s = \begin{cases} [\lambda_s(0), \lambda_s(\pi)], & s = 2l - 1; \\ [\lambda_s(\pi), \lambda_s(0)], & s = 2l, \quad l \geqslant 1, \end{cases}$$

coincides with the spectrum of h_0. Without loss of generality we may assume $\lambda_1(0) = 0$. The intervals σ_s and

$$l_s = \begin{cases} (\lambda_s(\pi), \lambda_{s+1}(\pi)), & s = 2l - 1; \\ (\lambda_s(0), \lambda_{s+1}(0)), & s = 2l, \end{cases}$$

are referred to as *spectral bands* and *spectral gaps* respectively. The semi-infinite interval $(-\infty, 0) = l_0$ is considered as being the zeroth gap. In general the edges of some (at least two) bands σ_s and σ_{s+1} can merge with each other. In this case $l_s = \varnothing$.

Suppose l_s is a nonempty gap and $k_0 = 0$ if $s = 2l$, $k_0 = \pi$ if $s = 2l - 1$; let $\lambda(k)$ be any of the functions $\lambda_s(k)$, $\lambda_{s+1}(k)$; and $\lambda^0 = \lambda(k_0)$ be the edge of the gap. The function $\lambda(k)$ is real-analytic in a neighborhood of k_0, has a nondegenerate extremum at k_0, and satisfies the relation

$$\lambda(k) - \lambda^0 = (2m)^{-1}(k - k_0)^2(1 + o(1)), \qquad k \to k_0, \tag{2.11}$$

where $m = [d^2\lambda(k_0)/dk^2]^{-1}$ is the *effective mass* associated with the edge λ^0. Here $m > 0$ (resp. $m < 0$) if λ^0 is the right (resp. left) edge.

By (2.10) for any $s \geqslant 1$ there exists a function $k_s(\lambda)$, $\lambda \in \sigma_s$, which is inverse to $\lambda_s(k)$ and infinitely differentiable inside σ_s. We define via $k_s(\lambda)$ the *global quasimomentum* $k = k(\lambda)$, $\lambda \in \mathbb{R}$ (see [Sh]). Namely, for $\lambda \in l_s$ we set

$$k(\lambda) = \pi s, \tag{2.12}$$

and for $\lambda \in \sigma_s$

$$k(\lambda) = \begin{cases} k_s(\lambda) + \pi(s-1), & s = 2l-1, \\ -k_s(\lambda) + \pi s, & s = 2l. \end{cases}$$

The function $k(\lambda)$ is obviously continuous and by virtue of (2.10) nondecreasing. By (2.11), near the edge λ^0 of a nonempty gap, the quasimomentum $k(\lambda)$ satisfies the bound

$$|k(\lambda) - k(\lambda^0)| \leqslant C|\lambda - \lambda^0|^{\frac{1}{2}}, \qquad |\lambda - \lambda^0| \leqslant 1. \tag{2.13}$$

In what follows we shall use the relationship of $k(\lambda)$ with the (*integrated*) *density of states* $\rho(\lambda)$ for the operator h_0. This quantity is defined for all $\lambda \in \mathbb{R}$ as the limit

$$\rho(\lambda) = \lim \frac{N(\lambda, 0; \Delta)}{|\Delta|}, \qquad |\Delta| \to \infty \tag{2.14}$$

($|\Delta|$ is the length of Δ). This limit is known to exist (see [Sh]) and to satisfy

$$\rho(\lambda) = \pi^{-1} k(\lambda). \tag{2.15}$$

Moreover,

$$\rho(\lambda) = \pi^{-1}\lambda^{\frac{1}{2}}[1 + O(\lambda^{-1})], \qquad \lambda \to \infty. \tag{2.16}$$

Now let us describe the eigenfunction expansion for the operator h_0. It is clear that the functions $\psi_s(x, k)$ can be presented as

$$\psi_s(x, k) = e^{ikx}\omega_s(x, k), \tag{2.17}$$

$\omega_s(\cdot, k)$ being periodic functions, so we may extend $\psi_s(x, k)$ to all $x \in \mathbb{R}$. Let P_s be the spectral projection of h_0 associated with the band σ_s. The map

$$(U_s u)(k) = (2\pi)^{-\frac{1}{2}} \int_{-\infty}^{\infty} \overline{\psi_s(x, k)} u(x)\, dx$$

defines a unitary operator from $P_s L^2(\mathbb{R})$ to $L^2(-\pi, \pi)$, which diagonalizes $h_0 P_s$. Namely,

$$(U_s h_0 P_s u)(k) = \lambda_s(k)(U_s P_s u)(k).$$

Let l_s be a nonempty gap and let $\omega(x, k)$ denote any of the functions $\omega_s(x, k)$ or $\omega_{s+1}(x, k)$. In the next section we shall use the following bounds for $\omega(x, k) - \omega(x, k_0)$:

$$\begin{gathered} \int_0^1 |\omega(x, k) - \omega(x, k_0)|^2\, dx \leqslant C(k - k_0)^2, \\ \int_0^1 |\omega'(x, k) - \omega'(x, k_0)|^2\, dx \leqslant C(k - k_0)^2, \end{gathered} \tag{2.18}$$

where C depends on s only (prime means differentiation with respect to x). These bounds are consequences of the analyticity of $\omega(x, k)$ in k near k_0.

In conclusion we make a remark about the values of $M(\lambda, q; \Delta)$ for $\Delta = \mathbb{R}_+$ or $\Delta = \mathbb{R}$ at the endpoints of a nonempty gap l_s (recall that the essential spectrum of the operator $h_0(\mathbb{R}_+)$ coincides with σ). Let λ_2^0 be its right edge. If $s \geqslant 1$ (i.e. the gap is bounded), λ_1^0 denotes its left edge. If $s = 0$, then λ_1^0 is an arbitrary number, $-\infty < \lambda_1^0 < \lambda_2^0$. By (2.7), the discrete spectrum of $h_0(\mathbb{R}_+)$ in every gap may consist of at most one eigenvalue. Thus, by virtue of (2.4), the function $M(\lambda, q; \Delta)$ is monotone near the edge of the gap for both $\mathbb{R}_+$ and $\mathbb{R}$. Hence the following limits (finite or infinite) exist:

$$M(\lambda_i^0, gq) = \lim_{\lambda \to \lambda_i^0} M(\lambda, gq), \qquad i = 1, 2, \quad \lambda \in (\lambda_1^0, \lambda_2^0).$$

We consider these equalities as a definition of $M(\cdot, q)$ at the endpoints.

§3. Results

Our aim is to study the behavior of $M(\lambda, gq; \Delta)$ and $N(\lambda_1, \lambda_2, gq; \Delta)$ as $g \to \infty$ and $\lambda, \lambda_1, \lambda_2 \in [\lambda_1^0, \lambda_2^0]$, when $\Delta = \mathbb{R}_+$ or $\mathbb{R}$. As we mentioned in §1, we may restrict ourselves to the case $\Delta = \mathbb{R}_+$. Results for $\Delta = \mathbb{R}$ follow immediately by (2.7), (2.8). Unless otherwise stated, we assume $\Delta = \mathbb{R}_+$ and omit Δ from notation.

To start with we make

ASSUMPTION 3.1. *There exists a nondecreasing function $\varphi = \varphi(x) > 0$, such that*

$$\int (\varphi|q| + \varphi^{-1})\, dx < \infty.$$

This assumption first appeared in [BB] in connection with the (negative) discrete spectrum of the Schrödinger operator.

For an arbitrary interval $\Delta \subseteq \mathbb{R}$, we introduce the notation

$$I(q; \Delta) = \int_\Delta |q(x)| \varphi(x)\, dx, \qquad J(\Delta) = \int_\Delta (\varphi(x))^{-1}\, dx,$$
$$\Theta(q; \Delta) = \sum_{l=0}^{\infty} \left[\int_l^{l+1} |q(x)| \chi_\Delta(x)\, dx \right]^{\frac{1}{2}},$$

where $\chi_\Delta(\cdot)$ is the characteristic function of Δ. One can easily check that $\Theta(q; \Delta)$ is finite for an arbitrary Δ if Assumption 3.1 is satisfied.

We shall use the following estimate for the discrete spectrum of $h_{00} + q$ obtained in [BB]:

$$\mathring{N}(0, q; \Delta) = \mathring{M}(0, q; \Delta) \leqslant C I^{\frac{1}{2}}(q, \Delta) J^{\frac{1}{2}}(\Delta), \qquad q \leqslant 0. \tag{3.1}$$

Assumption 3.1 is clearly fulfilled if, for example,

$$\int (1 + |x|)^{1+\beta} |q(x)|\, dx < \infty \tag{3.2}$$

with some $\beta > 0$. The other sufficient condition is that $|q(\cdot)|$ does not increase and

$$\int |q(x)|^{\frac{1}{2}}\, dx < \infty.$$

According to [Kh], [BB], Assumption 3.1 guarantees the finiteness of the discrete spectrum of $h(q)$ in the gap. However, this follows independently from the next theorem, which yields an estimate for $M(\lambda, q)$ similar to (3.1).

THEOREM 3.2. *Let q be a function with fixed sign satisfying Assumption 3.1. Then given an interval $\Delta \subseteq \mathbb{R}_+$ (compact or not) the estimate*

$$M(\lambda, q; \Delta) \leqslant C\left[I^{\frac{1}{2}}(q; \Delta)J^{\frac{1}{2}}(\Delta) + \Theta(q; \Delta) + 1\right] \tag{3.3}$$

holds, where the constant C does not depend on $\lambda \in [\lambda_1^0, \lambda_2^0]$, q, and φ.

Theorems 3.3–3.6 give asymptotics of the function $M(\lambda, gq)$. First we treat the case $q \leqslant 0$.

THEOREM 3.3. *Let $q \leqslant 0$ and Assumption 3.1 be fulfilled. Then the asymptotics (1.2) holds uniformly in $\lambda \in [\lambda_1^0, \lambda_2^0]$.*

Further results depend explicitly on the behavior of q at infinity. Let condition (1.3) be fulfilled, $A(\lambda; q_0, \alpha)$ and $A(\lambda_1, \lambda_2; q_0, \alpha)$ be defined by (1.4) and (1.7) respectively. If $\alpha > 2$, then the function q satisfies (3.2) (with $\beta < \alpha - 2$), so that the discrete spectrum in the gap is finite. When $\alpha \leqslant 2$, the discrete spectrum in $[\lambda_1^0, \lambda_2^0]$ is in general infinite. A closed interval $\delta \subseteq [\lambda_1^0, \lambda_2^0]$ is said to be *admissible* if it contains finitely many discrete eigenvalues for all g. It is clear that for $q \geqslant 0$ (resp. $q \leqslant 0$) and $\alpha \leqslant 2$ an arbitrary interval $\delta \subseteq (\lambda_1^0, \lambda_2^0]$ (resp. $\delta \subseteq [\lambda_1^0, \lambda_2^0)$) is admissible.

In what follows q is assumed to satisfy (1.3). Asymptotic relations in the following theorems are uniform in $\lambda, \lambda_1, \lambda_2 \in \delta$, where δ is an admissible interval.

THEOREM 3.4. *If $q \leqslant 0$ and $\alpha = 2$, then*

$$\lim (g^{\frac{1}{2}} \ln g)^{-1} M(\lambda, gq) = (2\pi)^{-1} |q_0|^{\frac{1}{2}}, \qquad g \to \infty. \tag{3.4}$$

THEOREM 3.5. *If $q \leqslant 0$ and $\alpha < 2$, then the asymptotics (1.5) holds.*

The asymptotics (3.4) may be regarded as an intermediary one between $\alpha > 2$ and $\alpha < 2$. On one hand, as in the case $\alpha > 2$ (when (1.2) holds), the right-hand side of (3.4) does not depend on λ. On the other hand, as in the case $\alpha < 2$, it is determined by q_0 only.

In the case $q \geqslant 0$, the behavior of $M(\lambda, gq)$ is described by a common formula for all $\alpha > 0$:

THEOREM 3.6. *If $q \geqslant 0$, then the asymptotics (1.5) holds for all $\alpha > 0$.*

The next theorems give the asymptotics of $N(\lambda_1, \lambda_2, gq)$.

THEOREM 3.7. *Let* $\alpha \geqslant 2$. *Suppose that* $q \leqslant 0$ *and the following conditions are fulfilled*:

(1) q *is a differentiable function with uniformly bounded derivative*;
(2) $\operatorname{supp} q$ *contains finitely many roots* $x_1, x_2, \dots, x_m$, $m < \infty$, *of* q;
(3) *the derivative* $q'(x)$ *obeys the estimates*

$$|q'(x)| \leqslant C|q(x)|, \qquad x \geqslant r > 0; \tag{3.5}$$

$$|q'(x)| \leqslant C|x - x_l|^{-1}|q(x)|, \qquad |x - x_l| \leqslant 1, \quad l = 1, 2, \dots, m. \tag{3.6}$$

Then the asymptotics (1.6) *holds.*

In the first assertion of the following theorem q is allowed to change its sign.

THEOREM 3.8. *If* $\alpha < 2$ *and* $q_0 \neq 0$, *then the asymptotics* (1.6) *holds. If* $q \geqslant 0$, *then* (1.6) *is valid for all* $\alpha > 0$.

It is trivially seen from (2.6) and (1.4), (1.7) that for $q \geqslant 0$ (or $q \leqslant 0$, $\alpha < 2$) relation (1.6) is a direct consequence of (1.5). On the contrary, in the case $q \leqslant 0$, $\alpha > 2$, the asymptotics (1.2) implies only that $N(\lambda_1, \lambda_2, g) = o(g^{\frac{1}{2}})$, $g \to \infty$. In fact, as we shall see in §5, the verification of (1.6) for $q \leqslant 0$, $\alpha > 2$ is similar to the calculation of the second term in the asymptotics of $M(\lambda, gq)$.

REMARK 3.9. Under the assumptions of Theorems 3.5, 3.6 or 3.7, 3.8, the right-hand sides of (1.5) or (1.6) are finite. Let us prove for example that $A(\lambda_1, \lambda_2; q_0, \alpha) < \infty$ if q satisfies the conditions of Theorem 3.7. By (2.15), (2.16)

$$k(\lambda_i + |q_0|t^{-\alpha}) = |q_0|^{\frac{1}{2}} t^{-\frac{\alpha}{2}} + O(t^{\frac{\alpha}{2}}), \qquad t \to 0, \tag{3.7}$$

so that

$$k(\lambda_2 + |q_0|t^{-\alpha}) - k(\lambda_1 + |q_0|t^{-\alpha}) = O(t^{\frac{\alpha}{2}}), \qquad t \to 0.$$

Thus, the integrand in (1.7) is an L^1-function in a neighborhood of the origin. If $\lambda_2 < \lambda_2^0$, then (2.12) implies

$$k(\lambda_1 + |q_0|t^{-\alpha}) = k(\lambda_2) = k(\lambda_2 + |q_0|t^{-\alpha}) \tag{3.8}$$

for t sufficiently large. Thus, the integral is trivially finite. When $\lambda_2 = \lambda_2^0$ and $\alpha > 2$, the left equality in (3.8) and (2.13) ensure that

$$k(\lambda_2^0 + |q_0|t^{-\alpha}) - k(\lambda_1 + |q_0|t^{-\alpha}) = k(\lambda_2^0 + |q_0|t^{-\alpha}) - k(\lambda_2^0) = O(t^{-\frac{\alpha}{2}})$$

as $t \to \infty$. Hence the integral (1.7) converges in this case as well.

REMARK 3.10. Though $A(\lambda_1, \lambda_2; q_0, \alpha) < \infty$, the asymptotic coefficients $A(\lambda_i; q_0, \alpha)$, $i = 1, 2$, are infinite under the assumptions of Theorem 3.7. Indeed, for $\alpha > 2$ the right-hand side of (3.7) is not integrable near the origin. This agrees with the fact that for $q \leqslant 0$, $\alpha \geqslant 2$ the behavior of $M(\lambda, gq)$

is governed by the relations (1.2) or (3.4) but not by (1.5), although the asymptotics (1.6) of $N(\lambda_1, \lambda_2, gq)$ holds for all $\alpha > 0$.

As we mentioned in the Introduction, relations (1.2), (1.5), and (1.6) can be thought of as Weyl type asymptotics. In particular, (1.6) was reduced to (1.8). Similarly, the asymptotics (1.5) can be presented as

$$M(\lambda, gq) \approx \pi^{-1} \iint dk(\lambda')\, dx, \qquad g \to \infty, \tag{3.9}$$

where the double integral is taken over the domain

$$\{\lambda' \in \mathbb{R}, x \geqslant 0 \mid \lambda - gq(x) \leqslant \lambda' \leqslant \lambda\}, \qquad \text{if } q \geqslant 0,$$
$$\{\lambda' \in \mathbb{R}, x \geqslant 0 \mid \lambda \leqslant \lambda' \leqslant \lambda - gq(x)\}, \qquad \text{if } q \leqslant 0.$$

Note that the asymptotics (1.2) can also be rewritten in the form (3.9). We do not give a complete proof here but show this for a finite range potential $q(x) \leqslant 0$, assuming for simplicity that $q(x) \leqslant -C$, $0 \leqslant x \leqslant a$, and $q(x) = 0$, $x > a$. Then the right-hand side of (3.9) equals

$$\pi^{-1} \int_0^a dx\, [k(\lambda + g|q(x)|) - k(\lambda)].$$

Using (2.15), (2.16), we obtain (1.2).

§4. Proofs of Theorems 3.2 and 3.3

1. Proof of Theorem 3.2. By (2.8) it suffices to prove the bound (3.3) for $M(gq, \lambda) = M(gq, \lambda; \mathbb{R})$, where $q = q\chi_\Delta$. To be definite, we suppose that $q \leqslant 0$ and the gap under consideration is even, i.e. $s = 2l$, $l \geqslant 0$, so that $k_0 = 0$. Set

$$\lambda(k) = \lambda_{s+1}(k), \qquad \lambda^0 = \lambda_2^0, \qquad P = P_{s+1},$$
$$P^+ = \sum_{l=s+1}^{\infty} \oplus P_l, \qquad Q = P^+ - P, \qquad U = U_{s+1}.$$

Since the Hill operator has no discrete spectrum in gaps, the relation (2.4_+) holds for any $\lambda \in (\lambda_1, \lambda_2)$. The right-hand side of (2.4_+) will only increase if the resolvent $r_0(\lambda)$ is replaced by its positive part $r_0(\lambda)P^+ = r_0(\lambda)P + r_0(\lambda)Q$. Set

$$T_1 = T_1(\lambda) = |q|^{\frac{1}{2}} r_0(\lambda) P |q|^{\frac{1}{2}} \geqslant 0,$$
$$T_2 = T_2(\lambda) = |q|^{\frac{1}{2}} r_0(\lambda) Q |q|^{\frac{1}{2}} \geqslant 0.$$

According to (2.2), it is sufficient to prove the bound (3.3) for the functions $n_+(s, T_i)$, $i = 1, 2$, with $s > 0$ fixed, separately.

At first let $i = 2$. The obvious estimates $\|(h_0 + 1) r_0(\lambda) Q\| \leqslant C$ and $\|(h_{00} + 1)^{\frac{1}{2}} (h_0 + 1)^{-\frac{1}{2}}\| \leqslant C$ lead to the inequality $T_2(\lambda) \leqslant C\mathring{T}(-1)$ and consequently

$$n_+(s, T_2(\lambda)) \leqslant n_+(s, C\mathring{T}(-1)). \tag{4.1}$$

Using (2.4_+) again, we conclude that the right-hand side of inequality (4.1) equals $\mathring{N}(-1, -s^{-1}C|q|)$. Since $q = q\chi_\Delta$, the latter quantity is bounded from above by $\mathring{N}(-1, -s^{-1}C|q|; \Delta) + 2$. By (3.1), this implies

$$n_+(s, T_2(\lambda)) \leqslant CI^{\frac{1}{2}}(q; \Delta)J^{\frac{1}{2}}(\Delta) + 2. \qquad (4.2)$$

Let $i = 1$. Using the operator U, one can rewrite $T_1(\lambda)$ in the form

$$T_1(\lambda) = |q|^{\frac{1}{2}} U^* (\lambda(k) - \lambda)^{-1} U |q|^{\frac{1}{2}}.$$

Thus by virtue of (2.1) it suffices to estimate the s-values of the operator

$$|q|^{\frac{1}{2}} U^* (\lambda(k) - \lambda)^{-\frac{1}{2}}, \qquad (4.3)$$

acting from $L^2(-\pi, \pi)$ to $L^2(\mathbb{R})$. The relation (2.11) implies that $\lambda(k) \geqslant \varkappa k^2 + \lambda^0$ with some $\varkappa > 0$. Therefore s-values of (4.3) do not exceed (up to a constant factor) those of the operator

$$B = B(\eta) = |q|^{\frac{1}{2}} U^*(k^2 + \eta)^{-\frac{1}{2}}, \qquad \eta = \varkappa^{-1}(\lambda^0 - \lambda) > 0.$$

By (2.17) B can be represented as the sum of two operators

$$B_1(\eta), B_2(\eta) \colon L^2(-\pi, \pi) \to L^2(\mathbb{R})$$

with kernels

$$(2\pi)^{-\frac{1}{2}} |q(x)|^{\frac{1}{2}} \omega(x, 0) e^{ikx} (k^2 + \eta)^{-\frac{1}{2}},$$
$$(2\pi)^{-\frac{1}{2}} |q(x)|^{\frac{1}{2}} f(x, k)(k^2 + \eta)^{-\frac{1}{2}},$$

where $f(x, k) = [\omega(x, k) - \omega(x, 0)]e^{ikx}$. Since $\omega(\cdot, 0)$ is bounded, we have

$$n(s, B_1(\eta)) \leqslant n_+(s^2, C\mathring{T}(-\eta))$$

by (2.1). Now relation (2.4_+) and inequality (3.1) ensure that

$$n(s, B_1(\eta)) \leqslant CI^{\frac{1}{2}}(q; \Delta)J^{\frac{1}{2}}(\Delta) + 2. \qquad (4.4)$$

Let us show that $B_2(\eta) \in \mathfrak{S}_1$ and

$$n(s, B_2(\eta)) \leqslant C\Theta(q; \Delta). \qquad (4.5)$$

To that end, for the operators $B_2^{(l)} = \chi_{[l, l+1)} B_2(\eta)$, $l \in \mathbb{Z}$, we establish the upper bound

$$\|B_2^{(l)}\|_1 \leqslant C\left[\int_l^{l+1} |q(x)|\, dx\right]^{\frac{1}{2}}, \qquad (4.6)$$

where C does not depend on l. The desired estimate (4.4) follows from (4.6) by the triangle inequality for the trace norm and (2.3). By virtue of (2.18), the norm of the function $f(\cdot, k)$ in the Sobolev space $\mathscr{H}_1 = \mathscr{H}_1(l, l+1)$ satisfies the estimate

$$\|f(\cdot, k)\|^2_{\mathscr{H}_1} \leqslant Ck^2 \quad (C \text{ does not depend on } l).$$

According to [BS2] (Proposition 2.1),

$$s_n(B_2^{(l)}) \leqslant Cn^{-\frac{3}{2}}\left[\int_l^{l+1}|q(x)|\,dx\right]^{\frac{1}{2}}\left[\int_{-\pi}^{\pi}(k^2+\eta)^{-1}\|f(\cdot\,,k)\|_{\mathscr{H}_1}^2\,dk\right]^{\frac{1}{2}}$$

$$\leqslant Cn^{-\frac{3}{2}}\left[\int_l^{l+1}|q(x)|\,dx\right]^{\frac{1}{2}}, \qquad l\in\mathbb{Z}.$$

This leads to (4.6) and, consequently, to (4.5). Combining (4.2), (4.4), and (4.5), we get (3.3). The proof is complete. □

2. Proof of Theorem 3.3. Let $R>0$ be an arbitrary number and $\Delta_1 = \Delta_1(R) = (0\,,R)$, $\Delta_2 = \Delta_2(R) = [R\,,\infty)$. By (2.8) the problem reduces to a separate examination of $M(\lambda\,,gq\,;\Delta_1)$ and $M(\lambda\,,gq\,;\Delta_2)$. Theorem 3.2 says that

$$M(\lambda\,,gq\,;\Delta_2) \leqslant C_1 g^{\frac{1}{2}}[I^{\frac{1}{2}}(q\,;\Delta_2)J^{\frac{1}{2}}(\Delta_2)+\Theta(q\,;\Delta_2)]+C_2.$$

Here the coefficient by $g^{\frac{1}{2}}$ tends to zero as $R\to\infty$. Thus,

$$\limsup_{g\to\infty} g^{-\frac{1}{2}}M(\lambda\,,gq\,;\Delta_2)=o(1)\,, \qquad R\to\infty. \tag{4.7}$$

Let us evaluate the asymptotics of $M(\lambda\,,gq\,;\Delta_1)$. We have by (2.5)

$$M(\lambda\,,gq\,;\Delta_1) = N(\lambda\,,gq\,;\Delta_1) - N(\lambda\,,0\,;\Delta_1).$$

Since the second summand does not depend on g, it remains to study the first one. The asymptotics of $N(\lambda\,,gq\,;\Delta_1) = \mathring{N}(0\,,p-\lambda+gq\,;\Delta_1)$ coincides with that of $\mathring{N}(0\,,gq\,;\Delta_1)$ because $p-\lambda$ is a relatively compact perturbation of $h_{00}(\Delta_1)$. This allows us to apply a result from [BS1] which says that

$$\lim g^{-\frac{1}{2}}\mathring{N}(0\,,gq\,;\Delta_1) = \pi^{-1}\int_{\Delta_1}|q(x)|^{\frac{1}{2}}\,dx\,, \qquad g\to\infty. \tag{4.8}$$

Passing to the limit as $R\to\infty$ and taking into account (4.7), we arrive at (1.2). □

§5. Estimate for $N(\lambda_1\,,\lambda_2\,,gq\,;\Delta)$

1. Here we obtain an efficient bound on the number $N(\lambda_1\,,\lambda_2\,,gq\,;\Delta)$ of eigenvalues for a compact interval $\Delta=[0\,,R)$ as $q\leqslant 0$. In what follows we suppose that the function q satisfies conditions (1)–(3) of Theorem 3.7 and that λ, λ_1, λ_2 are arbitrary real numbers. The asymptotics (1.3) is not assumed.

Theorem 5.1. *The estimate*

$$N(\lambda_1\,,\lambda_2\,,gq\,;\Delta)\leqslant C(\log g+R+1) \tag{5.1}$$

holds, where C *does not depend on* g, R, q *and* λ_1, λ_2, *on a compact interval.*

We shall deduce this result from the following estimate for the distribution function $N(\lambda\,,gq\,;\Delta)$.

LEMMA 5.2. *The estimate*

$$\left| N(\lambda, gq; \Delta) - g^{\frac{1}{2}}\pi^{-1}\int_0^R |q(x)|^{\frac{1}{2}}\,dx \right| \leqslant C(\log g + R + 1) \qquad (5.2)$$

holds, where C *does not depend on* g, R, q *and* λ, *on a compact interval.*

2. The proof of Lemma 5.2 uses the relationship between $N(\lambda, q; \Delta)$ and properties of the solution of the problem

$$-u'' + (p+q)u = \lambda u, \qquad u(0) = 0, \quad u'(0) = 1. \qquad (5.3)$$

Namely, it follows from the theory of ordinary differential equations (see [T]) that $N(\lambda, q; \Delta)$ equals the number of roots $\mu = \mu(p, q, \lambda)$ of the function u lying strictly inside Δ. To calculate μ, we present $u(x)$ in "polar" form. Denote $f(x) = (|q(x)| + 1)^{1/2}$. The equalities

$$u(x) = \beta(x)\sin\xi(x), \qquad u'(x) = \beta(x)f(x)\cos\xi(x) \qquad (5.4)$$

and equation (5.3) with initial conditions define the functions β (amplitude) and ξ (phase) uniquely, provided that ξ is assumed to be continuous. Using (5.3), one can easily derive the equation

$$\xi' = f + f'(2f)^{-1}\sin(2\xi) + (\lambda - 1 - p)f^{-1}\sin^2\xi, \qquad \xi(0) = 0. \qquad (5.5)$$

The function β can be evaluated via ξ:

$$\beta(x) = [f(0)]^{-1}\exp\left\{\int_0^x [(p+1-\lambda)\sin\xi - f'\cos\xi]f^{-1}\cos\xi\,dt\right\}.$$

However, we do not need this explicit formula for β. It is important for us to know only that β does not vanish anywhere. Therefore, by (5.4), the value $\mu(p, q, \lambda)$ equals the number of points x where $\sin\xi(x) = 0$. In other words, μ coincides with the total number of roots of the equation $\xi(x) = 0 \pmod{\pi}$. According to (5.5), we have $\xi'(x) = f(x) > 0$ for those x, so that for any fixed integer l the function ξ can take the value πl once only. Thus

$$\pi^{-1}\xi(R) - 1 \leqslant \mu \leqslant \pi^{-1}\xi(R).$$

Since $N(\lambda, q; \Delta) = \mu(p, q, \lambda)$, it follows from (5.5) that

$$\begin{aligned} &\left| N(\lambda, q; \Delta) - \pi^{-1}\int_0^R |q(x)|^{\frac{1}{2}}\,dx \right| \\ &\leqslant (2\pi)^{-1}\int_0^R |f'(x)[f(x)]^{-1}|\,dx + \pi^{-1}\int_0^R (|\lambda| + 2 + |p(x)|)\,dx + 1 \\ &\leqslant (4\pi)^{-1}\int_0^R |q'(x)|(|q(x)|+1)^{-1}dx + \pi^{-1}R(|\lambda| + 2 + \sup_x |p(x)|) + 1. \end{aligned} \qquad (5.6)$$

3. PROOF OF LEMMA 5.2. By (2.7) it suffices to establish (5.2) for q having only one root $x_1 = 0$. Together with (3.5), this implies that $|q'(x)||q(x)|^{-1}$

is bounded from above uniformly in $x \geqslant 1$. Thus,

$$g \int_1^R |q'(x)|(g|q(x)|+1)^{-1}\,dx \leqslant CR. \tag{5.7}$$

Let us estimate a similar integral over the interval $(0, 1)$. It follows from the boundedness of $q'(x)$ and (3.6) that

$$g|q'(x)|(g|q(x)|+1)^{-1} \leqslant C_1 g, \qquad x \leqslant \varepsilon,$$
$$g|q'(x)|(g|q(x)|+1)^{-1} \leqslant C_2 x^{-1}, \qquad \varepsilon < x \leqslant 1,$$

for an arbitrary $\varepsilon \in (0, 1)$ with C_1, C_2 not depending on ε and q. These inequalities lead to the estimate

$$g \int_0^1 |q'(x)|(g|q(x)|+1)^{-1}\,dx \leqslant C_1 g\varepsilon \sup_{x\leqslant 1}|q'(x)| + C_2(|\log\varepsilon|+1).$$

Setting $\varepsilon = g$ and combining the last inequality with (5.6), (5.7), we obtain (5.2). □

To complete the proof of (5.1), it remains to take into account the obvious inequality

$$\left|N(\lambda_2, gq; \Delta) - N(\lambda_1, gq; \Delta) - N(\lambda_1, \lambda_2, gq; \Delta)\right| \leqslant 1$$

and use (5.2).

§6. Proofs of Theorems 3.4–3.8

1. Throughout this section we assume that q satisfies condition (1.3) for some $\alpha > 0$ and the numbers λ, λ_1, λ_2 belong to an admissible interval δ. Again, by the splitting principle, we may separately consider three intervals

$$\Delta_1 = \Delta_1(\varepsilon) = \{x \leqslant \varepsilon g^{\frac{1}{\alpha}}\},$$
$$\Delta_2 = \Delta_2(\varepsilon, A) = \{\varepsilon g^{\frac{1}{\alpha}} \leqslant x \leqslant A g^{\frac{1}{\alpha}}\},$$
$$\Delta_3 = \Delta_3(A) = \{x \geqslant A g^{\frac{1}{\alpha}}\},$$

with arbitrary ε, A, $0 < \varepsilon < A$.

2. First of all, we prove that Δ_1 and Δ_2 do not contribute to the asymptotics of $M(\lambda, gq)$. Namely, we verify that

$$\limsup_{g\to\infty} g^{-\frac{1}{\alpha}} M(\lambda, gq; \Delta_1(\varepsilon)) = o(1), \qquad \varepsilon \to 0, \tag{6.1}$$

if $q \geqslant 0$ or $q \leqslant 0$, $\alpha < 2$ and

$$\limsup_{g\to\infty} g^{-\frac{1}{\alpha}} M(\lambda, gq; \Delta_3(A)) = o(1), \qquad A \to \infty, \tag{6.2}$$

for any $\alpha > 0$ independently of the sign of q.

As to the interval Δ_2, we check that for any $\Delta \subseteq \Delta_2(\varepsilon, A)$ and $\alpha > 0$, the estimate

$$M(\lambda, gq; \Delta) \leqslant C_1 g^{\frac{1}{\alpha}} + C_2, \qquad C_1 = C_1(\varepsilon, A), \tag{6.3}$$

holds.

We start by proving (6.1). First let $q \geqslant 0$. It follows from (2.5) and from the obvious inequality

$$N(\lambda, 0; \Delta_1) \leqslant C|\Delta_1| \tag{6.4}$$

that $M(\lambda, gq; \Delta_1) \leqslant C\varepsilon g^{\frac{1}{\alpha}}$. Let $q \leqslant 0$, $\alpha < 2$. Using Theorem 3.2 with $\varphi(x) = (1+x)^{\frac{\alpha}{2}}$, we obtain the bound

$$\begin{aligned} M(\lambda, gq; \Delta_1) &\leqslant C\left\{g^{\frac{1}{2}} \int_{\Delta_1} (1+x)^{-\frac{\alpha}{2}} dx + \sum_{0 \leqslant l \leqslant \varepsilon g^{\frac{1}{\alpha}}+1} \left[\int_l^{l+1} |q(x)|\, dx\right]^{\frac{1}{2}} + 1\right\} \\ &\leqslant C(\varepsilon^{1-\frac{\alpha}{2}} g^{\frac{1}{\alpha}} + 1), \qquad q \leqslant 0, \quad \alpha < 2. \end{aligned} \tag{6.5}$$

This gives (6.1).

Now let us verify (6.2). To be definite, we assume $q \leqslant 0$. For $\alpha > 2$, we apply Theorem 3.2 again. It ensures us that

$$\begin{aligned} M(\lambda, gq; \Delta_3) &\leqslant C\left\{g^{\frac{1}{2}} \int_{\Delta_3} (1+x)^{-\frac{\alpha}{2}} dx + \sum_{l \geqslant A g^{\frac{1}{\alpha}}-1} \left[\int_l^{l+1} |q(x)|\, dx\right]^{\frac{1}{2}} + 1\right\} \\ &\leqslant C(A^{1-\frac{\alpha}{2}} g^{\frac{1}{\alpha}} + 1), \end{aligned}$$

so relation (6.2) is valid. In the case $\alpha \leqslant 2$, the interval δ is bounded away from λ_2^0, so that $|\lambda - \lambda_2^0| \geqslant \mu$, $\lambda \in \delta$ for some $\mu > 0$. Now our argument is simpler than in the previous case, since we do not use the upper bound (3.3). Namely, by (2.4_+) and (2.8)

$$M(\lambda, gq; \Delta_3) \leqslant M(\lambda, gq\chi_{\Delta_3}; \mathbb{R}) + 1 = n_+(1, T(\lambda, gq\chi_{\Delta_3}; \mathbb{R})) + 1.$$

Since $r_0(\lambda) \leqslant \mu^{-1}$, the operator $T = T(\lambda, gq\chi_{\Delta_3}; \mathbb{R})$ has the upper bound $g \max_x |q(x)\chi_{\Delta_3}(x)|\mu^{-1} \leqslant CA^{-\alpha}\mu^{-1}$. Taking A sufficiently large, we see that $n_+(1, T) = 0$. Hence, $M(\lambda, gq; \Delta_3) \leqslant 1$ and, certainly, (6.2) holds.

The proof of (6.3) is similar to that of (6.5), and we omit it.

Relations similar to (6.1) and (6.2) are valid for $N(\lambda_1, \lambda_2, gq; \Delta_j)$, $j = 1, 3$ under the conditions of Theorems 3.7 or 3.8:

$$\limsup_{g \to \infty} g^{-\frac{1}{\alpha}} N(\lambda_1, \lambda_2, gq; \Delta_1(\varepsilon)) = o(1), \qquad \varepsilon \to 0, \tag{6.6}$$

$$\limsup_{g \to \infty} g^{-\frac{1}{\alpha}} N(\lambda_1, \lambda_2, gq; \Delta_3(A)) = o(1), \qquad A \to \infty. \tag{6.7}$$

Equality (6.7) follows from (6.2) by (2.6), since the potential q does not change its sign at $x \in \Delta_3$. To verify (6.6) for $q \geqslant 0$, one can use (2.6) and (6.1). If q satisfies the conditions of Theorem 3.7, then (6.6) is a direct consequence of Theorem 5.1 with $R = \varepsilon g^{\frac{1}{\alpha}}$.

Let $\alpha < 2$. We cannot apply (6.1), since q may change its sign. However by (2.5) we have

$$N(\lambda_1, \lambda_2, gq; \Delta_1) \leqslant N(\lambda_2, -g|q|; \Delta_1) \leqslant M(\lambda_2, -g|q|; \Delta_1) + N(\lambda_2, 0; \Delta_1).$$

Taking into account (6.1) and (6.4), we get (6.6).

3. PROOF OF THEOREM 3.4. Let $q \leqslant 0$, $\alpha = 2$. First of all we check that it suffices to establish the following relation:

$$\lim_{g\to\infty} (g^{\frac{1}{2}} \log g)^{-1} M(\lambda, gq_{as}; [1, \infty)) = |q_0|^{\frac{1}{2}} (2\pi)^{-1}, \tag{6.8}$$

where $q_{as}(x) = q_0 x^{-2}$.

Indeed, by (2.8) and Theorem 3.2, the asymptotics of $M(\lambda, gq; \mathbb{R}_+)$ coincides with that of $M(\lambda, gq; \Delta)$ for $\Delta = \Delta(R) = [R, \infty)$, $R > 0$ being an arbitrary number. Since q obeys (1.3), we have

$$\begin{aligned}&\limsup (g^{\frac{1}{2}} \log g)^{-1} M(\lambda, gq; \mathbb{R}_+) \\ &\quad\leqslant \limsup (g^{\frac{1}{2}} \log g)^{-1} M(\lambda, -g(|q_0| + \varkappa)x^{-2}; \Delta) \\ &\quad- \lim (g^{\frac{1}{2}} \log g)^{-1} M(\lambda, -g(|q_0| + \varkappa)x^{-2}; [1, \infty)), \qquad g \to \infty,\end{aligned}$$

and similarly

$$\begin{aligned}&\liminf (g^{\frac{1}{2}} \log g)^{-1} M(\lambda, gq; \mathbb{R}_+) \\ &\quad\geqslant \lim (g^{\frac{1}{2}} \log g)^{-1} M(\lambda, -g(|q_0| - \varkappa)x^{-2}; [1, \infty)), \qquad g \to \infty,\end{aligned}$$

where $\varkappa = \varkappa(R) \to +0$ as $R \to \infty$. Since $\varkappa$ may be chosen arbitrarily small by appropriately choosing R, (6.8) yields (3.4).

Now let us prove (6.8). Note that (6.8) is obvious if $q_0 = 0$, so that we assume $q_0 < 0$. By (6.3) and (6.2), the intervals $\Delta_2(\varepsilon, A)$ and $\Delta_3(A)$ do not contribute to the asymptotics (3.4) for any ε and A. Moreover, by virtue of (2.5) and (6.4), it remains to check that

$$\lim_{g\to\infty} (g^{\frac{1}{2}} \log g)^{-1} N(\lambda, gq_{as}; \overline{\Delta}_1) = (2\pi)^{-1} |q_0|^{\frac{1}{2}}, \tag{6.9}$$

where $\overline{\Delta}_1 = \overline{\Delta}_1(\varepsilon) = [1, \varepsilon g^{\frac{1}{2}})$. Taking into account the boundedness of p, the obvious inequality

$$g|q_{as}(x)| \geqslant |q_0| \varepsilon^{-2},$$

and the equality

$$N(\lambda, gq_{as}; \overline{\Delta}_1) = \mathring{N}(0, p + gq_{as} - \lambda; \overline{\Delta}_1),$$

we get

$$\mathring{N}(0, gq_{as}(1-\mu); \overline{\Delta}_1) \leqslant N(\lambda, gq_{as}; \overline{\Delta}_1) \leqslant \mathring{N}(0, gq_{as}(1+\mu); \overline{\Delta}_1),$$

where $\mu = \mu(\varepsilon) \to +0$ as $\varepsilon \to 0$. Thus, the problem reduces to the calculation of the asymptotics (6.9) for $\mathring{N}(0, gq_{as}; \Delta_1)$. Again, taking into account (6.2) and (6.3), we may assume that $\overline{\Delta}_1 = [1, 2^n)$, where $n = [(\log g)/(2\log 2)]$ ($[\cdot]$ means the integral part). We divide the interval $\overline{\Delta}_1$ by the points $2, \dots, 2^{n-1}$ into n subintervals $\Delta^{(j)} = [2^j, 2^{j+1})$, $j = 0, 1, \dots, n-1$. By virtue of (2.8), the asymptotics will not change if we replace $\mathring{N}(0, gq_{as}; \overline{\Delta}_1)$ by $\sum \mathring{N}(0, gq_{as}; \Delta^{(j)})$. Then, by means of scaling, one can easily verify that $\mathring{N}(0, gq_{as}; \Delta^{(j)}) = \mathring{N}(0, gq_{as}; \Delta^{(0)})$ for any j. Now, using (4.8) for $\Delta^{(0)}$, we get

$$\lim_{g\to\infty} g^{-\frac{1}{2}} \mathring{N}(0, gq_{as}; \Delta^{(0)}) = \pi^{-1}|q_0|^{\frac{1}{2}} \int_1^2 x^{-1}\, dx = \pi^{-1}|q_0|^{\frac{1}{2}} \log 2. \tag{6.10}$$

Eventually we obtain

$$\begin{aligned} &\lim (g^{\frac{1}{2}} \log g)^{-1} \mathring{N}(0, gq_{as}; \overline{\Delta}_1) \\ &= \lim (g^{\frac{1}{2}} \log g)^{-1} \sum_{j=0}^{n-1} \mathring{N}(0, gq_{as}; \Delta^{(j)}) \\ &= \lim (g^{\frac{1}{2}} \log g)^{-1} \frac{\log g}{2\log 2} \mathring{N}(0, gq_{as}; \Delta^{(0)}), \qquad g \to \infty. \end{aligned}$$

Substituting (6.10) into the latter equality, we get (6.9) for $\mathring{N}(0, gq_{as}; \overline{\Delta}_1)$, which yields (6.9). In turn, this leads to (6.8) and, consequently, to (3.4). $\square$

4. Here we evaluate the asymptotics of $M(\lambda, gq; \Delta_2)$ needed for Theorems 3.5–3.8. Our approach basically follows that of [ADH]. We present it here for the sake of completeness. First we shall prove that if q satisfies condition (1.3), then

$$\lim_{g\to\infty} g^{-\frac{1}{\alpha}} N(\lambda, gq; \Delta_2(\varepsilon, A)) = \int_\varepsilon^A \rho(\lambda - q_0 t^{-\alpha})\, dt. \tag{6.11}$$

Just as in the proof of Theorem 3.4, one can check that it suffices to establish (6.11) for $q(x) = q_{as}(x) = q_0 x^{-\alpha}$, $x \geqslant 1$. To be definite, we assume $q_0 \leqslant 0$. We divide the interval $[\varepsilon, A)$ into L identical parts by the points $t_0 = \varepsilon$, $\{t_i\}$, $i = 1, 2, \dots, L-1$, $t_L = A$, so that $t_i - t_{i-1} = (A-\varepsilon)L^{-1}$. For a fixed L, the asymptotics of $N(\lambda, gq; \Delta_2)$ coincides with that of the sum

$$N^L(\lambda, gq) = \sum_{i=1}^{L} N(\lambda, gq; \Delta_2^{(i)}),$$

where $\Delta_2^{(i)} = [t_{i-1} g^{\frac{1}{\alpha}}, t_i g^{\frac{1}{\alpha}})$, $i = 1, 2, \dots, L$. Let us introduce the step functions q_1 and q_2:

$$\begin{aligned} q_1(x) &= q(t_{i-1} g^{\frac{1}{\alpha}}) = q_0 t_{i-1}^{-\alpha} g^{-1}, \qquad && x \in \Delta_2^{(i)}; \\ q_2(x) &= q(t_i g^{\frac{1}{\alpha}}) = q_0 t_i^{-\alpha} g^{-1}, && x \in \Delta_2^{(i)}. \end{aligned}$$

Clearly, $q_1(x) \leqslant q(x) \leqslant q_2(x)$, $x \in \Delta_2$. Hence,

$$N^L(\lambda, gq_2) \leqslant N^L(\lambda, gq) \leqslant N^L(\lambda, gq_1). \tag{6.12}$$

Now we evaluate the asymptotics of $N^L(\lambda, gq_1)$. By definition (2.14)

$$\begin{aligned}\lim_{g\to\infty} g^{-\frac{1}{\alpha}} N(\lambda, gq_1; \Delta_2^{(i)}) &= (t_i - t_{i-1}) \lim_{g\to\infty} |\Delta_2^{(i)}|^{-1} N(\lambda - q_0 t_{i-1}^{-\alpha}, 0; \Delta_2^{(i)}) \\ &= (t_i - t_{i-1})\rho(\lambda - q_0 t_{i-1}^{-\alpha}), \qquad i = 1, 2, \dots, L.\end{aligned}$$

Thus,

$$\lim_{g\to\infty} g^{-\frac{1}{\alpha}} N^L(\lambda, gq_1) = \sum_{i=1}^{L} (t_i - t_{i-1})\rho(\lambda - q_0 t_{i-1}^{-\alpha}).$$

Together with (6.12), this implies

$$\limsup_{g\to\infty} g^{-\frac{1}{\alpha}} N(\lambda, gq, \Delta_2) \leqslant \sum_{i=1}^{L} (t_i - t_{i-1})\rho(\lambda - q_0 t_{i-1}^{-\alpha}) \tag{6.13}$$

for any L. Passing to the limit as $L \to \infty$ and taking into account the continuity of $\rho(\lambda)$ in λ, we find that the left-hand side of (6.13) does not exceed the right-hand side of (6.11).

Using the left inequality (6.12), one can verify in the same way that a similar lower bound for

$$\liminf_{g\to\infty} g^{-\frac{1}{\alpha}} N(\lambda, gq, \Delta_2)$$

is valid. This completes the proof of (6.11).

Now it follows from (6.11) and (2.5), (2.15) that

$$\lim_{g\to\infty} g^{-\frac{1}{\alpha}} M(\lambda, gq; \Delta_2(\varepsilon, A)) = \pi^{-1} \int_\varepsilon^A \{k(\lambda - q_0 t^{-\alpha}) - k(\lambda)\} dt, \qquad q \leqslant 0, \tag{6.14}$$

and

$$\lim_{g\to\infty} g^{-\frac{1}{\alpha}} M(\lambda, gq; \Delta_2(\varepsilon, A)) = \pi^{-1} \int_\varepsilon^A \{k(\lambda) - k(\lambda - q_0 t^{-\alpha})\} dt, \qquad q \geqslant 0. \tag{6.15}$$

Moreover,

$$\lim_{g\to\infty} g^{-\frac{1}{\alpha}} N(\lambda_1, \lambda_2, gq; \Delta_2(\varepsilon, A)) = \pi^{-1} \int_\varepsilon^A \{k(\lambda_2 - q_0 t^{-\alpha}) - k(\lambda_1 - q_0 t^{-\alpha})\} dt. \tag{6.16}$$

5. Proof of Theorems 3.5–3.8. According to Remark 3.9, the integrals (6.14)–(6.16) converge as $\varepsilon \to 0$ and $A \to \infty$. To complete the proof of Theorems 3.5–3.8, it remains to take into account relations (6.1), (6.2) and (6.6), (6.7). □

Acknowledgment

The author is grateful to professor M. Sh. Birman for valuable discussions and his kind attention to this work.

References

[ADH] S. Alama, P. A. Deift, and R. Hempel, *Eigenvalue branches of the Schrödinger operator* $H - \lambda W$ *in a gap of* $\sigma(H)$, Comm. Math. Phys. **121** (1989), 291–321.

[B1] M. Sh. Birman, *Discrete spectrum in the gaps of the continuous one in the large-coupling-constant limit*, Oper. Theory: Adv. Appl. **46** (1990), 17–25.

[B2] ———, *The discrete spectrum in a gap of the perturbed periodic operator at large coupling constants*, Proc. Conf. Rigorous Results in Quantum Dynamics. Liblice Castle, June 11–15, 1990 (to appear).

[BB] M. Sh. Birman and V. V. Borzov, *On the asymptotics of the discrete spectrum for certain singular differential operators*, Probl. Mat. Fiz. **5** (1971), 24–38; English transl. in Topics in Math. Phys. **5** (1972).

[BS1] M. Sh. Birman and M. Z. Solomyak, *On the principal term of the spectral asymptotics for "non-smooth" elliptic problems*, Funktsional. Anal. i Prilozhen. **4** (1970), no. 4, 1–13; English transl. in Functional Anal. Appl. **4** (1970).

[BS2] ———, *Estimates of singular numbers of integral operators*, Uspekhi Mat. Nauk **32** (1977), no. 1, 17–84; English transl. in Russian Math. Surveys **32** (1977), no. 1, 15–89.

[BS3] ———, *Spectral theory of self-adjoint operators in Hilbert space*, Izdat. Leningrad. Univ., Leningrad, 1980; English transl., Reidel, Dordrecht, 1987.

[DH] P. A. Deift and R. Hempel, *On the existence of eigenvalues of the Schrödinger operator* $H - \lambda W$ *in a gap of* $\sigma(H)$, Comm. Math. Phys. **103** (1986), 461–490.

[H] R. Hempel, *On the asymptotic distribution of the eigenvalue branches of a Schrödinger operator* $H - \lambda W$ *in a spectral gap of* H, J. Reine Angew. Math. **399** (1989), 38–59.

[Kh] S. V. Khryashchev, *Discrete spectrum for a periodic Schrödinger operator perturbed by a decreasing potential*, Oper. Theory: Adv. Appl. **46** (1990), 109–114.

[RS] M. Reed and B. Simon, *Methods of modern mathematical physics*, IV. *Analysis of operators*, Academic Press, New York, 1979.

[Sh] M. A. Shubin, *The spectral theory and the index of elliptic operators with almost periodic coefficients*, Uspekhi Mat. Nauk **34** (1979), no. 2, 95–135; English transl. in Russian Math. Surveys **34** (1979), no. 2, 109–157.

[T] F. Trikomi, *Equazioni differenziale*, Einaudi, Torino, 1948; 2nd ed., 1953; 3rd ed., Paolo Boringhieri, Torino, 1961; English transl., Hafner, New York, 1961; Russian transl., IL, Moscow, 1962.

Leningrad Branch
Steklov Mathematical Institute (LOMI)
Fontanka 27
Leningrad 191011 USSR

Translated by THE AUTHOR

ADVANCES IN SOVIET MATHEMATICS
Volume 7, 1991

On Solutions of the Schrödinger Equation with Radiation Conditions at Infinity

D. R. YAFAEV

ABSTRACT. We consider the Schrödinger equation with a potential $q(x)$ vanishing as $|x|^{-1-\varepsilon}$, $\varepsilon > 0$, $|x| \to \infty$. It is shown that every solution of this equation satisfying some a priori bound at infinity is asymptotically expressed as a linear combination of outgoing and incoming spherical waves. Coefficients of these waves are related by the scattering matrix. An expansion theorem in terms of such solutions is established.

§1. Introduction

There are several equivalent definitions of the scattering matrix for the Schrödinger operator with a potential decaying sufficiently rapidly at infinity. Roughly speaking, they can be gathered into time-dependent and stationary versions. Stationary definitions can in turn be introduced either in terms of the resolvent of the Schrödinger operator or in terms of certain solutions of the homogeneous differential equation

$$-\Delta u + q(x)u = \lambda u, \qquad x \in \mathbb{R}^d, \ \lambda > 0. \tag{1.1}$$

Under the assumption that

$$|q(x)| \leqslant C(1+|x|)^{-\alpha}, \qquad q = \bar{q}, \tag{1.2}$$

where $2\alpha > d+1$, for every $\omega \in \mathbb{S}^{d-1}$, the equation (1.1) has the solution $u(x, \omega)$ with asymptotics

$$\begin{aligned} u(x, \omega) = {} & \exp(i\lambda^{1/2}\langle x, \omega\rangle) \\ & + ar^{-(d-1)/2}\exp(i\lambda^{1/2}r) + o(r^{-(d-1)/2}), \qquad r = |x|, \end{aligned} \tag{1.3}$$

at infinity. The coefficient $a = a(\varphi, \omega; \lambda)$ of the outgoing spherical wave depends on the direction ω of the incident plane wave and on the direction $\varphi = r^{-1}x$ of observation and is called the scattering amplitude. The scattering matrix $\mathscr{S} = \mathscr{S}(\lambda)$ is an operator in the space $\mathscr{G} = L_2(\mathbb{S}^{d-1})$. It differs

1991 *Mathematics Subject Classification.* Primary 35J10; Secondary 35B40.

from the identity operator by an integral operator whose kernel is equal (up to some numerical factor) to the scattering amplitude.

On the other hand, in the physics literature (see, e.g., [LL]) one can find the definition of the scattering matrix in terms of solutions of equation (1.1) which are asymptotically expressed as linear combinations of incoming and outgoing spherical waves:

$$u(x) = r^{-(d-1)/2}[\gamma b_+(\varphi)\exp(i\lambda^{1/2}r) - \bar{\gamma} b_-(-\varphi)\exp(-i\lambda^{1/2}r)] + o(r^{-(d-1)/2}), \qquad \gamma = \exp\left(-\frac{\pi i(d-3)}{4}\right). \tag{1.4}$$

The existence of such solutions can be established by integration over $\omega \in \mathbb{S}^{d-1}$ of the product $b(\omega)u(x, \omega)$, where b is a smooth function. The stationary phase method shows that this integral has (up to some numerical factor) the asymptotics (1.4), where $b_- = b$ and $b_+ = \mathcal{S}b$. This definition of the scattering matrix is quite similar to that in the one-dimensional case (or to the problem on the half-axis), where every solution of (1.1) asymptotically equals a linear combination of exponentials $\exp(\pm i\lambda^{1/2}x)$. In the multidimensional case, the solution (1.4) determines the action of the scattering matrix, while its kernel is found with the help of solutions (1.3).

Our aim is to ascribe a precise meaning to the definition of the scattering matrix in terms of solutions of the equation (1.1) with asymptotics (1.4) and to justify such a definition. It turns out that the solutions (1.4) exist under the natural assumption $\alpha > 1$. This requires, however, that the relation $v(x) = o(r^{-(d-1)/2})$ be understood in an appropriate averaged sense. The coefficient b_- (and b_+) in (1.4) can be an arbitrary function from $\mathcal{G}$ which determines the action of $\mathcal{S}$ on the whole space $\mathcal{G}$.

Our main result is that in the class of functions obeying the condition

$$\int_{|x|<\rho} |u(x)|^2\, dx < C\rho \tag{1.5}$$

every solution of equation (1.1) has asymptotics (1.4) at infinity. Thus such solutions are equal to linear combinations of functions satisfying outgoing and incoming radiation conditions. Moreover, the coefficients $b_\pm$ in (1.4) are related by $b_+ = \mathcal{S}b_-$. Therefore all solutions of (1.1) obeying (1.5) are parametrized by the functions $b_- \in \mathcal{G}$.

It is convenient for us to introduce, together with the scattering matrix $\mathcal{S}$, the operator $\Sigma = \mathcal{S}\mathcal{J}$, where $\mathcal{J}$, $(\mathcal{J}f)(\omega) = f(-\omega)$, is the reflection operator in $\mathcal{G}$. The operator $\Sigma = \Sigma(\lambda)$ is unitary in the space $\mathcal{G}$. It has a purely point spectrum, so that linear combinations of its eigenvectors are dense in $\mathcal{G}$. The asymptotics (1.4) is particularly simple if $b = \mathcal{J}b_-$ is an eigenvector of Σ. Indeed, if $\Sigma b = \exp(-2i\eta)b$, then (1.4) reduces to

$$u(x) = r^{-(d-1)/2}b(\varphi)\sin(\lambda^{1/2}r - (d-3)\pi/4 - \eta) + o(r^{-(d-1)/2}), \tag{1.6}$$

that is u has the asymptotics of a standing wave at infinity. In the spherically symmetric case $q(x) = q(r)$, solutions with asymptotics (1.6) are naturally parametrized by the orbital quantum number $l = 0, 1, 2, \ldots$. In this case $b = b_l$ is a spherical function and $\eta - \pi l/2$ is the corresponding scattering phase.

Our construction of solutions of equation (1.1) with asymptotics (1.4) relies on the analytical machinery of stationary scattering theory. In particular, we need sharp estimates for the resolvent of the Schrödinger operator obtained by S. Agmon and L. Hörmander [AH] and expounded in the monograph [H1]. We also use well-known results (see the paper [H2] by L. Hörmander and references therein) about solutions of equation (1.1) for $q = 0$. For the sake of completeness, we give proofs of all necessary auxiliary results. Note that our proofs are somewhat different from those in [H1]. We always assume the bound (1.2) with $\alpha > 1$. As in [AH, H1], this condition can be relaxed a little bit, but we shall not be concerned with this here.

The paper is organized as follows. In §§2 and 3 different results of stationary scattering theory are discussed. Solutions of equation (1.1) with asymptotics (1.4) at infinity are studied in §4. In particular, the main result is formulated as Theorem 4.7. Finally, in §5 we establish an expansion theorem in terms of solutions of (1.1) satisfying condition (1.4).

§2. Estimates for the resolvent

1. Denote by $H_0 = -\Delta$ a selfadjoint operator in the space $\mathscr{H} = L_2(\mathbb{R}^d)$ defined on the Sobolev class $W_2^2(\mathbb{R}^d)$. Let Q be the operator of multiplication by a function $q(x)$ which is assumed to be real and bounded. Then the operator $H = H_0 + Q$ is selfadjoint on the domain $W_2^2(\mathbb{R}^d)$.

Let $\Phi\colon L_2(\mathbb{R}^d) \to L_2(\Xi^d)$ be the Fourier transform, i.e.,

$$(\Phi f) = (2\pi)^{-d/2} \int_{\mathbb{R}^d} \exp(-i\langle x, \xi\rangle) f(x)\,dx, \qquad \hat{f} = \Phi f.$$

Sometimes we do not distinguish the operators $L\colon L_2(\mathbb{R}^d) \to L_2(\mathbb{R}^d)$ and $\Phi L \Phi^*$, indicating only that the latter acts in the momentum representation. For example, H_0 acts as multiplication by ξ^2 in the momentum representation. We denote by $\mathbb{S}_\lambda^{d-1}$ the sphere of radius $\lambda^{1/2}$ in $\mathbb{R}^d$ or Ξ^d; $\mathbb{S}^{d-1} = \mathbb{S}_1^{d-1}$ and $\mathscr{G} = L_2(\mathbb{S}^{d-1})$. All adjoint operators are constructed either with respect to the scalar product in $\mathscr{H}$ or in $\mathscr{G}$. By C and c we denote positive constants whose precise values are immaterial.

Let

$$(\Gamma_0(\lambda) f)(\omega) = 2^{-1/2} \lambda^{(d-2)/4} (2\pi)^{-d/2} \int_{\mathbb{R}^d} \exp(-i\lambda^{1/2}\langle x, \omega\rangle) f(x)\,dx,$$

that is (up to a numerical factor) $\Gamma_0(\lambda)$ acts as the restriction of the Fourier transform $\hat{f}$ of f on the sphere $\mathbb{S}_\lambda^{d-1}$. Then the adjoint operator is defined by

$$(\Gamma_0^*(\lambda)a)(x) = 2^{-1/2}\lambda^{(d-2)/4}(2\pi)^{-d/2}\int_{\mathbb{S}^{d-1}} \exp(i\lambda^{1/2}\langle x, \omega\rangle)a(\omega)\,d\omega. \quad (2.1)$$

Clearly, the resolvent $R_0(z) = (H_0 - z)^{-1}$ of H_0 can be represented as the Cauchy integral

$$(R_0(z)f, g) = \int_0^\infty (\lambda - z)^{-1}(\Gamma_0(\lambda)f, \Gamma_0(\lambda)g)_{\mathcal{G}}\,d\lambda, \qquad \operatorname{Im} z \neq 0. \quad (2.2)$$

Let $\mathcal{H}^\beta = L_2^{(\beta)}(\mathbb{R}^d)$ be the space of functions with the norm

$$\|f\|_\beta^2 = \int_{\mathbb{R}^d} (x^2+1)^\beta |f(x)|^2\,dx.$$

By Sobolev's trace theorem for $\beta > 1/2$, the operator $\Gamma_0(\lambda)\colon \mathcal{H}^\beta \to \mathcal{G}$ is bounded (and compact) and Hölder continuous in $\lambda \in (0, \infty)$. It follows that the operator $\Gamma_0^*(\lambda)\colon \mathcal{G} \to \mathcal{H}^{-\beta}$, $\beta > 1/2$, has the same properties. According to the Privalov theorem, the representation (2.2) ensures that the operator $R_0(z)\colon \mathcal{H}^\beta \to \mathcal{H}^{-\beta}$ for any $\beta > 1/2$ is continuous in norm with respect to z in the complex plane cut along $[0, \infty)$, possibly with exception of the point $z = 0$. Moreover, the relation

$$\begin{aligned}(R_0(\lambda + i0)f, g) &- (R_0(\lambda - i0)f, g)\\ &= 2\pi i(\Gamma_0(\lambda)f, \Gamma_0(\lambda)g)_{\mathcal{G}}, \qquad f, g \in \mathcal{H}^\beta,\end{aligned} \quad (2.3)$$

holds. Proofs of these results can be found in, say, [K].

We note that $R_0(z)$ is an integral operator in $L_2(\mathbb{R}^d)$ with kernel

$$\begin{aligned}R_0(x, x'; z) &= i4^{-1}(2\pi)^{-\nu}(z^{1/2}|x - x'|^{-1})^\nu \mathbf{H}_\nu^{(1)}(z^{1/2}|x - x'|),\\ \nu &= 2^{-1}d - 1, \quad \operatorname{Im} z^{1/2} > 0,\end{aligned} \quad (2.4)$$

where $\mathbf{H}_\nu^{(1)}$ is the first Hankel function of order ν.

2. We need a precise result on the boundedness of $R_0(z)$ which supplements assertions of the previous part. Let the space B consist of functions with the finite norm

$$\|f\|_B = \left(\int_{|x|\leqslant 1} |f(x)|^2\,dx\right)^{1/2} + \sum_{n=0}^\infty \left(2^n \int_{2^n \leqslant |x| \leqslant 2^{n+1}} |f(x)|^2\,dx\right)^{1/2}. \quad (2.5)$$

This norm is very close to but somewhat larger than the norm $\|f\|_{1/2}$. The difference is that (2.5) is constructed in terms of the l_1-norm of the sequence

$$\left(\int_{2^n \leqslant |x| \leqslant 2^{n+1}} |x||f(x)|^2\,dx\right)^{1/2},$$

whereas $\|f\|_{1/2}$ corresponds to the l_2-norm of this sequence. The space B^* dual to B (with respect to $\mathscr{H}$) consists of all functions for which the norm

$$\|f\|_{B^*} = \sup\left\{\left(\int_{|x|\leqslant 1} |f(x)|^2\,dx\right)^{1/2}, \sup_{n=0,1,2,\dots}\left(2^{-n}\int_{2^n\leqslant|x|\leqslant 2^{n+1}} |f(x)|^2\,dx\right)^{1/2}\right\} \quad (2.6)$$

is finite. By the Schwarz inequality $\mathscr{H}^\beta \subset B$ if $\beta > 1/2$, and hence

$$\mathscr{H}^\beta \subset B \subset \mathscr{H}^{1/2} \subset \mathscr{H} \subset \mathscr{H}^{-1/2} \subset B^* \subset \mathscr{H}^{-\beta}, \qquad \beta > 1/2.$$

We recall that, by definition, $f_n \in B^*$ is $*$-weak convergent as $n \to \infty$ to f if $(f_n, g) \to (f, g)$ for every $g \in B$.

The space B^* of course is not separable. The closure of $\mathscr{H}$ in the norm $\|\cdot\|_{B^*}$ gives the space B_0^* of functions $f(x)$ such that the integrals of $|f(x)|^2$ over spherical layers $\{x \in \mathbb{R}^d : 2^n \leqslant |x| \leqslant 2^{n+1}\}$ are $o(2^n)$ as $n \to \infty$. It is not difficult to see that the norm

$$\|f\|_{B^*} = \sup_{\rho\geqslant 1} \rho^{-1} \int_{|x|\leqslant\rho} |f(x)|^2\,dx \quad (2.7)$$

is equivalent to (2.6). Moreover, $f \in B_0^*$ if and only if

$$\lim_{\rho\to\infty} \rho^{-1}\int_{|x|\leqslant\rho} |f(x)|^2\,dx = 0.$$

It can be shown (but we do not need this information) that $f \in B$ ($f \in B^*$) if and only if its Fourier transform $\hat{f}$ belongs to the Besov space $\mathbf{B}_{2,1}^{1/2}$ (to the Besov space $\mathbf{B}_{2,\infty}^{-1/2}$). Here the standard notation for Besov spaces (see, e.g., [BL, T]) is used.

The following two properties of the class B are quite elementary. The first of them implies that an operator acting in the momentum representation as multiplication by a function $\hat{\zeta}$ from the Schwartz class $\mathfrak{S}(\Xi^d)$ is bounded in B.

LEMMA 2.1. *Let* $\zeta \in \mathfrak{S}(\mathbb{R}^d)$, $f \in B$, *and*

$$g(x) = \int_{\mathbb{R}^d} \zeta(x-y) f(y)\,dy.$$

Then $g \in B$ *and* $\|g\|_B \leqslant C\|f\|_B$.

Let $\mathbb{R}^{d-1}_{e_1}$ be the orthogonal complement in $\mathbb{R}^d$ to a given unit vector $e_1 \in \mathbb{R}^d$, and $x = x_1 e_1 + x^1$, where $x^1 \in \mathbb{R}^{d-1}_{e_1}$. We write briefly $x = (x_1, x^1)$, $f(x) = f(x_1, x^1)$.

LEMMA 2.2. *We have the bound*

$$\int_{-\infty}^{\infty} \|f(x_1, \cdot)\|_{L_2(\mathbb{R}^{d-1})} dx_1 \leqslant 2^{1/2} \|f\|_B. \tag{2.8}$$

PROOF. With the help of the Schwarz inequality, the bound (2.8) can be directly obtained separately for each of the functions $\chi_n f$, where χ_n is the indicator of the set $\{x \in \mathbb{R}^d : 2^n \leqslant |x| \leqslant 2^{n+1}\}$. Then one can sum up over n all such bounds. □

Let us now prove the boundedness (uniform in z) of the resolvent $R_0(z)$: $B \to B^*$. This result was obtained in [AH]. Our proof is apparently simpler than that given in [AH] or [H1] where, however, a rather large class of differential operators with constant coefficients H_0 was considered. It is convenient for us to consider products $LR_0(z)$, where L is the operator of multiplication in the momentum representation by a C^∞-function $l(\xi)$ obeying the bound $|l(\xi)| \leqslant C\xi^2$ as $|\xi| \to \infty$. This situation is (formally) more general and the results for $R_0(z)$ are obtained if $L = I$.

THEOREM 2.3. *We have the bound*

$$\|LR_0(\lambda \pm i\varepsilon)\|_{B, B^*} \leqslant C, \tag{2.9}$$

where the constant C *does not depend on* $\varepsilon > 0$ *and* $\lambda \in [\lambda_0, \lambda_1]$ *if* $0 < \lambda_0 < \lambda_1 < \infty$.

PROOF. We introduce a partition of unity ζ_n,

$$\sum_{n=0}^{N} \zeta_n(\xi) = 1, \qquad \zeta_n \in C^\infty(\Xi^d),$$

with the following properties. A function $\zeta_0(\xi) = 0$ in some neighborhood of the sphere $\mathbb{S}_\lambda^{d-1}$ and $\zeta_0(\xi) = 1$ sufficiently far from it (for example, for $|\xi^2 - \lambda| \geqslant \lambda/2$). Functions $\zeta_n \in C_0^\infty(\Xi^d)$ are chosen in such a way that in some Cartesian coordinate system (ξ_n, ξ^n), $\dim \xi_n = 1$, $\dim \xi^n = d-1$, the equality $\zeta_n(\xi_n, \xi^n) = 0$ holds for a sufficiently small quantity $(\xi^n)^2 - \lambda$. Such a partition of unity can be constructed for any d.

The bound (2.9) is equivalent to the inequality

$$\left| \int_{\Xi^d} (\xi^2 - \lambda \mp i\varepsilon)^{-1} l(\xi) \hat{f}(\xi) \overline{\hat{g}(\xi)}\, d\xi \right| \leqslant C \|f\|_B \|g\|_B. \tag{2.10}$$

We set $l_n = \zeta_n l$ and estimate integrals (2.10) for each l_n separately. If $n = 0$, then the function $(\xi^2 - \lambda \mp i\varepsilon)^{-1} l_0(\xi)$ is uniformly bounded so that the integral (2.10) does not exceed $C\|f\|\,\|g\|$. If $n \geqslant 1$, we write the integral (2.10) with $l = l_n$ in the variables (ξ_n, ξ^n). The dual variables will be denoted by (x_n, x^n). First we estimate the integral over the one-dimensional variable ξ_n. Let $\tilde{v}_n(x_n, \xi^n)$ be the (partial) inverse Fourier transform of the function $\hat{v}_n(\xi) = l_n(\xi)\hat{f}(\xi)$ in the variable ξ_n. Then according to (2.4) for $d = 1$ the integral (2.10) with $l = l_n$ equals

$$i2^{-1}\int_{\Xi^{d-1}} d\xi^n \int_{-\infty}^{\infty}\int_{-\infty}^{\infty} dx_n\, dx_n'\ \tilde{v}_n(x_n, \xi^n)\overline{\tilde{g}(x_n', \xi^n)}$$
$$\times\left(\lambda - (\xi^n)^2\right)^{-1/2}\exp\left(i\left(\lambda - (\xi^n)^2\right)^{1/2}|x_n - x_n'|\right).$$

By the construction of l_n, the integration over ξ^n is actually restricted to the region $|(\xi^n)^2 - \lambda| \geqslant c > 0$ so that the last integral does not exceed

$$C\int_{\Xi^{d-1}}\int_{-\infty}^{\infty}\int_{-\infty}^{\infty}|\tilde{v}_n(x_n, \xi^n)||\tilde{g}(x_n', \xi^n)|\,d\xi^n\,dx_n\,dx_n'.$$

By the Schwarz inequality, this integral is bounded by

$$\int_{-\infty}^{\infty}dx_n\left(\int_{\Xi^{d-1}}|\tilde{v}_n(x_n, \xi^n)|^2\,d\xi^n\right)^{1/2}\times\int_{-\infty}^{\infty}dx_n\left(\int_{\Xi^{d-1}}|\tilde{g}(x_n, \xi^n)|^2\,d\xi^n\right)^{1/2}.$$

According to the Parseval relation, this product equals

$$\int_{-\infty}^{\infty}dx_n\left(\int_{\mathbb{R}^{d-1}}|v_n(x_n, x^n)|^2\,dx^n\right)^{1/2}\times\int_{-\infty}^{\infty}dx_n\left(\int_{\mathbb{R}^{d-1}}|g(x_n, x^n)|^2\,dx^n\right)^{1/2}.$$

Lemma 2.2 shows that the last term is bounded by $2\|v_n\|_B\|g\|_B$, which by Lemma 2.1 does not exceed $C\|f\|_B\|g\|_B$. □

COROLLARY 2.4. *The operator $LR_0(\lambda \pm i\varepsilon)\colon B \to B^*$ is weakly* continuous in $\lambda > 0$, $\varepsilon \geqslant 0$. The operator $LR_0(\lambda \pm i\varepsilon)\colon B \to \mathscr{H}^{-\beta}$ for $\beta > 1/2$ is strongly continuous in $\lambda > 0$, $\varepsilon \geqslant 0$.*

PROOF. It follows from the results of subsection 1 that the function $(LR_0(\lambda \pm i\varepsilon)f, g)$ is continuous in $\lambda > 0$ and $\varepsilon \geqslant 0$ if $f, g \in \mathscr{H}^\beta$, $\beta > 1/2$. Since $\mathscr{H}^\beta$ is dense in B, this result extends to arbitrary $f, g \in B$ according to (2.9). Similarly, if $f \in \mathscr{H}^\beta$, $\beta > 1/2$, the function $LR_0(\lambda \pm i\varepsilon)f$ is strongly continuous in $\mathscr{H}^{-\beta}$. The bound (2.9) again allows us to extend this result to arbitrary $f \in B$. □

According to Corollary 2.4, the operator $R_0(\lambda \pm i0)\colon B \to B^*$ is well-defined and, moreover, $R_0(\lambda \pm i0)f \in W^2_{2,\mathrm{loc}}$.

COROLLARY 2.5. *Both functions $\psi = R_0(\lambda \pm i0)f$, $f \in B$, satisfy the equation $-\Delta\psi - \lambda\psi = f$.*

PROOF. Let $\psi_\varepsilon = R_0(\lambda \pm i\varepsilon)f$. Then $\psi_\varepsilon \in W_2^2(\mathbb{R}^d)$ and $(-\Delta - \lambda \mp i\varepsilon)\psi_\varepsilon = f$. By Theorem 2.3 $\|\psi_\varepsilon\|_{B^*} \leqslant C < \infty$ so that $(-\Delta - \lambda)\psi_\varepsilon \to f$ as $\varepsilon \to 0$ in B^*. By Corollary 2.4 $((-\Delta - \lambda)\psi_\varepsilon, g) \to ((-\Delta - \lambda)\psi, g)$ as $\varepsilon \to 0$ if $g \in B$. It follows that $((-\Delta - \lambda)\psi, g) = (f, g)$ for every $g \in B$. □

According to relation (2.3), results about the resolvent $R_0(z)$ induce results about the operator $\Gamma_0(\lambda)$.

COROLLARY 2.6. *The operators* $\Gamma_0(\lambda): B \to \mathcal{G}$ *are bounded and strongly continuous in* $\lambda > 0$. *The operators* $\Gamma_0^*(\lambda): \mathcal{G} \to B^*$ *and* $D_j \Gamma_0^*(\lambda): \mathcal{G} \to B^*$, $D_j = -i\partial/\partial x_j$, $j = 1, \dots, d$, *are bounded and weakly** *continuous in* $\lambda > 0$.

3. In this subsection we shall continue our study of the operators Γ_0 and Γ_0^*. Let us introduce special notation for spherical waves

$$w_\pm(r) = \gamma^{\pm 1} r^{-(d-1)/2} \exp(\pm i\lambda^{1/2} r), \qquad \gamma = \exp(-\pi i(d-3)/4).$$

LEMMA 2.7. *If* $a \in C^\infty(\mathbb{S}^{d-1})$, *then, as* $r \to \infty$,

$$\begin{aligned}(\Gamma_0^*(\lambda)a)(x) = &- i2^{-1}\pi^{-1/2}\lambda^{-1/4}(a(\varphi)w_+(r) - a(-\varphi)w_-(r)) \\ &+ o(r^{-(d-1)/2}), \qquad x = r\varphi.\end{aligned} \tag{2.11}$$

PROOF. It is sufficient to apply the stationary phase method (see, e.g., [V]) to the integral (2.1), which has two stationary points $\omega = \varphi$ and $\omega = -\varphi$. □

In the next section the asymptotics (2.11) is extended to arbitrary $a \in \mathcal{G}$. Now we shall need the following

THEOREM 2.8. *If* $a \in \mathcal{G}$, *then there exists*

$$\lim_{\rho \to \infty} \rho^{-1} \int_{|x| \leqslant \rho} |(\Gamma_0^*(\lambda)a)(x)|^2 \, dx = (2\pi)^{-1}\lambda^{-1/2}\|a\|_{\mathcal{G}}^2. \tag{2.12}$$

PROOF. If $a \in C^\infty(\mathbb{S}^{d-1})$, then (2.12) is a direct consequence of (2.11). It is taken into account here that the contribution of the cross-terms is proportional to

$$\operatorname{Re}\left(\gamma^2 \rho^{-1} \int_0^\rho \exp(2i\lambda^{1/2} r)\, dr\right) \cdot \int_{|\varphi|=1} a(\varphi)\overline{a(-\varphi)}\, d\varphi$$

and hence vanishes as $\rho \to \infty$. If $a \in \mathcal{G}$ is arbitrary, then

$$\rho^{-1} \int_{|x| \leqslant \rho} |(\Gamma_0^*(\lambda)a)(x)|^2 \, dx \leqslant \|\Gamma_0^*(\lambda)a\|_{B^*}^2 \leqslant C\|a\|_{\mathcal{G}}^2, \qquad \rho \geqslant 1,$$

by (2.7) and Corollary 2.6. Therefore approximation by smooth functions allows us to extend (2.12) to all $a \in \mathcal{G}$. □

COROLLARY 2.9. *We have the two-sided estimate*

$$c\|a\|_{\mathcal{G}} \leqslant \|\Gamma_0^*(\lambda)a\|_{B^*} \leqslant C\|a\|_{\mathcal{G}}, \tag{2.13}$$

so that the kernel $N(\Gamma_0^*)$ *of the operator* $\Gamma_0^*: \mathcal{G} \to B^*$ *is trivial and its image* $R(\Gamma_0^*)$ *is closed.*

Thus the result of Corollary 2.6 about the boundedness of the operator $\Gamma_0^*: \mathcal{G} \to B^*$ cannot be improved further.

According to Corollary 2.9, $\overline{R(\Gamma_0)} = \mathcal{G}$ and, moreover, $R(\Gamma_0) = \overline{R(\Gamma_0)}$. The last equality would have been quite obvious if B were reflexive. Then it would have been sufficient to take into account that if some operator has a closed image, then its adjoint has the same property. Since B is not reflexive, it is necessary to use a fairly subtle theorem due to H. Hahn (see, e.g., [DS, Chapter VI, §6]). This theorem asserts that an operator has closed image if its adjoint has this property. This justifies the following

COROLLARY 2.10. *The image of the operator* $\Gamma_0 : B \to \mathcal{G}$ *coincides with the entire space* $\mathcal{G}$.

Denote by $K_\rho(B) = \{ f \in B : \|f\|_B < \rho \}$ the ball of radius ρ in the space B. The notation $K_\rho(\mathcal{G})$ has a similar meaning. For an arbitrary set $M \subset B$, denote by $\Gamma_0(M)$ the image of M under the mapping Γ_0.

COROLLARY 2.11. *For sufficiently large (but finite)* ρ

$$K_1(\mathcal{G}) \subset \Gamma_0(K_\rho(B)), \tag{2.14}$$

and hence for an arbitrary $a \in \mathcal{G}$ *an element* f *can be chosen so that* $\Gamma_0 f = a$ *and*

$$\|f\|_B \leqslant \rho \|a\|_{\mathcal{G}}.$$

In particular, the set $\Gamma_0(\mathfrak{S}(\mathbb{R}^d) \cap K_\rho(B))$ *is dense in* $K_1(\mathcal{G})$ *in the norm of* $\mathcal{G}$.

PROOF. Since the operator $\Gamma_0 : B \to \mathcal{G}$ is bounded and $R(\Gamma_0) = \mathcal{G}$, the set $\Gamma_0(K_1(B))$ is open (see, e.g., [DS, Chapter II, §2]) and therefore contains a certain ball $K_\varepsilon(\mathcal{G})$, $\varepsilon > 0$. This ensures (2.14). □

§3. Radiation condition. Uniqueness theorem

1. It is usually agreed that a function $u(x)$, $x \in \mathbb{R}^d$, satisfies the radiation condition if $u(x) = b(\varphi) w_\pm(r) + o(r^{-(d-1)/2})$, $x = r\varphi$, as $r = |x| \to \infty$. The upper sign here corresponds to the outgoing and the lower to the incoming radiation condition. The precise meaning of the radiation condition depends on the sense in which the relation $v(x) = o(r^{-(d-1)/2})$ is understood. If not specified otherwise, we understand it in an averaged sense. Namely, $v(x) = o_{\mathrm{av}}(r^{-(d-1)/2})$ means that

$$\rho^{-1} \int_{|x| \leqslant \rho} |v(x)|^2 \, dx = o(1), \quad \rho \to \infty, \tag{3.1}$$

which is equivalent to the inclusion $v \in B_0^*$.

Let us show that functions $R_0(\lambda \pm i0) f$ satisfy the radiation condition.

LEMMA 3.1. *If* $f \in B$, *then*

$$\begin{aligned} (R_0(\lambda \pm i0) f)(x) = \pi^{1/2} \lambda^{-1/4} (\Gamma_0(\lambda) f)(\pm\varphi) w_\pm(r) \\ + o_{\mathrm{av}}(r^{-(d-1)/2}), \qquad r \to \infty. \end{aligned} \tag{3.2}$$

PROOF. Let us proceed from the explicit expression (2.4) for the kernel of the resolvent $R_0(z)$. Using the asymptotics of the Hankel functions, we find that

$$R_0(x, x'; \lambda \pm i0) = 2^{-1}(2\pi)^{-(d-1)/2}\lambda^{(d-3)/2}w_\pm(|x-x'|) + O(|x-x'|^{-(d+1)/2}) \tag{3.3}$$

as $|x-x'| \to \infty$, where the remainder term is estimated pointwise. Moreover, if $|x'| \leqslant C$ and $|x| \to \infty$, then

$$w_\pm(|x-x'|) = \exp(\mp i\lambda^{1/2}|x|^{-1}\langle x, x'\rangle)w_\pm(|x|) + O(|x|^{-(d+1)/2}). \tag{3.4}$$

It follows that the asymptotics (3.2) is valid for $f \in C_0^\infty(\mathbb{R}^d)$ (or, more generally, for compactly supported bounded functions) with the remainder term estimated pointwise by $r^{-(d+1)/2}$. Relation (3.2) is extended to arbitrary $f \in B$ by its approximation in the norm of B by $f_n \in C_0^\infty(\mathbb{R}^d)$. Indeed, considering that the left-hand side of (3.1) is bounded for $\rho \geqslant 1$ by $\|v\|_{B^*}^2$ and

$$\|\Gamma_0^{(\pm)} f w_\pm\|_{B^*} = \|\Gamma_0^{(\pm)} f\|_{\mathscr{G}}, \qquad (\Gamma_0^{(\pm)} f)(\varphi) = (\Gamma_0(\lambda)f)(\pm\varphi),$$

we obtain the inequality

$$\begin{aligned}\rho^{-1}\|R_0(\lambda \pm i0)f - \pi^{1/2}\lambda^{-1/4}\Gamma_0^{(\pm)} f w_\pm\|_{L_2(K_\rho)}^2 \\ \leqslant C(\|R_0(\lambda \pm i0)(f-f_n)\|_{B^*}^2 + \|\Gamma_0^{(\pm)}(f-f_n)\|_{\mathscr{G}}^2 \\ + \rho^{-1}\|R_0(\lambda \pm i0)f_n - \pi^{1/2}\lambda^{-1/4}\Gamma_0^{(\pm)} f_n w_\pm\|_{L_2(K_\rho)}^2).\end{aligned}$$

By virtue of Theorem 2.3 and Corollary 2.6, the first two terms on the right-hand side vanish as $n \to \infty$. The last term tends to zero for fixed n as $\rho \to \infty$, since the pointwise convergence is stronger than the averaged one. □

Quite similarly, the closure of the asymptotics (2.11) can be carried out with the help of Corollary 2.6.

LEMMA 3.2. *If $a \in \mathscr{G}$ and $r \to \infty$, then*

$$(\Gamma_0^*(\lambda)a)(x) = -i2^{-1}\pi^{-1/2}\lambda^{-1/4}(a(\varphi)w_+(r) - a(-\varphi)w_-(r)) + o_{\mathrm{av}}(r^{-(d-1)/2}). \tag{3.5}$$

We note that the results about $\Gamma_0^* a$ could be deduced from those about $R_0 f$ on the basis of the relation

$$2\pi i\Gamma_0^*(\lambda)a = R_0(\lambda + i0)f - R_0(\lambda - i0)f, \qquad a = \Gamma_0(\lambda)f,$$

which is a consequence of (2.3). However, this requires the equality $R(\Gamma_0) = \mathscr{G}$, whose proof relies on the asymptotics (2.11) for $a \in C^\infty(\mathbb{S}^{d-1})$. Moreover, our justification of Corollary 2.10 uses a rather subtle theorem of Banach space theory.

Now we discuss asymptotics of derivatives of the functions $R_0(\lambda \pm i0)f$ and $\Gamma_0^*(\lambda)a$ with respect to radial and angular variables. Denote

$$D_r = r^{-1}\sum_{j=1}^{d} x_j \partial/\partial x_j, \qquad \nabla_\varphi = \nabla - \varphi D_r, \quad x = r\varphi.$$

LEMMA 3.3. *If $f \in B$, $a \in \mathcal{G}$, and $r \to \infty$, then*

$$\begin{aligned}(D_r R_0(\lambda \pm i0)f)(x) = {} & \pm i\pi^{1/2}\lambda^{1/4}(\Gamma_0(\lambda)f)(\pm\varphi)w_\pm(r) \\ & + o_{\mathrm{av}}(r^{-(d-1)/2}),\end{aligned} \tag{3.6}$$

$$\begin{aligned}(D_r\Gamma_0^*(\lambda)a)(x) = {} & 2^{-1}\pi^{-1/2}\lambda^{1/4}(a(\varphi)w_+(r) + a(-\varphi)w_-(r)) \\ & + o_{\mathrm{av}}(r^{-(d-1)/2}).\end{aligned} \tag{3.7}$$

PROOF. Asymptotics (3.3) and (3.4) admit differentiation with respect to r. Similarly to Lemma 3.1, this yields (3.6) (with the remainder term estimated by $O(r^{-(d+1)/2})$) for $f \in C_0^\infty(\mathbb{R}^d)$. Relation (3.6) is extended to arbitrary $f \in B$ on the basis of the boundedness (see Theorem 2.3) of the operator $D_r R_0(\lambda \pm i0)$: $B \to B^*$.

Differentiating (2.1) with respect to r, we get the integral representation for $D_r\Gamma_0^* a$. Then, similarly to Lemma 2.7, the asymptotics (3.7) for $a \in C^\infty(\mathbb{S}^{d-1})$ is obtained by the stationary phase method. Relation (3.7) is extended to arbitrary $a \in \mathcal{G}$ on the basis of the boundedness (see Corollary 2.9) of the operator $D_r\Gamma_0^*\colon \mathcal{G} \to B^*$. □

The angular part of the gradient improves the fall-off of the functions $R_0(\lambda \pm i0)f$ and $\Gamma_0^*(\lambda)a$ at infinity.

LEMMA 3.4. *If $f \in B$, $a \in \mathcal{G}$, and $r \to \infty$, then*

$$(\nabla_\varphi R_0(\lambda \pm i0)f)(x) = o_{\mathrm{av}}(r^{-(d-1)/2}), \tag{3.8}$$

$$(\nabla_\varphi \Gamma_0^*(\lambda)a)(x) = o_{\mathrm{av}}(r^{-(d-1)/2}). \tag{3.9}$$

PROOF. Boundedness of the operators $\nabla_\varphi R_0\colon B \to B^*$ and $\nabla_\varphi\Gamma_0^*\colon \mathcal{G} \to B^*$ ensures (cf. Lemma 3.3) that it is sufficient to verify (3.8) for $f \in C_0^\infty(\mathbb{R}^d)$ and (3.9) for $a \in C^\infty(\mathcal{S}^{d-1})$. The asymptotics (3.3) and (3.4) admit differentiation with respect to x_j. Taking into account the equality

$$\nabla_\varphi \exp(\pm i\lambda^{1/2}\langle\varphi, x'\rangle) = \pm i\lambda^{1/2}r^{-1}(x' - \varphi\langle\varphi, x'\rangle)\exp(\pm i\lambda^{1/2}\langle\varphi, x'\rangle),$$

we find that the application of the operator ∇_φ gives rise to the additional factor $\mp i\lambda^{1/2}r^{-1}(x' - \varphi\langle\varphi, x'\rangle)$ in the asymptotics of $R_0(x, x'; \lambda \pm i0)$. It follows that for $f \in C_0^\infty(\mathbb{R}^d)$ the right-hand side of (3.8) equals $O(r^{-(d+1)/2})$. To obtain (3.9), we differentiate (2.1) and use the equality

$$\nabla_\varphi \exp(i\lambda^{1/2}\langle x, \omega\rangle) = i\lambda^{1/2}(\omega - \varphi\langle\varphi, \omega\rangle)\exp(i\lambda^{1/2}\langle x, \omega\rangle).$$

PROOF. To study $R(z)\colon B \to B^*$, we notice that according to (3.14)

$$R(z) = R_0(z) - R_0(z)QR_0(z) + R_0(z)QR(z)QR_0(z).$$

We shall obtain the desired result separately for each of the three terms on the right-hand side. The first term was already considered in Corollary 2.4. The same assertion ensures that the function $R_0(z)f$ is continuous, where $f \in B$ is strongly continuous in the space $\mathscr{H}^{-\beta}$ if $\beta > 1/2$. Since the operator $Q\colon \mathscr{H}^{-\beta} \to \mathscr{H}^{\beta}$, $2\beta \leqslant \alpha$, is bounded, it follows that the sesquilinear form of the second term $\big(QR_0(z)f, R_0(\bar{z})g\big)$ is continuous in z if $f, g \in B$. To treat the last term it should be additionally taken into account that by Theorem 3.8 the operator $R(z)\colon \mathscr{H}^{\beta} \to \mathscr{H}^{-\beta}$ is continuous in norm with respect to z.

The result about $R(z)\colon B \to \mathscr{H}^{-\beta}$ can be obtained similarly on the basis of the left equality (3.14). Thus we see that according to Corollary 2.4 and Theorem 3.8 the operators $R_0(z)\colon B \to \mathscr{H}^{-\beta}$ and $R(z)\colon \mathscr{H}^{\beta} \to \mathscr{H}^{-\beta}$ are strongly continuous. □

Finally we note that the uniqueness Theorem 3.5 was used only to exclude the sets $\mathscr{N}_{\pm}$. It can be shown without this theorem that $\mathscr{N}_+ = \mathscr{N}_-$ and that all the results of Theorem 3.8 are valid on the open set $\mathbb{R}_+ \setminus \mathscr{N}_{\pm}$ of full measure.

§4. Main theorem

1. Let us consider first the "free" problem when $q = 0$. By definition (2.1), every function $u \in R\big(\Gamma_0^*(\lambda)\big)$ belongs to $C^\infty(\mathbb{R}^d)$ and satisfies the equation

$$-\Delta u = \lambda u. \tag{4.1}$$

It turns out that all solutions of (4.1) from B^* belong to $R\big(\Gamma_0^*(\lambda)\big)$. This is a consequence of the following more general assertion.

LEMMA 4.1. *Let $u \in B^*$ and $(u, f) = 0$ if $f \in \mathfrak{S}(\mathbb{R}^d)$ and $\hat{f}(\xi) = 0$ in some neighborhood of the sphere $\mathbb{S}_\lambda^{d-1}$. Then $u \in R\big(\Gamma_0^*(\lambda)\big)$.*

PROOF. Every function from $\mathfrak{S}(\Xi^d)$ which equals zero on $\mathbb{S}_\lambda^{d-1}$ can be approximated (see, e.g., [LM]) in the metric of the Sobolev space $W_2^1(\Xi^d)$ by functions which belong to $\mathfrak{S}(\Xi^d)$ and are equal to zero in some neighborhoods of $\mathbb{S}_\lambda^{d-1}$. Therefore the equality $(u, f) = 0$ extends to all $f \in \mathfrak{S}(\mathbb{R}^d)$ such that $\Gamma_0 f = 0$, $\Gamma_0 = \Gamma_0(\lambda)$. It follows that the functional (u, f), defined on $f \in \mathfrak{S}(\mathbb{R}^d)$, depends actually only on functions $\Gamma_0 f$.

Now on the set $\Gamma_0\big(\mathfrak{S}(\mathbb{R}^d)\big)$ let us introduce the functional $l(g) = (u, f)$, where $g = \Gamma_0 f$, $f \in \mathfrak{S}(\mathbb{R}^d)$. Since the value of $l(g)$ does not depend on the choice of f, this definition is correct. By Corollary 2.11,

$$|l(g)| \leqslant \|u\|_{B^*} \|f\| \leqslant \rho \|u\|_{B^*}.$$

on the set $\Gamma_0(\mathfrak{S}(\mathbb{R}^d) \cap K_p(B))$, which is dense in $K_1(\mathcal{G})$. Therefore $l(g)$ extends by continuity to a bounded functional on the whole space $\mathcal{G}$. This implies $l(g) = (a, g)_{\mathcal{G}}$ for some $a \in \mathcal{G}$. By definition of $l(g)$, it follows that

$$(u, f) = (a, \Gamma_0 f)_{\mathcal{G}}$$

for every $f \in \mathfrak{S}(\mathbb{R}^d)$. Thus $u = \Gamma_0^* a$. □

REMARK 4.2. The class of those $f \in \mathfrak{S}(\mathbb{R}^d)$ for which $\hat{f}(\xi) = 0$ in some neighborhood of $\mathbb{S}_\lambda^{d-1}$ is dense in the metric of B in the class of those $f \in B$ for which $\Gamma_0 f = 0$. Therefore in the proof of Lemma 4.1 the functional l can from the very beginning be defined for all $g \in \mathcal{G}$ by the equality $l(g) = (u, f)$, where $g = \Gamma_0 f$ with $f \in B$. We stick to the definition of l on the dense set $\Gamma_0(\mathfrak{S}(\mathbb{R}^d))$ only because we did not find the approximation result in the space B formulated above in the literature.

Let us apply now Lemma 4.1 to the study of the equation (4.1).

THEOREM 4.3. *In order that a function u belong to $W_{2,\mathrm{loc}}^2(\mathbb{R}^d) \cap B^*$ and satisfy equation* (4.1), *it is necessary and sufficient to have $u \in R(\Gamma_0^*(\lambda))$.*

PROOF. The sufficiency follows from definition (2.1) and Corollary 2.6. To prove the necessity, suppose that $u \in W_{2,\mathrm{loc}}^2 \cap B^*$ is a solution of (4.1). Let $f \in \mathfrak{S}(\mathbb{R}^d)$, $\hat{f}(\xi) = 0$ in some neighborhood of the sphere $\mathbb{S}_\lambda^{d-1}$, and $\hat{g}(\xi) = (\xi^2 - \lambda)^{-1}\hat{f}(\xi)$. Then $g \in \mathfrak{S}(\mathbb{R}^d)$ and $(u, f) = (u, (-\Delta - \lambda)g) = 0$ by equation (4.1). According to Lemma 4.1, it follows that $u \in R(\Gamma_0^*)$. □

2. Now we are able to study equation (1.1) with a potential q satisfying the bound (1.2) with $\alpha > 1$. We emphasize that solutions of (1.1) are always assumed to belong to $W_{2,\mathrm{loc}}^2(\mathbb{R}^d)$. The generalization of the operator $\Gamma_0(\lambda)$ is introduced by the formula

$$\Gamma(\lambda) = \Gamma_0(\lambda)\big(I - QR(\lambda - i0)\big). \tag{4.2}$$

Properties of this operator are summarized in the two following assertions.

LEMMA 4.4. *The operators $\Gamma(\lambda)\colon B \to \mathcal{G}$ are bounded and strongly continuous in $\lambda > 0$. The adjoint operators*

$$\Gamma^*(\lambda) = \big(I - R(\lambda + i0)Q\big)\Gamma_0^*(\lambda)\colon \mathcal{G} \to B^* \tag{4.3}$$

are bounded and weakly continuous in $\lambda > 0$.*

PROOF. It suffices to combine Corollaries 2.6 and 3.9. □

Now we shall need the relation

$$\begin{aligned}&\big(I - R(\lambda + i0)Q\big)\big(I + R_0(\lambda + i0)Q\big)\\ &\quad = \big(I + R_0(\lambda + i0)Q\big)\big(I - R(\lambda + i0)Q\big) = I,\end{aligned} \tag{4.4}$$

which follows from the resolvent identity (3.14) and Corollary 3.9.

LEMMA 4.5. *The kernel* $N(\Gamma^*)$ *of the operator* (4.3) *is trivial and its image* $R(\Gamma^*)$ *is closed. The image of* Γ *coincides with* $\mathcal{G}$.

PROOF. According to (4.4) the operator $I-R(\lambda+i0)Q$ is invertible. Therefore assertions about Γ^* are consequences of Corollary 2.9. Similarly, for the proof of the equality $R(\Gamma)=\mathcal{G}$, one can use Corollary 2.10. □

The following statement generalizes Theorem 4.3.

THEOREM 4.6. *Let the estimate* (1.2) *hold for some* $\alpha>1$. *In order that a function* $u\in B^*$ *satisfy the equation* (1.1) *it is necessary and sufficient to have* $u\in R(\Gamma^*(\lambda))$.

PROOF. Let $u\in R(\Gamma^*)\subset B^*$, i.e.,

$$u=(I-RQ)u_0, \qquad u_0=\Gamma_0^*a, \qquad R=R(\lambda+i0).$$

Then according to (4.4),

$$u=u_0-R_0Qu, \qquad R_0=R_0(\lambda+i0). \tag{4.5}$$

By Corollary 2.5, $R_0Qu\in W^2_{2,\mathrm{loc}}$ and $(-\Delta-\lambda)R_0Qu=qu$. Since, moreover, $(-\Delta-\lambda)u_0=0$, it follows from (4.5) that $u\in W^2_{2,\mathrm{loc}}$ and u satisfies (1.1).

Conversely, let $u\in B^*$ satisfy (1.1). We define u_0 by (4.5) and apply the operator $-\Delta-\lambda$ to this equality. Then, again on the basis of Corollary 2.5, we find that $(-\Delta-\lambda)u_0=0$. By Theorem 4.3 this implies $u_0=\Gamma_0^*a$ for some $a\in\mathcal{G}$. Finally applying the operator $I-RQ$ to (4.5) we find that $u=\Gamma^*a$. □

3. Now it is easy to establish our main result which describes solutions of the equation (1.1) from the class B^* in terms of their asymptotic behavior at infinity. We recall that the scattering matrix $\mathcal{S}(\lambda)\colon\mathcal{G}\to\mathcal{G}$ for the Schrödinger operator $H=H_0+Q$ is defined by the relation

$$\mathcal{S}(\lambda)=I-2\pi i\Gamma_0(\lambda)\big(Q-QR(\lambda+i0)Q\big)\Gamma_0^*(\lambda). \tag{4.6}$$

The right-hand side of (4.6) can be considered as a sum of products of bounded operators. Indeed, $\Gamma_0^*(\lambda)\colon\mathcal{G}\to B^*$ is bounded by Corollary 2.6, $Q\colon B^*\to B$ is bounded by (1.2), where $\alpha>1$, $R(\lambda+i0)\colon B\to B^*$ is bounded by Theorem 2.3 and, finally, $\Gamma_0(\lambda)\colon B\to\mathcal{G}$ is bounded again by Corollary 2.6. It is possible to replace here B and B^* by $\mathcal{H}^{\beta}$ and $\mathcal{H}^{-\beta}$ (where $\beta>1/2$), respectively. According to identities (2.3) and (3.14), the operator (4.6) is unitary. Using notation (4.3), we can rewrite (4.6) as

$$\mathcal{S}(\lambda)=I-2\pi i\Gamma_0(\lambda)Q\Gamma^*(\lambda). \tag{4.7}$$

Denote by $\mathcal{J}$, $(\mathcal{J}f)(\omega)=f(-\omega)$, the reflection operator in $\mathcal{G}$ and set $\Sigma(\lambda)=\mathcal{S}(\lambda)\mathcal{J}$. These operators are also unitary in $\mathcal{G}$.

THEOREM 4.7. *Let the estimate* (1.2) *hold for some* $\alpha > 1$. *Suppose that* $u \in B^*$ *satisfies equation* (1.1). *Then there exists an element* $b \in \mathscr{G}$ *such that*

$$u(x) = (\Sigma(\lambda)b)(\varphi)w_+(r) - b(\varphi)w_-(r) + o_{\text{av}}(r^{-(d-1)/2}), \quad r \to \infty. \tag{4.8}$$

Conversely, for every $b \in \mathscr{G}$ *there exists a unique solution of the equation* (1.1) *with asymptotics* (4.8) (*such a function obviously belongs to* B^*). *Moreover, the estimate*

$$\|u\|_{B^*} \leqslant C\|b\| \tag{4.9}$$

holds and the following limit exists

$$\lim_{\rho\to\infty} \rho^{-1} \int_{|x|\leqslant\rho} |u(x)|^2\, dx = 2\|b\|^2. \tag{4.10}$$

PROOF. We proceed from Theorem 4.6. Every solution $u \in B^*$ of (1.1) has the representation $u = \Gamma^* a$, where $a \in \mathscr{G}$. This implies that u satisfies (4.5) with $u_0 = \Gamma_0^* a$. The asymptotics of the right-hand side of (4.5) is given by Lemmas 3.1 and 3.2, so that

$$\begin{aligned} u(x) = &- i2^{-1}\pi^{-1/2}\lambda^{-1/4}[a(\varphi)w_+(r) - a(-\varphi)w_-(r)] \\ &- \pi^{1/2}\lambda^{-1/4}(\Gamma_0 Qu)(\varphi)w_+(r) + o_{\text{av}}(r^{-(d-1)/2}). \end{aligned} \tag{4.11}$$

According to (4.7),

$$\Gamma_0 Qu = \Gamma_0 Q\Gamma^* a = (2\pi i)^{-1}(I - \mathscr{S})a.$$

Thus the coefficient of $w_+(r)$ in (4.11) equals $-i2^{-1}\pi^{-1/2}\lambda^{-1/4}(\mathscr{S}a)(\varphi)$. It follows that (4.11) can be rewritten in the form (4.8) if we set

$$b = -i2^{-1}\pi^{-1/2}\lambda^{-1/4}\mathscr{S}a. \tag{4.12}$$

For every $b \in \mathscr{G}$, the solution of (1.1) with asymptotics (4.8) can be defined by the relation $u = \Gamma^* a$, where a and b are connected by (4.12). The estimate (4.9) is a consequence of the boundedness of the operator $\Gamma^* : \mathscr{G} \to B^*$. Finally, according to (4.8),

$$\begin{aligned} &\rho^{-1} \int_{|x|\leqslant\rho} |u(x)|^2\, dx \\ &\quad = \|\Sigma b\|_{\mathscr{G}}^2 + \|b\|_{\mathscr{G}}^2 - 2\,\text{Re}\left[\gamma^2\rho^{-1}\int_0^\rho \exp(2i\lambda^{1/2}r)\, dr (\Sigma b, b)_{\mathscr{G}}\right] + o(1). \end{aligned}$$

This yields (4.10). □

REMARK 4.8. We have shown that as $r \to \infty$,

$$\begin{aligned} &(\Gamma^*(\lambda)a)(x) \\ &\quad = i2^{-1}\pi^{-1/2}\lambda^{-1/4}((\mathscr{S}a)(\varphi)w_+(r) - a(-\varphi)w_-(r)) + o_{\text{av}}(r^{-(d-1)/2}). \end{aligned}$$

As an easy consequence of Theorem 4.7, we obtain

THEOREM 4.9. *In order that the equation* (1.1) *have a solution with asymptotics*

$$u(x) = b_+(\varphi)w_+(r) - b_-(\varphi)w_-(r) + o_{\mathrm{av}}(r^{-(d-1)/2}), \quad b_\pm \in \mathcal{G},$$

at infinity, it is necessary and sufficient to have $b_+ = \Sigma(\lambda)b_-$.

COROLLARY 4.10. *If* u *satisfies the equation* (1.1) *and*

$$u(x) = b(\varphi)w_\pm(r) + o_{\mathrm{av}}(r^{-(d-1)/2}), \quad b \in \mathcal{G}, \tag{4.13}$$

as $r \to \infty$, *for one of the signs, then* $u = 0$.

Since relation (4.13) is not *a priori* assumed to be differentiable with respect to r, the last assertion improves Corollary 3.6. On the other hand, all asymptotics of this section can be actually differentiated with respect to radial and angular variables.

THEOREM 4.11. *Under the assumptions of Theorem* 4.7, *the following asymptotics, as* $r \to \infty$, *are true*

$$\begin{aligned}(D_r u)(x) &= i\lambda^{1/2}\left[(\Sigma(\lambda)b)(\varphi)w_+(r) + b(\varphi)w_-(r)\right] \\ &\quad + o_{\mathrm{av}}(r^{-(d-1)/2}), \\ (\nabla_\varphi u)(x) &= o_{\mathrm{av}}(r^{-(d-1)/2}).\end{aligned} \tag{4.14}$$

PROOF. A solution $u \in B^*$ of (1.1) satisfies relation (4.5), where $u_0 = \Gamma_0^* a$ and a is connected by (4.12) with the coefficient b in (4.8). The asymptotics of the derivatives of u_0 and $R_0 Q u$ are given by Lemmas 3.3 and 3.4. □

4. Let us return to the problem mentioned in the Introduction—that of constructing solutions of the equation (1.1) with asymptotics (1.6) of a standing wave at infinity. We note first of all that according to (4.6) or (4.7) the compactness of the operator $\Gamma_0 \colon \mathcal{H}^\beta \to \mathcal{G}$, $\beta > 1/2$, ensures that $\mathcal{S} - I$ is compact in $\mathcal{G}$. It follows that $\Sigma - \mathcal{J}$ is also compact, so that the spectrum of the unitary operator Σ consists of eigenvalues lying on the unit circle and accumulating only at the points $+1$ and -1 (from above and below). Moreover, eigenvalues of Σ which do not coincide with ± 1 have finite multiplicities. As a particular case of Theorem 4.9 we have

THEOREM 4.12. *In order that the equation* (1.1) *have a solution with asymptotics*

$$\begin{aligned}u(x) &= b(\varphi)r^{-(d-1)/2}\sin(\lambda^{1/2}r - (d-3)\pi/4 - \eta) \\ &\quad + o_{\mathrm{av}}(r^{-(d-1)/2}), \qquad b \in \mathcal{G}, \quad r \to \infty,\end{aligned} \tag{4.15}$$

it is necessary and sufficient that $\mu = \exp(-2i\eta)$ *be an eigenvalue of the operator* $\Sigma(\lambda)$ *and* b *be its eigenvector,* $\Sigma(\lambda)b = \mu b$.

§5. Expansion theorem

1. Under the assumption (1.2), where $2\alpha > d+1$, an arbitrary function (orthogonal to the eigenvectors of the operator H) can be expanded (see, e.g., [AJS]) as the integral over solutions of the Schrödinger equation (1.1) with asymptotics (1.3) at infinity. This expansion generalizes the Fourier integral and reduces to it if $q=0$. On the other hand, in the spherically symmetric case $q(x) = q(r)$, the Schrödinger operator H can be decomposed into the orthogonal sum of operators corresponding to fixed values of the orbital quantum number $l = 0, 1, 2, \dots$. For every l there exists an expansion in eigenfunctions of such an ordinary differential operator. This expansion requires only the condition $\alpha > 1$ in (1.2) and reduces to the Fourier-Bessel integral in the case $q=0$.

In this section we construct an expansion in terms of solutions of the equation (1.1) which have asymptotics (1.4) as $r \to \infty$. This expansion holds for arbitrary $\alpha > 1$ and includes as a particular case the "spherically symmetric" result mentioned above. Our expansion theorem is actually different from the familiar (see, e.g., [H1]) expansion theorem formulated in operator terms only in that it is directly connected with solutions of the Schrödinger equation. The information collected in previous sections allows us to give a complete proof of this theorem quite easily.

We start with a generalization of relation (2.3). According to Theorem 3.8, the positive part of the operator H is absolutely continuous. Its spectral family $E(\lambda)$ satisfies the equality

$$(R(\lambda+i0)f, g) - (R(\lambda-i0)f, g) = 2\pi i \frac{(E(\lambda)f, g)}{d\lambda}, \qquad f, g \in B, \tag{5.1}$$

where both sides are continuous in $\lambda > 0$. We recall that the operator $\Gamma(\lambda): B \to \mathcal{G}$ is defined by (4.2).

LEMMA 5.1. *For $f, g \in B$ and $\lambda > 0$,*

$$(\Gamma(\lambda)f, \Gamma(\lambda)g)_{\mathcal{G}} = \frac{(E(\lambda)f, g)}{d\lambda}. \tag{5.2}$$

PROOF. According to (2.3) and (3.14),

$$\begin{aligned} &2\pi i(\Gamma(\lambda)f, \Gamma(\lambda)g)_{\mathcal{G}} \\ &\quad = \Big((R_0(\lambda+i0) - R_0(\lambda-i0))(I - QR(\lambda-i0))f, (I-QR(\lambda-i0))g\Big) \\ &\quad = (f, R(\lambda-i0)g) - (R(\lambda-i0)f, g). \end{aligned}$$

It now suffices to take (5.1) into account. □

Let us denote by $\tilde{\mathcal{H}} = L_2(\mathbb{R}_+; \mathcal{G})$ the space of vector-functions $\tilde{f}(\lambda)$, $\lambda \in \mathbb{R}_+$, with values in $\mathcal{G}$. The mapping $F: \mathcal{H} \to \tilde{\mathcal{H}}$ is defined on the set B dense in $\mathcal{H}$ by the formula

$$(Ff)(\lambda) = \Gamma(\lambda)f. \tag{5.3}$$

Integrating (5.2) over $\mathbb{R}_+$, we find by the spectral theorem that

$$\|Ff\|^2 = \int_0^\infty \|\Gamma(\lambda)f\|_{\mathscr{G}}^2 \, d\lambda = \|Pf\|^2, \quad P = E((0, \infty)), \quad f \in B.$$

Thus the operator F extends by continuity to a bounded operator on the entire space $\mathscr{H}$. This operator is isometric on the absolutely continuous subspace $\mathscr{H}^{(a)} = P\mathscr{H}$ of the operator H and vanishes on its orthogonal complement. In other words,

$$F^* F = P. \tag{5.4}$$

Denote by Λ multiplication by the independent variable (λ) in the space $\tilde{\mathscr{H}}$ and by $E_\Lambda(X)$ its spectral projection acting as multiplication by the characteristic function of the Borel set $X \cap \mathbb{R}_+$. We shall establish the intertwining property

$$FE(X) = E_\Lambda(X)F. \tag{5.5}$$

It is sufficient to verify that

$$(FHf, \tilde{g}) = (Ff, \Lambda \tilde{g}) \tag{5.6}$$

for $f \in \mathfrak{S}(\mathbb{R}^d)$ and $\tilde{g}(\lambda) = \zeta(\lambda)a$, where $\zeta \in C_0^\infty(\mathbb{R}_+)$ and $a \in \mathscr{G}$. By definition of F, (5.6) is equivalent to the equality

$$\int_0^\infty (Hf, \Gamma^*(\lambda)a)\zeta(\lambda)\, d\lambda = \int_0^\infty \lambda(f, \Gamma^*(\lambda)a)\zeta(\lambda)\, d\lambda.$$

Now it suffices to take into account the fact that, according to Theorem 4.6, the function $\Gamma^*(\lambda)a$ belongs to B^* and satisfies equation (1.1).

Let us show further that $R(F) = \tilde{\mathscr{H}}$, i.e.,

$$FF^* = I. \tag{5.7}$$

This property of F is the most difficult to prove in the scheme which we use. We start with a preliminary assertion.

LEMMA 5.2. *Let* $\tilde{g} \in \tilde{\mathscr{H}}$ *be arbitrary and* $\tilde{h} = FF^* \tilde{g}$. *Then*

$$\Gamma^*(\lambda)(\tilde{h}(\lambda) - \tilde{g}(\lambda)) = 0$$

for almost all (a.a.) $\lambda \in \mathbb{R}_+$.

PROOF. By (5.4), (5.5), for any $f \in \mathscr{H}$ and any $X \subset \mathbb{R}_+$,

$$(E_\Lambda(X)FF^*\tilde{g}, Ff) = (E_\Lambda(X)\tilde{g}, Ff).$$

If $f \in B$, this relation can be written as

$$\int_X (\tilde{h}(\lambda), \Gamma(\lambda)f)_{\mathscr{G}}\, d\lambda = \int_X (\tilde{g}(\lambda), \Gamma(\lambda)f)_{\mathscr{G}}\, d\lambda.$$

Since $f \in B$ is arbitrary, it follows that

$$\int_X \Gamma^*(\lambda)(\tilde{h}(\lambda) - \tilde{g}(\lambda))\, d\lambda = 0.$$

It suffices to take into account the fact that the set X is also arbitrary. □

We deduce (5.7) from a similar equality related to the case $q = 0$. Indeed, let $F_0 \colon \mathscr{H} \to \tilde{\mathscr{H}}$ be defined on B by $(F_0 f)(\lambda) = \Gamma_0(\lambda) f$. Since the Fourier transform is unitary, we have $R(F_0) = \tilde{\mathscr{H}}$ and therefore the image $F_0(B)$ of B is dense in $\tilde{\mathscr{H}}$. Set $\tilde{f} = F_0 f_0$, where $f_0 \in B$, i.e. $\tilde{f}(\lambda) = \Gamma_0(\lambda) f_0$. Clearly, for arbitrary $h \in \mathscr{H}$ and $\tilde{h} = Fh$,

$$(\tilde{f}, Fh) = \int_0^\infty (\Gamma_0(\lambda) f_0, \tilde{h}(\lambda))_{\mathscr{G}} \, d\lambda. \tag{5.8}$$

According to (4.2), (4.4),

$$\Gamma_0(\lambda) f_0 = \Gamma(\lambda)\big(I + QR_0(\lambda - i0)\big) f_0, \tag{5.9}$$

where by Theorem 2.3 and Corollary 2.4, $f_0 + QR_0(\lambda - i0) f_0 \in B$ and this function is weakly continuous in $\lambda > 0$. Therefore

$$(\Gamma_0(\lambda) f_0, \tilde{h}(\lambda))_{\mathscr{G}} = \Big(\big(I + QR_0(\lambda - i0)\big) f_0, \Gamma^*(\lambda)\, \tilde{h}(\lambda)\Big).$$

Suppose now that $h = F^* \tilde{g}$ for arbitrary $\tilde{g} \in \tilde{\mathscr{H}}$. Then by Lemma 5.2 and (5.9),

$$\begin{aligned}(\Gamma_0(\lambda) f_0, \tilde{h}(\lambda))_{\mathscr{G}} &= \Big(\big(I + QR_0(\lambda - i0)\big) f_0, \Gamma^*(\lambda)\, \tilde{g}(\lambda)\Big) \\ &= (\Gamma_0(\lambda) f_0, \tilde{g}(\lambda))_{\mathscr{G}} \\ &= (\tilde{f}(\lambda), \tilde{g}(\lambda))_{\mathscr{G}}.\end{aligned}$$

It follows from (5.8) that

$$(\tilde{f}, FF^* \tilde{g}) = (\tilde{f}, \tilde{g}).$$

This concludes the proof of (5.7). We summarize the results obtained in the following

THEOREM 5.3. *For every $f \in B$ the function $\Gamma(\lambda) f$ belongs to the space $\tilde{\mathscr{H}}$. The operator F defined by* (5.3) *extends by continuity to a bounded operator on the entire space $\mathscr{H}$ and satisfies relations* (5.4), (5.5), *and* (5.7).

We note that (5.7) is equivalent to the formal relation

$$(\Gamma^*(\lambda) a_1, \Gamma^*(\mu) a_2) = \delta(\lambda - \mu)(a_1, a_2)_{\mathscr{G}}, \quad a_j \in \mathscr{G}, \tag{5.10}$$

where $\delta(\cdot)$ is the Dirac function. It is not difficult to give a direct proof of a similar but somewhat weaker result. Indeed, let $a_j \in \mathscr{G}$, $j = 1, 2$, be arbitrary and $u_1(\lambda) = \Gamma^*(\lambda) a_1$, $u_2(\mu) = \Gamma^*(\mu) a_2$. Then we have the asymptotic relation

$$\begin{aligned}\int_{|x|\leqslant r} & u_1(x, \lambda)\overline{u_2(x, \mu)}\,dx \\ &= i(\lambda^{1/2}-\mu^{1/2})^{-1}\Big[(b_1, b_2)_{\mathscr{G}}\exp(-i(\lambda^{1/2}-\mu^{1/2})r) \\ &\qquad - (\Sigma(\lambda)b_1, \Sigma(\mu)b_2)_{\mathscr{G}}\exp(i(\lambda^{1/2}-\mu^{1/2})r)\Big] \\ &+ i(\lambda^{1/2}+\mu^{1/2})^{-1}\Big[\gamma^2(\Sigma(\lambda)b_1, b_2)_{\mathscr{G}}\exp(i(\lambda^{1/2}+\mu^{1/2})r) \\ &\qquad - \bar{\gamma}^2(b_1, \Sigma(\mu)b_2)_{\mathscr{G}}\exp(-i(\lambda^{1/2}+\mu^{1/2})r)\Big] \\ &+ o_{\mathrm{av}}(1),\end{aligned} \tag{5.11}$$

where $\lambda \neq \mu$ and

$$b_1 = -i2^{-1}\pi^{-1/2}\lambda^{-1/4}\mathscr{F}a_1, \qquad b_2 = -i2^{-1}\pi^{-1/2}\mu^{-1/4}\mathscr{F}a_2.$$

For the proof of (5.11), we can consider that the functions u_j satisfy the equation (1.1) and hence

$$\begin{aligned}(\lambda-\mu)\int_{|x|\leqslant r} & u_1(x, \lambda)\overline{u_2(x, \mu)}\,dx \\ &= \int_{|x|\leqslant r}[(-\Delta u_1 + qu_1)\bar{u}_2 - u_1(-\Delta\bar{u}_2 + q\bar{u}_2)]\,dx \\ &= r^{d-1}\int_{\mathbb{S}^{d-1}}\left(u_1\frac{\partial\bar{u}_2}{\partial r} - \bar{u}_2\frac{\partial u_1}{\partial r}\right)d\varphi, \qquad x = r\varphi.\end{aligned}$$

Replacing (on the right-hand side) u_j and $\partial u_j/\partial r$ by their asymptotics (4.8) and (4.14), we arrive at (5.11).

Relation (5.11) shows that the functions $u_1(x, \lambda)$ and $u_2(x, \mu)$ are orthogonal (in the sense of distributions) if $\lambda \neq \mu$. However, for the proof of (5.10) it is necessary to show additionally that the remainder term $o_{\mathrm{av}}(1)$ in (5.11) does not give any contribution as $\lambda - \mu \to 0$.

2. Theorem 5.3 is of course the standard expansion theorem. Results of §4 allow us to reformulate it in terms of solutions of equation (1.1) with special asymptotics at infinity. By (5.3), (5.4) the equality

$$(Pf, g) = \int_0^\infty (\Gamma^*(\lambda)\tilde{f}(\lambda), g)\,d\lambda, \qquad \tilde{f} = Ff, \tag{5.12}$$

holds for arbitrary $f \in \mathscr{H}$ and $g \in B$. Theorem 4.6 and Remark 4.8 ensure that for a.a. $\lambda \in \mathbb{R}_+$ the function $u_f(\lambda) = \Gamma^*(\lambda)\tilde{f}(\lambda) \in B^*$ satisfies (1.1) and

$$\begin{aligned}u_f(x;\lambda) = &-i2^{-1}\pi^{-1/2}\lambda^{-1/4}\Big[(\mathscr{S}(\lambda)\tilde{f}(\lambda))(\varphi)w_+(r) - \tilde{f}(-\varphi;\lambda)w_-(r)\Big] \\ &+ o_{\mathrm{av}}(r^{-(d-1)/2})\end{aligned} \tag{5.13}$$

as $r \to \infty$. Moreover, the following limit

$$\lim_{\rho \to \infty} \rho^{-1} \int_{|x| \leqslant \rho} |u_f(x, \lambda)|^2 \, dx = (2\pi)^{-1} \lambda^{-1/2} \| \tilde{f}(\lambda) \|_{\mathcal{G}}^2 \tag{5.14}$$

exists. Now suppose additionally that $f \in E((\mu, \nu))\mathcal{H}$ for some $0 < \mu < \nu < \infty$. Then the integral (5.12) is actually taken only over the interval (μ, ν). According to Lemma 4.4 and (5.4),

$$\begin{aligned} \int_\mu^\nu d\lambda \int_{\mathbb{R}^d} |u_f(x, \lambda)| |g(x)| \, dx &\leqslant \int_\mu^\nu d\lambda \|u_f(\lambda)\|_{B^*} \|g\|_B \\ &\leqslant C(\mu, \nu) \int_\mu^\nu \| \tilde{f}(\lambda)\|_{\mathcal{G}} \, d\lambda \cdot \|g\|_B \\ &\leqslant C_1(\mu, \nu) \|f\| \, \|g\|_B. \end{aligned} \tag{5.15}$$

Since $(u_f(\lambda), g)$ is an integral in x, (5.12) can be regarded as a double integral. By the Fubini theorem, it is possible to interchange the order of integration in the variables λ and x there. It follows that

$$(f, g) = \left(\int_0^\infty u_f(\lambda) \, d\lambda, g \right)$$

and therefore

$$f(x) = \int_0^\infty u_f(x, \lambda) \, d\lambda. \tag{5.16}$$

The integral (5.16) converges (absolutely) for a.a. $x \in \mathbb{R}^d$. This is a consequence of (5.15) if the g are chosen as the characteristic functions of the balls $\{ x \in \mathbb{R}^d : |x| \leqslant \rho \}$. We emphasize that the integral (5.16) belongs to $\mathcal{H}$ although its integrand is from B^* only.

For arbitrary $f \in \mathcal{H}$

$$Pf = \operatorname*{s-lim}_{\mu \to 0, \nu \to \infty} f_{\mu,\nu}, \qquad f_{\mu,\nu} = E((\mu, \nu)) f,$$

where s-lim means the strong limit in $\mathcal{H}$. By (5.5) $\tilde{f}_{\mu,\nu}(\lambda) = \tilde{f}(\lambda)$ if $\lambda \in (\mu, \nu)$, and $\tilde{f}_{\mu,\nu}(\lambda) = 0$ if $\lambda \notin (\mu, \nu)$. Now using (5.16) for $f_{\mu,\nu}$, we obtain the following version of the expansion theorem.

THEOREM 5.4. *Let $f \in \mathcal{H}$ be arbitrary, $\tilde{f} = Ff$, and let $u_f(x, \lambda)$ be the solution of equation* (1.1) *with asymptotics* (5.13) *at infinity. Then*

$$Pf = \operatorname*{s-lim}_{\mu \to 0, \nu \to \infty} \int_\mu^\nu u_f(\lambda) \, d\lambda \tag{5.17}$$

and the generalized Parseval identity

$$\|Pf\|^2 = 2\pi \int_0^\infty \lim_{\rho \to \infty} \left(\rho^{-1} \int_{|x| \leqslant \rho} |u_f(x, \lambda)|^2 \, dx \right) \lambda^{1/2} \, d\lambda$$

holds. On the other hand, for arbitrary $\tilde{f} \in \tilde{\mathscr{H}}$, *the right-hand side of* (5.17) *is well defined, belongs to* $\mathscr{H}^{(a)}$, *and by applying the operator* F *to this element one recovers* $\tilde{f}$.

Theorem 5.4 can be formulated in a more concrete form if the functions $\tilde{f}(\lambda)$ are decomposed with respect to some basis in $\mathscr{G}$. Indeed, suppose that for every (or almost every) $\lambda > 0$ some orthonormal basis $a_n(\cdot, \lambda)$, $n = 1, 2, \dots$, is chosen in the space $\mathscr{G}$. We stress that the choice of this basis may depend on the spectral parameter $\lambda > 0$. The functions $u_n(\lambda) = \Gamma^*(\lambda)a_n(\lambda)$ satisfy equation (1.1) and have the asymptotics

$$u_n(x, \lambda) = -i2^{-1}\pi^{-1/2}\lambda^{-1/4}\left[(\mathscr{S}(\lambda)a_n(\lambda))(\varphi)w_+(r) - a_n(-\varphi, \lambda)w_-(r)\right] + o_{\mathrm{av}}(r^{-(d-1)/2}) \tag{5.18}$$

as $r \to \infty$. For every $f \in \mathscr{H}$, the function $\tilde{f}(\lambda) = (Ff)(\lambda)$ can be represented as the series

$$\tilde{f}(\lambda) = \sum_{n=1}^{\infty} \psi_n(\lambda)a_n(\lambda), \qquad \psi_n(\lambda) = (\tilde{f}(\lambda), a_n(\lambda))_{\mathscr{G}}, \qquad \text{for a.a. } \lambda \in \mathbb{R}_+.$$

Applying the operator $\Gamma^*(\lambda)$ to this equality, we get

$$u_f(x, \lambda) = \sum_{n=1}^{\infty} \psi_n(\lambda)u_n(x, \lambda),$$

where the series converges in the space B^* for a.a. $\lambda \in \mathbb{R}_+$. Therefore Theorem 5.4 ensures

THEOREM 5.5. *Let* u_n *be solutions of the equation* (1.1) *with the asymptotics* (5.18) *at infinity. Then for every* $f \in \mathscr{H}$

$$Pf = \operatorname*{s-lim}_{\mu \to 0, \nu \to \infty} \int_\mu^\nu \sum_{n=1}^{\infty} \psi_n(\lambda)u_n(\lambda)\, d\lambda, \tag{5.19}$$

where

$$\psi_n(\lambda) = ((Ff)(\lambda), a_n(\lambda))_{\mathscr{G}} \tag{5.20}$$

and

$$\|Pf\|^2 = \int_0^\infty \sum_{n=1}^{\infty} |\psi_n(\lambda)|^2\, d\lambda. \tag{5.21}$$

In the case $f \in B$, *the coefficient* $\psi_n(\lambda)$ *has the standard form*

$$\psi_n(\lambda) = (\Gamma(\lambda)f, a_n(\lambda))_{\mathscr{G}} = (f, u_n(\lambda)) = \int_{\mathbb{R}^d} f(x)\overline{u_n(x, \lambda)}\, dx.$$

On the other hand, suppose that $\psi_n(\lambda)$ *is an arbitrary family of functions for which the right-hand side of* (5.21) *is finite. Then the right-hand side of* (5.19) *is well defined and belongs to* $\mathscr{H}^{(a)}$. *Denoting this element by* f, *we can recover* $\psi_n(\lambda)$ *via* (5.20).

In particular, if $a_n(\lambda)$ are eigenvectors of the operator $\Sigma(\lambda)$, then (5.19) yields an expansion in terms of solutions of the Schrödinger equation with the asymptotics (4.15) of a standing wave at infinity.

3. We have constructed the operator F in subsection 1 proceeding from the operator $\Gamma(\lambda)$ defined by equality (4.2). In fact, we can replace there $R(\lambda - i0)$ by $R(\lambda + i0)$. Then we obtain two operators

$$\Gamma_{\pm}(\lambda) = \Gamma_0(\lambda)\big(I - QR(\lambda \pm i0)\big) : B \to \mathscr{G}$$

and two mappings $(F_{\pm}f)(\lambda) = \Gamma_{\pm}(\lambda)f$. Clearly, $\Gamma(\lambda) = \Gamma_{-}(\lambda)$ and $F = F_{-}$. The mapping F_{+} has the same properties as F_{-}. Relations (2.3) and (3.14) ensure that

$$\Gamma_{+}(\lambda) = \mathscr{S}(\lambda)\Gamma_{-}(\lambda), \tag{5.22}$$

where the unitary operator (the scattering matrix) $\mathscr{S}(\lambda)$ is defined by (4.6).

We could have formulated an expansion theorem of the type of Theorem 5.4, starting from F_{+} instead of F_{-}. It is not, however, difficult to see that, due to relation (5.22), these two versions actually coincide.

Finally we note that the mappings $F_{\pm}$ are intimately connected with the so-called wave operators introduced by the relation

$$W_{\pm} = \operatorname*{s-lim}_{t \to \pm\infty} \exp(iHt)\exp(-iH_0t).$$

We do not dwell neither on the proof of the existence of these limits nor on the verification of the formula

$$W_{\pm} = F_{\pm}^{*}F_0.$$

Both of them can be found, for example, in [H1]. Theorem 5.4 gives the following representation for the wave operators. Let $f \in \mathscr{H}$ be arbitrary, $\tilde{f}_0 = F_0 f$,

$$f_{+}^{(+)}(\lambda) = \tilde{f}_0(\lambda), \qquad f_{-}^{(+)}(\lambda) = \mathscr{J}\mathscr{S}^{*}(\lambda)\,\tilde{f}_0(\lambda),$$
$$f_{+}^{(-)}(\lambda) = \mathscr{S}(\lambda)\,\tilde{f}_0(\lambda), \qquad f_{-}^{(-)}(\lambda) = \mathscr{J}\tilde{f}_0(\lambda),$$

and let $u_f^{(\pm)}(x, \lambda)$ be the solution of equation (1.1) with asymptotics

$$u_f^{(\pm)}(x, \lambda) = -i2^{-1}\pi^{-1/2}\lambda^{-1/4}\Big[f_{+}^{(\pm)}(\lambda, \varphi)w_{+}(r) - f_{-}^{(\pm)}(\lambda, \varphi)w_{-}(r)\Big] + o_{\mathrm{av}}\big(r^{-(d-1)/2}\big)$$

at infinity. Then

$$W_{\pm}f = \operatorname*{s-lim}_{\mu \to 0,\, \nu \to \infty} \int_{\mu}^{\nu} u_f^{(\pm)}(\lambda)\, d\lambda. \tag{5.23$_\pm$}$$

Since $F_{\pm}W_{\pm}f = F_0 f$, the relation (5.23$_-$) is an immediate consequence of Theorem 5.4 applied to $W_{-}f$. For the proof of (5.23$_+$) one must, moreover, take (5.22) into account.

REFERENCES

[AJS] W. O. Amrein, J. M. Jauch, and K. B. Sinha, *Scattering theory in quantum mechanics*, W. A. Benjamin, Inc., Reading, MA, 1977.

[AH] S. Agmon and L. Hörmander, *Asymptotic properties of solutions of differential equations with simple characteristics*, J. Analyse Math. **30** (1976), 1–38.

[BL] J. Bergh and J. Löfström, *Interpolation spaces. An Introduction*, Springer-Verlag, Berlin and New York, 1976.

[DS] N. Dunford and J. T. Schwartz, *Linear operators, I, General Theory*, Interscience, New York, 1958.

[H1] L. Hörmander, *The analysis of linear partial differential operators*, vols. I, II, Springer-Verlag, Berlin and New York, 1983.

[H2] ———, *Lower bounds at infinity for solutions of differential equations with constant coefficients*, Israel J. Math. **16** (1973), no. 1, 103–116.

[K] S. T. Kuroda, *Scattering theory for differential operators*, J. Math. Soc. Japan **25** (1973), 75–104, 222–234.

[LL] L. D. Landau and E. M. Lifshits, *Quantum mechanics: nonrelativistic theory*, Fizmatgiz, Moscow, 1963; English transl., Pergamon Press, Oxford, and Addison-Wesley, Reading, MA, 1965.

[LM] J.-L. Lions and E. Magenes, *Problèmes aux limites non homogènes et applications*, vol. 1, Dunod, Paris, 1968.

[S] Y. Saitō, *Spectral representations for Schrödinger operators with long-range potentials*, Lecture Notes in Math., vol. 727, Springer, Berlin, 1979.

[T] H. Triebel, *Theory of function spaces*, Akademische Verlagsgesellschaft, Leipzig, 1983.

[V] B. R. Vaĭnberg, *Asymptotic methods in equations of mathematical physics*, Izdat. Moskov. Univ., Moscow, 1982. (Russian)

Translated by THE AUTHOR